People and
Environment
A Global Approach

WITHDRAWN

PEARSON
Education

We work with leading authors to develop the strongest
educational materials in geography, bringing cutting-edge
thinking and best learning practice to a global market.

Under a range of well-known imprints, including
Prentice Hall, we craft high-quality print and
electronic publications which help readers to understand
and apply their content, whether studying or at work.

To find out more about the complete range of our
publishing, please visit us on the World Wide Web at:
www.pearsoned.co.uk

People and Environment

A Global Approach

GARETH JONES

PEARSON
Prentice
Hall

Harlow, England • London • New York • Boston • San Francisco • Toronto • Sydney • Singapore • Hong Kong
Tokyo • Seoul • Taipei • New Delhi • Cape Town • Madrid • Mexico City • Amsterdam • Munich • Paris • Milan

Pearson Education Limited
Edinburgh Gate
Harlow
Essex CM20 2JE
England

and Associated Companies throughout the world

Visit us on the World Wide Web at:
www.pearsoned.co.uk

First published 2004

ISBN 978-0-582-41412-9

British Library Cataloguing-in-Publication Data
A catalogue record for this book is available from the British Library

Library of Congress Cataloging-in-Publication Data
A catalog record for this book is available from the Library of Congress

10   9   8   7   6   5   4
09   08

Typeset by 35 in 10/12pt Sabon
Printed and bound in Malaysia (CTP-VVP)

The publisher's policy is to use paper manufactured from sustainable forests.

Contents

List of black and white plates

List of colour plates

A   Hong Kong. A modern-day city

B   Third World village. Botan, western Nepal in 1995. This mountain settlement had no road access, no electricity, no water supply or sewage system. It was totally lacking in any modern infrastructure

C   Deforestation resulting in severe soil erosion on steep ground. The white area at the top of the main slope and behind the trees is bare rock, exposed by soil erosion. Natal Province, South Africa

D   Peaty-gley soil. Permanent water-logging of the soil has resulted in the production of ferrous iron compounds characterised by predominantly grey colours in the soil profile. The darker coloured bands are layers of un-decomposed organic material, or peat. There are two layers of peat, the lower one representing the former surface of the soil, above which more mineral soil has been deposited, and finally a new layer of peat is currently forming

E   Multiple use of an urban waterway. The River Avon as it passes through the centre of Christchurch, New Zealand, is used for recreation, wildlife and amenity purposes. The quality and quantity of the water flow are unaltered by the uses

F   An extensive agricultural terrace system, Banyalbufar on the north of the island of Mallorca in the Mediterranean Basin. Irrigation is possible from water stored in subterranean tanks in the mountains

G   View down the McKinnon Valley, Fiordland National Park, New Zealand. This region now forms part of the Southwest New Zealand World Heritage Area, covering 2.6 million hectares. Permanent human residents are, for the most part, not allowed in the area

H   (a) Purple saxifrage, a late-glacial relict species growing on Ben Lawers, Perthshire, Scotland; (b) *Scilla peruviana*, growing in a fragment of soil on bare limestone, Mallorca, Spain

List of figures

List of tables

List of boxes

Preface

This book examines the complex relationship between our own species, *Homo sapiens*, and all the other living and non-living components of the planet. Learning how to manage the environment has proved a difficult task, but even harder is finding ways of managing ourselves! Our species is the most innovative, competitive and opportunistic species to inhabit the biosphere. When environmental issues began to emerge in the 1970s they were of concern only to a tiny fragment of the population – to the so-called 'eco-freaks' or the hippy communities. Wider acceptance by society of the need to take care of the environment developed quite quickly, taking as little as ten years from the mid-1980s. Since that time, our views on how to achieve successful environmental management have matured and it is unlikely that any inhabitant, certainly of the developed world, remains oblivious to the changes that have occurred in the way we behave towards the environment.

Despite the progress, for the most part, our skills in managing the environment on a day-to-day basis remain poorly developed. We remain uncertain about the criteria we should use as 'standards' on which to base our assessment of the environment. As our knowledge of the environment has improved, so we have been required to change the standards to which we operate. For example in 1997, participants at the Kyoto Climate Conference agreed a reduction of carbon dioxide to the atmosphere of 5 per cent but even before the Conference had drawn to a close, Member States of the European Union were proposing that a reduction of 12 per cent in CO_2 output was justified. On the basis of new scientific findings some EU countries raised their target to 20 per cent by 2002 and a small group of EU countries began planning for a 60 per cent reduction by 2050 in order to contain global warming. The consequences of such a huge reduction in CO_2 for industry and, in turn, society, will be immense but the repercussions of *not* taking action are recognised as being even more disruptive. A reduction of CO_2 by 60 per cent in 1997 would have been scientifically and socially unjustifiable and yet within five years a complete turnaround in thinking had been achieved. We have clearly made very considerable progress in facing up to environmental problems.

To sustain our ability to cope with environmental problems we need a better understanding of the ways in which all the many components of the environment fit together. No longer is it sufficient to be educated in the traditional sciences of biology, chemistry and physics. Understanding the environment requires that we cut across the traditional boundaries of knowledge. We need a knowledge of the basic sciences, so that we can understand how plants, animals and humans live within the context of a biosphere that comprises water, atmosphere and biogeochemical cycles. However, knowing how the components of the natural world function is no longer enough. We need specific information of how and why we behave towards the biosphere and the environment in the ways that we do. This information requires that we acquire skills in sociology, in politics and in economics. Finally, we need to be certain of the ethical and moral obligations to ourselves (or more correctly, to future generations of humans) as well as to the rights of all non-human species upon which we rely for the correct functioning of the biosphere. Located somewhere in this huge amalgam of information is the basis for the new science of environmental studies – part geography, biology, sociology and economics. The objective of this book is that it can contribute in a small way to providing some of the information necessary to achieve a new environmental awareness.

The first two chapters explain the way in which the biosphere and all its physical and biological components have evolved and how the different components are assembled into the complex forms we call ecosystems. Chapter 3 looks at the social and political organisation necessary to plan and legislate for the use of the biosphere. For much of the time, society has been concerned with reacting to problems that we have inherited from the past and less to the proactive management of the contemporary environment. Subsequent chapters use the information in the first three chapters to examine some of the major environmental problems that now face us. Chapter 4 deals with the management of the atmosphere. The air we breathe remains one of the very few 'free' natural resources. The immense size of the atmosphere combined with its ceaseless activity of winds and air movements made us believe that it had an infinite capacity to absorb our pollutants and at the same time continue to deliver clean air for us to breathe and provide solar radiation to warm our planet and rainfall to supply fresh water. Looking back only two decades to the 1980s, it is difficult to understand why so many scientists could not accept that human actions were capable of bringing about climate change and yet now, in the early years of the twenty-first century, there are few who would doubt the role of human activity as the cause of global warming.

Chapter 5 deals with the mismanagement of our supply of fresh water. Because water was seen as a free resource, we overlooked the need to manage it. Only when countries began to face a critical water shortage were steps taken to begin to manage this resource. By 2020, the average supply of water worldwide is expected to drop by one-third. From being a 'free good' water has become a valuable commodity. The cost of guaranteeing a reliable supply of good-quality water to the consumer is now a cost that developed countries are prepared to bear. While the supply of fresh water is increasingly under the control of the water engineer, the majority of the salt water, the oceans, remain in an unmanaged state. Chapter 6 is concerned with the oceans and assesses its role in making our planet unique. Partly because the majority of the ocean is considered 'empty' space by humans and partly because of its vast, inexhaustible size, we have treated the oceans as an infinite resource, over-harvesting its organic resources and dumping vast quantities of our waste materials into it. Chapter 6 also looks at the problems that have resulted as a consequence of overexploitation and lack of management of the oceans.

The next chapter focuses on management of the land, and in particular, looks at land degradation and compares natural degradation processes with degradation due to human processes. Chapter 8 deals with the topical issue of biodiversity, and its new-found relevance as a reservoir of genetic material that is being exploited by genetic engineers to create varieties of plants, especially food crops, that will allow an increase in food production. This chapter highlights the conflict that can possibly arise between the almost limitless capabilities of the geneticist to change the course of evolution and the ethical debate that surrounds such actions. The final chapter considers how our use of the management of the biosphere might evolve during the present century and what new natural and anthropogenic problems might emerge.

Throughout the book, the fundamental hypothesis is that people cause environmental 'problems' and therefore 'solutions' to the problems lie in our hands. We are used to facing challenges and, provided these are not too severe, we appear to thrive under conditions in which we are required to prove ourselves against the forces that threaten to subdue us. For example, the quest to produce sufficient food, of finding cures to illness and for providing resources with which to build a strong society, are all challenges that have been faced over the millennia. Yet having solved these problems in one era, they reappear in a new guise – new human diseases, further catastrophic famines, the shortages of key commodities that threaten our industrial processes – each catastrophe apparently more serious than the last.

There is no doubt that we are members of an incredibly competent species whose capabilities appear to know no bounds but it has only been in the most recent centuries, the most recent 200 years or so at most, that our species has acquired such complete dominance over other life forms and of the physical environment. It could all have been so different. During the Pleistocene era (the Ice Age), our numbers probably underwent a major decline and had the Ice Age lasted another 10,000 years or had pushed further towards the equator, it is possible that our species would have become extinct. The success of our species in using the resources of our planet and altering the biosphere to further our own well-being separates us from all other animals. Our success as a species in dominating the biosphere has become absolute. We are the victors in the long battle for supremacy over all other species and of the physical resources of the planet, but as victors we have an obligation to take care of the other less successful components.

Acknowledgements

A large number of people, mainly many generations of undergraduates, have helped in the preparation of this book, most of them without knowing of their contribution! Many of the ideas and themes have been tried out as lecture material or as tutorial topics in my 'home' university and at others where I have had opportunity to learn from the new environments provided by academic visits or attendance at conferences.

I owe specific gratitude to all members of staff in the Geography Department at the University of Strathclyde for their support and encouragement, especially to George Yule for technical assistance and to Lorraine Nelson for general assistance. The editorial staff at the publisher Pearson Education have been extremely patient with my numerous questions and delays concerning the manuscript. I especially thank Morten Fuglevand and his staff.

To my wife, Lucy, who often despaired at the slow progress of my writing, I owe a particular debt of gratitude. She read all the initial drafts of each chapter and was usually able to identify flaws in my arguments or in my written style. Any inaccuracies remain, however, entirely due to my own shortcomings.

For the first time in book format, I have made extensive use of web-based material. The full web addresses have been given in the text and at the end of each chapter along with the title of the organisation responsible for the material. Web addresses evolve rapidly and in the event of an address not working, the reader is advised to use a search engine such as Google and to make a search based on organisation name. Alternatively, the root of the web address may allow access to the home page of the organisation and links may exist from that page to other relevant pages.

Every effort has been made to give credit to copyright holders for material used in this book. If omissions remain or material is incorrectly credited it was unintentional and every effort will be made to correct it in future revisions.

Thanks are due to the following for allowing me to use data and/or figures:

Table 2.1 Jordan, A. and O'Riordan, T. (2000) Environmental politics and policy processes in O'Riordan, T. (ed.) *Environmental Science for Environmental Management*. Prentice Hall, Harlow, reproduced with permission from Pearson Education Ltd.

Table 4.4 IPCC (1999) *Climate Change 2001: The Scientific Basis*. Cambridge University Press, Cambridge.

Table 7.2 reprinted by permission of Kogan Page from Gardner, G. (1997) *State of the World*. Earthscan, London.

Figure 5.1 reprinted by permission of Sage Publications Ltd from Jones, G. E. and Hollier, G. P. (1997) *Resources, Society and Environmental Management*. Paul Chapman, London.

Figure 5.5 Arnell, N. (2002) *Hydrology and Global Environmental Change*. Prentice Hall, Harlow reproduced with permission from Pearson Education Ltd.

Figure 6.1 Bigg, G. R. (1996) *The Oceans and Climate*. Cambridge University Press, Cambridge.

1 The environment

We now know what was unknown to all the preceding caravan of generations, that men are only fellow voyagers with other creatures . . . This new knowledge should have given us, by this time, a sense of kinship . . . a wish to live and let live; a sense of wonder over the magnitude and duration of the biotic enterprise.

Leopold, 1949

The relationship between humans and the biosphere has been incredibly biased in favour of the latter for most of the time that our species has existed. Only in the last two and a half centuries has the balance begun to tip towards favouring ourselves, while in the most recent half century it seems that at times humans have become a new super-being, so competent have we become in harnessing the resources of the planet. There are some signs that we may be causing some of the great natural planetary cycles to move out of balance but we are uncertain of the significance of these changes. This book looks at some of these changes, not so much from the perspective of an environmental scientist but more from the standpoint of a social scientist. The underlying concept of this book is that human nature is such that we can always be expected to place the well-being of our species first, for it is this concern with ourselves and of our family and of the society in which we live that makes us unique. It is the socio-economic and political structures that place us apart from all other species and it will be up to society to decide how to manage the resources of our planet both at the present and in the near future. To make the correct decisions we will need the best scientific knowledge

and the most advanced technological skills. But the benefits of knowledge and skills can only be harnessed if society understands the problems it has caused and how they need to be managed. This book attempts to delve beneath the surface of some of the environmental problems that face us in the first years of a new century. It begins by examining the basic information required for an understanding of the environment. In particular, it looks at:

- the role played by the environment in supporting life on Earth
- changes in the meaning of the term 'environment'
- evolution of different biological components.

1.1 Why is the environment important?

It is difficult to accept that the species to which we belong, *Homo sapiens*, is the same species that existed 10,000 years previously or 1000 years ago or even 100 years ago, so rapid has been our ability to manage life on our planet. In comparison with our great grandparents who were born between 100 and 150 years ago, the achievements of present-day human beings are such that we appear to exist in a different world. Almost all the achievements attained during the twentieth century have been firmly based on science and technology. The century saw the development and perfection of electronic communication (first the telegraph, followed by radio, telephone, television and later still, digital communication using fibre-optic cables, satellites and culminating in the internet). Personal transport has been revolutionised as a result of the internal combustion engine, but the invention of the automobile has shown how overuse and misuse of an invention can result in pollution, congestion and the despoilment of our major cities. The twentieth century closed with considerable concerns being expressed over the virtues of genetically modified foods. It appears that despite the undisputable benefits that society has gained from technology, we tend to misuse or apply the benefits in an insensitive manner and, consequently, bring about new problems that require the application of further technological solutions. We seem trapped in a technology spiral, in which society avidly embraces the latest electronic wizardry or genetic modification with little thought for the implications the technology may bring. Our skills for predicting the impact of new technology on society and the natural environment are poorly developed. If only we had a fortune teller's crystal ball that could help predict where the next environmental problem would arise!

It is sometimes hard to realise that our concern for the well-being of the 'environment' is a comparatively recent phenomenon and has existed only since the mid-1930s. While there can be few people living on the Earth today who have not heard of the term 'environment', its significance will vary greatly depending on the type of society the individual inhabits. For the developed world, the term was primarily used by science and academia during the first half of the twentieth century and it only came into common usage in the press, general literature and especially television from the mid-1960s. Rachael Carson's book *Silent Spring* (Carson, 1963) did much to raise our awareness of 'environment' closely followed by the writings of Commoner (1971) and Ehrlich *et al.* (1973). For inhabitants of the developing world the term 'environment' was often subsumed into a much wider context involving the entire well-being of every living individual. Paradoxically, societies that are considered 'underdeveloped' or 'primitive' by western standards often show a far greater understanding and respect for the environment than that of western societies (Darling and Dasmann, 1972).

The widespread use of the term 'environment' has inevitably led to instances of its inappropriate use. This is particularly the case when the term is used by the popular press and general public. In a strict sense, the 'environment' comprises our surroundings, from which we receive all the stimuli and the resources necessary for life. In addition, we pass back all our pollutants into the environment in the expectation that they will be dispersed and diluted to safe levels. In reality, the environment does not exist. It is a 'virtual' layer that surrounds every living creature and provides an interface between the organism and the 'biosphere'. The biosphere provides the essential components of life: water, air, food, a place to live and a place in which other members of our species can be found and with whom relationships can be forged. In essence, there is only one biosphere to provide a living space for all organisms that inhabit the planet. By contrast, every single plant and animal requires its own highly specific environment so that individual species requirements can be met. Some species can modify the environment to ensure that they exist in an optimum environment. However, for all organisms apart from humankind, the extent to which modification of the environment is possible is minimal. By contrast, our species is extremely capable of modifying its surroundings to suit our individual needs. We achieve this by using science and technology to manipulate the amount and type of resources we extract from the biosphere. By using resources we are able to modify our environment.

Examples of environmental modification made by successive generations of human effort would fill many volumes of books. The impact of human ingenuity on the biosphere and environment can be seen in the development of settled agriculture. The development of agriculture involved several major technological breakthroughs that allowed small family groups to become settled in one place instead of roaming across large areas of land hunting and gathering food. As soon as our ancestors had become settled, they encountered a variety of new problems. They had to rely on the same small area of land to supply all their requirements for food and water and also to take back their wastes in such a way that dangerous levels of pollution did not occur. In the early period of settled agriculture many families probably died from starvation, or were killed by water-borne diseases transferred from animal and human sewage, or simply reverted to nomadism when the early agricultural system failed. By trial and error, the system of settled agriculture was gradually perfected but it took many generations to make it a reliable system.

The example of the new agricultural way of life illustrates the fundamental problems faced by our ancestors and by us at present: the use of technical skills allows us to become more settled within a specific physical location but in order to remain there we must resort to greater and greater 'management' of our environment. By resorting to the use of technology to support our preferred lifestyle, we have manoeuvred our species into a situation where, to make further advances along our chosen lifestyle, we are required to make more use of technology. Already, of the 6 billion human inhabitants of this planet, 2.7 billion live in urban areas (World Bank, 2000). These urban dwellers live in a highly modified environment in which heating, lighting, provision of fresh water, disposal of sewage and garbage, means of acquiring food, method of transport, contact with people and type of day-to-day employment are all totally different from those of a person living in a wilderness environment (see Colour Plates A and B).

In the confusion of everyday life it is sometimes difficult to see what we have achieved or, indeed, to know exactly where we want to go! It is worthwhile reflecting on what we have achieved in terms of managing our environment and also to look forward in an attempt to see how our new technological skills will allow us to manage the environment in totally new ways. Reflection is usually a much easier task than trying to predict what *may* happen in the future. However, we need to look to the future so that we can identify environmental problems as soon as they begin to appear. New technology is introduced so rapidly that it is often impossible to identify environmental problems until they have become well developed.

Why have we allowed human society to become so reliant on technology? The answer, quite simply, is that reliance on technology has proved to be a very successful strategy. It has enabled a dramatic growth in population during the twentieth century from about 1.5 billion in 1900 to 6 billion people in 1999 (Cutter and Renwick, 1999) but by so doing has created a demand for food and natural resources that could only be met by an ever-greater extraction of resources from the biosphere. As a consequence, some parts of the biosphere have become overused and show signs of deterioration, notably the loss of forests throughout the world, soil erosion and pollution of land, water and air. Alongside the negative impact of human society on the environment there are other signs that we have discovered how to manage the environment in a sympathetic and sustainable fashion. We have planted vast areas of new forest especially in mid- and high latitudes; we have designated extensive areas of land as national parks, heritage areas and wilderness land while most countries now have extensive planning and environmental protection legislation that ensures we remove the most harmful components of our pollution output before it is released to the environment.

Despite all the achievements that have taken place in support of the environment and its non-human species it appears that as fast as we solve one problem another, bigger problem appears on the horizon. In the 1970s we were confronted with the problem of precipitation that was becoming increasingly acidic in reaction (Mannion, 1992a). The 'acid rain' killed soil fauna, damaged the roots of plants and killed aquatic life in rivers and lakes especially in northern latitudes. Rigorous control on air and water pollution emanating from industrial sources gradually overcame the problem of acid precipitation but its place was taken by the damage caused to the atmospheric ozone layer from chlorofluorocarbons (CFCs) used as propellants in aerosol sprays (Farman *et al.*, 1985; Hofmann 1996). Once again, international action provided a rapid solution to this problem – although the replacement propellant remains a damaging agent to ozone. By the beginning of the 1990s a new problem had firmly established itself on the scientific, political and public agendas – global warming (Crutzen and Golitsyn, 1992). At present, there appears to be no 'quick fix' and despite the best intents of the Kyoto Protocol and subsequent meetings of the Conference of the Parties it is proving particularly difficult to find a way of bringing global warming under control (http://unfccc.int/resource/convkp.html).

The late twentieth century proved to be a time in our cultural evolution when politicians, industrialists and the general public accepted that such was the scale of change brought about by our technological capabilities that it was appropriate to look anew at our responsibility towards the biosphere, our environment, and the plants and animals that inhabited the Earth. The acceptance of the so-called 'environmental issues' by modern society raises many completely new debates, not least calling into question the basic principle on which our lifestyle is based – that of consumerism. While the popularisation of the term 'environment' has undoubtedly raised its profile among the general population it has also led to confusion. It has already been explained that the strict meaning of the term 'environment' is an invisible envelope that surrounds every living organism and through which are passed the stimuli necessary to fulfil our lives. However, because of the widespread, common use of the term, social scientists and natural scientists have also used 'environment' to mean the natural world that surrounds us, that provides natural resources, landscapes and the place where all other non-human organisms live. This broad definition of the environment will be used throughout this book.

1.2 Defining the 'environment' and how the meaning has changed

The term 'environment' is comparatively new and was rarely used by natural scientists working during the period 1850–1950. The leading British ecologist, Sir Arthur Tansley, made infrequent use of the word in his book *Practical Plant Ecology* (1923), while later, in his *Introduction to Plant Ecology* (1946), he preferred to use the term *habitat* to describe the surroundings from which an organism gained its resources. Gradually, through the 1970s the term gained widespread use and in *The Animal and the Environment* (Vernberg and Vernberg, 1970) the term 'environment' figures prominently. In its original use, biologists described the environment as a unique set of external conditions thought to influence the life of individual organisms. Used in this way, the environment consisted of two interrelated parts: a *physical* or *abiotic* environment comprising all non-living components (air, water, rocks and soil) and a *biological* or *biotic* environment that could be further subdivided into external and internal components. The external biotic environment included relationships between an organism and all other life forms while the internal biotic environment was made up primarily of biochemical control mechanisms (for example, hormones in animals), responsible for the overall well-being of the individual plant or animal. The environment was seen as an essential life support mechanism, nurturing, protecting and enabling life.

By the early years of the twentieth century, a new branch of biology, ecology, emerged as the scientific discipline responsible for the study of the environment. Initially, many of the ideas relating to the environment were only poorly understood and vaguely expressed and this contributed to the general uncertainty among scientists about whether the study of ecology was a genuine part of scientific study. Before ecology became recognised as a distinct branch of science, the more traditional branches of science were rapidly subdividing into sub-disciplines as a result of an ever-greater specialisation into the detail of the real world. Specialisation was considered to be the only way of providing the means of a fuller understanding of the ways in which our planet and all its life forms functioned. At first, the subject area embracing the natural world was called natural philosophy, but it soon became apparent that no one subject area could adequately cover the growing field of knowledge that eventually expanded to include geology, pedology, climatology, botany and zoology (Goudie, 2000: 1–12). The function of these distinct subject areas was to collect, in ever-greater detail, the specific knowledge related to their subject area. The individual specialist working within his own specific discipline used an analytical approach, breaking down the subject area into ever smaller and more specialised areas of knowledge. However, it was soon obvious that specialisation could never be an end to itself. Analysis must give way to synthesis because one of the findings that emerged from the analytical approach was that a close dependency existed, not only between the different sub-branches of a science, but also between one science and another (Reid, 1962).

There is little doubt that scientific knowledge has contributed greatly to the advancement of our species, enabling us to become the 'super-beings' we now are. Science has been seen as 'good' for society as it enabled improvements to be made in healthcare, agriculture, nutrition, communication and many other aspects of daily life that affected the general public. As Pepper (1996: 145–6) acknowledged, 'the professional scientist . . . was working for a universal good'. A consequence of learning about the detailed knowledge of our planet and the great natural processes that take place within its sphere and the millions of different species that are supported by those processes, has been the gradual understanding that no part of the planet, whether alive or dead, exists in isolation from any other part. We have learnt that there is a continuous movement of energy and matter between all the components of our planet. These movements have taken place almost since our planet was formed between 4000 and 4500 million years ago (Siever, 1975). For all but the last 250 years of this immense time period, these processes have been almost entirely beyond our influence. However, the Industrial Revolution heralded changes to the natural movement of energy and matter. Scientific knowledge gradually revealed that we have become responsible for altering some of the natural processes. Whether the alterations are sufficient to cause long-term change to our planet is, at this stage, conjecture. An alternative philosophy has emerged that proposes that until we gain a more thorough knowledge of how (or *if*) we have changed the natural environment, we should exercise caution in the way we use new technology (Wynne and Meyer, 1993).

Discovering how the meaning of the term 'environment' has changed is made more difficult because, like the make-up of the environment itself, the term means many different things to different people depending on their 'worldview'. The concept of the worldview in determining our envir-

onmental attitudes is considered in more detail in section 9.8. All that is necessary to know here is that our individual worldviews comprise a set of basic beliefs and values that influence the way in which we interact with one another and also with our surroundings. Few of us consciously pay much attention to our worldview. For most of us it is our instinctive 'character', learnt or inherited during our formative years and may be set by the time we are five years old (Cunningham and Saigo, 2001: 37–43).

Ironically, the reason why we now know far more about the environment of our planet is because of our quest to know more about 'space'. Space exploration and the sophisticated Earth surveillance satellites, provide a wealth of information that has allowed a reinterpretation of much of our previous scientific knowledge of our planet. The Landsat Program operated by NASA has provided the longest running sequence of imagery of the Earth from space. The first Landsat satellite was launched in 1972; the most recent, Landsat 7, was launched on 15 April 1999. The instruments on the Landsat satellites have acquired millions of images and provide a unique resource for research on the ecology and land use of the planet. Information contained in the Landsat website illustrates the way satellite-derived data has allowed a reanalysis of our impact on the natural world (http://geo.arc.nasa.gov/sge/landsat/landsat.html).

The modern use of the term 'environment' differs markedly from the earlier use of the word by ecologists working at the start of the twentieth century. Now, when we use 'environment' it is usually in the journalistic genre, used to summarise a complex group of factors that provide the essential natural resources we need to survive. For a subset of the population the term retains its deeper philosophical meaning similar to that of the original ecological sense in which the 'environment' is recognised as more than a provider of natural resources. The deeper connotation of the word signifies a safe place in which to live our lives and it is this meaning that indicates a clearer understanding of our position as a species within the protective web of the environment.

Until the 1970s human involvement with the environment had mainly been with the exploration of the surface of our planet, to find new land suitable for colonisation and for new resources. During the 1970s, the developed world became obsessed with the American and Soviet space programmes and the environment of our own planet was neglected. However, it was the finite nature of Earth as revealed by space exploration that gradually changed our concept of the planet from one of infinite richness to one of extremely limited resources. The concept of *Spaceship Earth* proposed by Ehrlich (1977) which likened our planet to a space mission with finite food and energy supplies enforced a new realism about the way in which we made use of the environment. We gradually realised that we understood very little about the very environment on which we were dependent. The growing discipline of environmental science attempted to remedy the lack of knowledge, but for some people, a scientific knowledge of the environment was not enough. It was, they argued, our reliance on science and technology that led to the belief that we could do whatever we wished to the environment. It was the 'cornucopian' approach that had caused a spate of environmental problems. The emergence of an environmental lobby in the 1970s resulted in a new consciousness – that humankind must accept the need to understand, protect and manage the environment in a responsible and sustainable way. Ironically, the environmental movement initially comprised young people, students, originating mainly from the west coast of the USA. Established society viewed these people with deep suspicion, believing them to be 'hippies', 'draft dodgers' wishing to escape the Vietnam War and general malcontents intent on destabilising the establishment. Given time, the establishment thought that environmentalists would be subsumed into the prevailing way of life in which consumerism was the dominant factor.

The environmental movement did not, however, disappear. Instead, concern with environmental issues gained support from the general public and the 'environment' gradually became more centrally positioned alongside a range of other problems such as international relations with other nations, the problem of drugs in society, and the availability of healthcare for old people. Only time will confirm whether our concern for the state of the environment is a genuine one and whether policies developed for the management of the environment reflect the needs of all the components and not only the specific needs of humans. It is the responsibility of our elected governments to ensure that we have the necessary environmental management systems in place to take account of the new levels of ethical, moral attitudes displayed by society towards the environment. Managing the environment in the twenty-first century involves far more than simply understanding the science of the environment. These issues will be examined in more detail in Chapters 2 and 3.

1.3 Environmental change

Changes to the physical and biological components of our planet is a natural and constant event. The environment responds to these changes, leading to innumerable new opportunities for evolution and development to occur. Most natural changes are relatively small scale and occur within controlled limits and, consequently, do not normally result in total destruction of habitats and their inhabitants. We need to know what constitutes a 'safe' magnitude of environmental change and what levels of change can be endured by the environment. An acceptable level of change in one location on the planet may be harmful in another and change brought about by anthropogenic effect may be less acceptable than natural change. Simmons (1996) suggested that any change brought about by human action should take place on 'a magnitude and at a rate that can be accommodated by the environment', but unfortunately, there is no absolute base line from which change can be measured and, therefore, we have no means by which to measure the benefits or dangers of environmental change. What ancillary information can we obtain that might help us determine the impact of change on the environment?

By observing the natural patterns of change we find that frequent, repetitive change appears to pose no threat to the environment. For example, the rotation of the Earth on its axis to give the diurnal change between night and day has been utilised by green plants in their respiration cycle into the dark and light phases (Keeton, 1980). Animals, too, have integrated the cycle of day and night into distinctive behavioural patterns. Change associated with the twice-daily tidal cycle poses no problems for life in the inter-tidal zones. The longer seasonal changes with variation in day length, temperature and moisture patterns have also been accommodated into the lifecycle of plants and animals.

The more erratic environmental changes associated with glacial and interglacial cycles that take place on much longer timescales (measured over time periods of many tens of thousands of years) may cause an adverse response with some species becoming extinct, usually due to the extreme cold conditions. However, we cannot be absolutely certain that it is the low temperatures that are responsible for the species response. It is almost certain that over the duration of an entire Ice Age lasting one million years, natural evolutionary processes will influence the fortunes of some species. Mannion (1992a) has compared the magnitude of natural changes measured over millions of years with that of anthropogenic change measured at most over 200,000 years. Could it have been coincidence that the emergence of modern humans began at about the same time as the world climate began to warm up at the end of the last Ice Age? Was this the opportunity that our ancestors had been waiting for? It is possible that we had developed sufficient skills to survive the harsh condition of the Ice Age and when climate conditions improved we were suddenly presented with new opportunities that allowed us to expand our geographical range.

From 200,000 years before present to about 10,000 years before present, changes caused by humans on the environment were relatively small scale. However, Leakey and Lewin (1992) suggested that in some parts of Africa our earliest ancestors had discovered the use of fire as a means of frightening wild animals, making them easier to kill. The use of fire was indiscriminate and much of the damage caused to vegetation was probably due to deliberately started fires getting out of control. People were probably killed by these 'wild fires' as well as animals!

Since those earliest days of human history, the magnitude and speed at which environmental change caused by humans has taken place increased to such an extent that it has overshadowed and obliterated many of the changes that occurred in the past (*New Scientist*, 2000). Now, many of the changes that occur are the result of major achievements in biotechnology. For example, sequencing the entire human genome will help us understand the way in which the human body functions as a biochemical unit. It will allow the diagnosis of disease both before and after birth. Gene therapy will allow babies to be designed, not just choosing between boys and girls, but height, eye colour, athletic prowess and academic ability could be selected. At present most people would consider such decision-making power to be detrimental, even inhuman. In another 50 years, however, attitudes could be very different. If the need to reduce the total world population meant that family size could not exceed one child, it might become acceptable practice that parents would have the right to select a customised set of attributes for their sole offspring.

The decisions we will be forced to take in the future are likely to place immense ethical and moral pressure and responsibility on every individual human being. We will need the best possible scientific information to help us make many of the choices. Many of the decisions will come on us quickly, with little time for debate. Our choices will inevitably have immense consequences for us at both individual and societal levels and if we make an incorrect decision we could find our future is bleak.

As the use of biotechnology becomes greater, it is likely that natural evolutionary processes will play a lesser role in determining the composition of species type and number. In our attempts to manage ourselves and the environment we will resort to a greater use of scientific and technical management. Moral and ethical assessment of the newly found techniques on the individual must be carried out. Already, bodies such as the US National Human Genome Research Institute gives financial support to the Ethical, Legal and Social Implications Program (*New Scientist*, 1999) to examine the implications of radical new scientific discoveries and their implication for society.

The remainder of this chapter examines the different components of the environment. Each component will be examined individually before finally reassembling them into the total environment we recognise around us at the present time.

1.4 The social environment

The emergence of a new and deeper appreciation of environmental matters since the 1970s has sometimes been associated with a change in the way in which humans interact with the environment. Cooper (1992) has been critical of the shallow acceptance of 'green ideas' by the general public and also by some politicians and suggests that real progress in achieving a human society more in balance with its environment will only come about if we appreciate our true place in the biosphere. But where is that 'place'?

A 'deep ecology' viewpoint would insist that *Homo sapiens* is merely another animal, with all the basic biological requirements of a living organism and as such, must be grouped with other animals. There can be no argument with this fundamental viewpoint. However, our species differs from all others in that we have evolved a consciousness associated with an enlarged brain that has allowed us to develop skills, cultures, religions and feelings, that, as far as we can tell, are unique to our species. The sociologist would argue that the unique features developed by *Homo sapiens* separates us from other animals and ensures that we have a different 'place' in the biosphere from other species. This viewpoint is also undisputed. The relevant question to ask would seem to be how far from the 'deep ecology' viewpoint can we move towards accepting humankind as an omnipotent power before we encounter real problems concerning our well-being? At present, we still require our food to be supplied from an agricultural system that is basically very similar to the natural food chain where crops (green plants) undergo photosynthesis to manufacture food which is subsequently transferred either directly to us or via farm animals to our food supply. Environmental problems seem to be created when we intensify our use of natural systems. For example,

the release of too much pollution into the atmosphere causes a pollution hazard (see Chapter 4), an excess consumption of fish stocks depletes fishing grounds (see Chapter 6) and poor agricultural systems cause soil erosion (see Chapter 7).

One of the unique human attributes mentioned earlier was the acquisition of a conscience, of an obligation to behave according to the laws of society (Des Jardins, 1993). The development of environmental issues and arguments has made it necessary to extend the human conscience to include environmental factors. As a result of this newly found environmental consciousness issues such as our 'right' to deforest tropical lands, to exterminate plant and animal species, to overconsume natural resources have come to the forefront of the environmental argument (Rolston, 1988: Chapter 3). Again, we encounter a dilemma not, however, over the basic correctness of killing animals or consuming natural resources. Most people accept that we must control the pests and predators that prey on humans and we must consume resources in order to survive. The question that must be asked is 'Do we need to kill *all* pests and predators?' and 'Can we consume *all* resources in the belief that new resources will be found in the future?' If the problem is that of insect-borne malaria, then the answer to the question is 'yes', we must totally eliminate the vector in order to totally prevent the occurrence of the disease. If, however, the problem is that of unpleasant insect bites typical of the northern taiga landscape during the short summer weeks then the answer would probably be 'no', because despite the considerable irritation caused by the biting insects their impact is not life threatening.

However, before a decision to eliminate the malaria-carrying mosquito can be put into effect there has to be agreement that such a decision is the 'correct' one under all the known circumstances. This involves interaction between scientists, medics, environmentalists, politicians, companies that make the insecticides that kill the mosquitoes and the general public (who in this case constitute the 'victims'). In the example used here, the elimination of malaria, the issues are unusually straightforward. More common-place is the situation where no absolutely clear case can be made for or against a particular course of action. In these cases we need to be able to take decisions that can be altered or 'fine tuned' at a later date if more scientific information becomes available or if public opinion changes. Combining our ecological know-how with our ability to make sound ethical judgement on behalf of society has proved to be one of the most difficult tasks confronting humankind (Cooper and Carling, 1996). The next section of this chapter examines three of the main areas in which environmental management decisions have proved most difficult.

1.4.1 Population pressure

All living organisms, plants and animals, live in an interconnected world. Some species have fewer lines of connection than others. The Albatross and the Coelacanth, for example, are solitary creatures for much of their life history, but at certain stages in their life history it becomes necessary to meet others of their own species if for no other reason than for the purpose of reproduction and the continuance of the species. All species form part of a feeding chain, in which one species is a food source for at least one other species. Only one species, humankind, has been able to take a small step out of this trophic relationship, giving us a unique position in the environment. Unfortunately, it also presents us with an ethical predicament. Devoid of a specific position in a feeding chain, and with no predator to keep our numbers in check, we have shown an inability to control our own numbers. In 1999 the world population of human beings exceeded six billion individuals (see Figure 1.1). The most up-to-date figures on world population can be obtained from the Zero Population Growth website at http://www.populationconnection.org.

In terms of population dynamics, our species displays the well-known feature of 'biotic potential' first identified by Chapman (1931) where, in the absence of a predator, population size will increase according to the laws of geometric growth (GP). The growth pattern depicted by a geometric rate of increase follows that of an *exponential curve* as shown in Figure 1.2. The consequences of a population increase that follows geometric growth

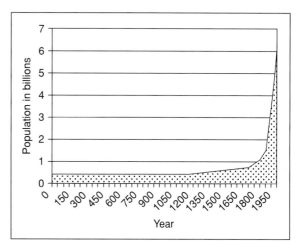

Figure 1.1 Growth in human population over the last 2000 years

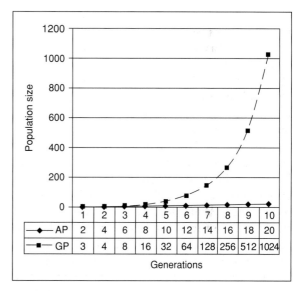

Figure 1.2 Comparison between arithmetic and geometric growth patterns

are clearly discernible when compared with arithmetic growth (AP).

The overall shape of an exponential curve is always the same and produces a line that eventually rises vertically on the graph. In reality, this is unsustainable, as the population would eventually deplete an essential resource such as food supply, oxygen or living space (Kormondy, 1969: Chapter 4). Exponential growth can end in one

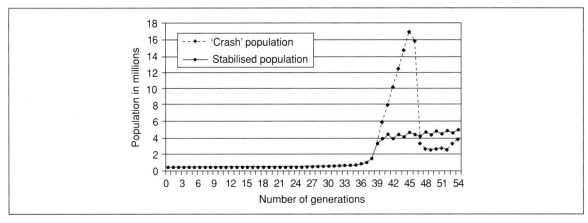

Figure 1.3 Comparison between a population growth curve that experiences exponential growth followed by a population 'crash' and a population that stabilises its size around a sustainable level

of two ways. In the most catastrophic cases, the population shows no ability to adjust to a sustainable level of numbers. The population continues to grow and eventually overexploits its resource base at which point an absolute scarcity of one critical resource occurs, causing a major reduction in the population size. This event is termed a 'population crash' and may reduce the population to as little as 5 or 10 per cent of its maximum population size (see Figure 1.3). Population crashes are highly undesirable events in the life history of a species. Under certain circumstances it could lead to an individual species becoming extinct. Most population crashes do not end in extinction. Instead, the population numbers recover, sometimes undergoing another phase of exponential growth that again is destined to end in a crash. More careful study of situations in which a species undergoes a succession of rapid growth periods followed by a population crash may reveal that the species was experiencing influences from outside its immediate environment. For example, species that follow an exponential growth curve are often those that have been introduced to an area (the so-called *exotic* species) and which have no local predators to help control their number. Alternatively, a native (or *endemic*) species may have had its predators removed and is no longer under the control of natural population dynamics.

Exponential growth does not always end in disaster. Many species show a population growth curve that, in its early stages, appears to follow a geometric growth sequence, but at some point in its development, the growth rate begins to decline and eventually becomes stable. This pattern is also shown in Figure 1.3 as the stabilised population curve. It is tempting to suggest that species that show an ability to stabilise its own number have, in some way, been able to calculate a 'sustainable' population size that can be supported by the environment. In reality, the population has probably attained its balance not from an internal population control but from external resistances received from the environment that limits the further growth of population. For example, within a finite living space, an increase in population size results in a decrease in the territorial space available for the population. The inability to find sufficient food and living space may begin to exert a positive feedback and result in a restriction in breeding rate. Overcrowding can lead to outbreak of disease while losses due to predation or competition may exceed the capacity to renew the population. Population numbers become stable because of a progressive increase in *environmental resistance* that restricts a further increase in numbers of the species (see Figure 1.4). From a human perspective, external limits on population growth may be viewed as disadvantageous, but for all other species, preventing

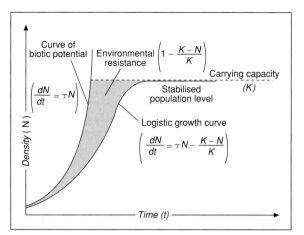

Figure 1.4 Comparison of two population models: a population growth curve that follows an exponential curve and a population that becomes stabilised due to environmental resistance

geometric growth from taking population numbers into an overshoot situation and the possibility of a population crash is highly advantageous.

The theories of population growth have been developed for simple non-human species and it would be incorrect to apply such simple models to even the most ancient of human communities. Considerable research on the statistical modelling of population growth has taken place, for example at the University of Minnesota. (Reports on this work can be found at the following website: http://www.geom.umn.edu/education/calc-init/population/.) Human culture which cares and protects the young, old and infirm, makes the application of simple population models impossible. Instead, we are faced with a far greater complexity in which the social structure has become modified. For example, organised agriculture and gainful employment leading to wealth generation enables a social support system to be placed in position to safeguard the population when hardship strikes. But, perhaps more than any other factor, the advent of modern medicine has resulted in infant mortality falling from 200 per 1000 live births to as few as four per 1000 live births (in Sweden and Japan) and life expectancy extending from 45 to 80 years (Jones and Hollier, 1997: Chapter 5; Lutz *et al.*, 2001). These changes have occurred in less than 100 years of our history and have given the

human population curve the appearance of an exponential curve.

1.4.2 Political issues

A full examination of the way in which concern over the environment has become incorporated into the political thinking of most of the world's major nations policy is made in Chapter 3 of this book. This section identifies the role that environmental pressures now place on society in general and, in turn, on our elected politicians.

The manner and speed in which the politicisation of environmental issues took place towards the end of the twentieth century will probably be ranked by future writers of human history alongside other great events that changed the direction of human society such as the Age of Enlightenment and the Agricultural Revolution. In the space of a few decades, political viewpoints that had prevailed throughout much of the industrial era were compelled to take account of the findings of environmental scientists and also of the somewhat more reactionary ideas that became prevalent within the general public. The popular press, and especially television, were responsible for making the findings of scientific research available to the public in a form that had been previously impossible. In some cases there has been an oversimplification of environmental problems leading to the wrong conclusion being transferred to the general public. For example, the public perception of an ozone 'hole' in the atmosphere over Antarctica is incorrect but is seemingly impossible to dispel, as is the belief that by eating genetically modified plants we will somehow poison ourselves!

The need for greater political control on environmental matters has come about for a number of reasons. First was the rise of environmental pressure groups or lobby groups. Their task was to identify elected politicians who were sympathetic to a specific cause, to inform and, in some cases, exert pressure on those persons to press for political recognition of the pressure group cause. The ultimate aim was to achieve a change in legislation so that, for example, an overused landscape or a species threatened with extinction would be given protection. Where members of a pressure group were highly organised or whenever progress in bringing about legislative change through the democratic system was very slow, pressure groups sometimes attempted to form their own political party. At times of election it became the objective to remove votes from established parties and cause political destabilisation. The pressure group party would never gain overall power, but hoped to gain a significant proportion of the vote to influence the elected party. Numerous examples of environmental political groupings exist: in California, the Big Green Party; in Britain, the Ecology Party (later renamed the Green Party) scored small local successes.

However, the main reason for a change of direction among the main political parties was the occurrence of major environmental 'scares' which, if left unattended, would destabilise society. It is sometimes forgotten that it is the prime task of politicians to create a structured, organised and stable society and legislation may be required to prevent labour unrest, shortage of resources or an environmental deterioration that could lead, for example, to unsafe living conditions. In the 1980s and 1990s, a series of 'scare stories' emerged about fish in the North American Great Lakes that could no longer reproduce or seals in the North Sea dying from a mysterious cause. In both cases, high levels of pesticides were thought to have been responsible for the problem. The scientific process normally followed would be to identify the problem; research its causes; find a solution; implement the solution. In a few cases this procedure was overtaken by scare-mongering publicity.

It was the occurrence of much larger scale problems that caused a change in political attitudes to the environment. The consequences to aquatic life of the acidification of the atmosphere by industrial pollutants, and to the defoliation of trees also from 'acid rain' led to legislation that set internationally accepted standards for the type, amount and concentration of pollutants that could be released into the atmosphere. Led by the United Nations, the majority of world nations and their political leaders participated in conferences that were intended to set new treaties, agreements and understandings whereby international standards of behaviour to the environment could be established. UN-organised conferences have taken place in 1972 in Stockholm (United Nations, 1973), in 1992 in Rio de Janeiro (the UN Conference on Environment and Development) (United Nations, 1992, 1993) and in June 2002 in Johannesburg, a 'World Summit on Sustainable Development'. A United Nations website dedicated to the Johannesburg conference can be found at http://www.un.org/events/wssd and this site has links to both 1972 and 1992 conference material.

While a start has been made in terms of legislating for an environment that is both safe and diverse in its range of habitats and life forms, much remains to be done. Not least, is the development of a system of monitoring to ensure that all signatories to international treaties actually fulfil their obligations. There has been strong criticism from environmental organisations that some governments are reneging on international agreements while governments who have fully implemented agreements try to ensure that other non-complying bodies play their part (Federal Ministry of the Environment, 1997). This problem is sometimes caused by the enthusiasm and persuasive powers of the environmental groups themselves. Their success in cajoling reticent politicians to sign agreements that make unrealistic demands on a nation often leads later to disappointment and disagreement. Whereas the objectives of environmental groups may be the creation of an ecologically sustainable environment, politicians are probably more concerned with establishing a 'safe' and economically sustainable environment.

Establishing an environment that provides a quality of life commensurate with twenty-first-century expectations, involves the provision of sufficient natural resources to sustain industrial growth and also sustain a rich and diverse range of plant and animal habitats. Achieving these aims presents a major challenge for politicians, scientists, industrialists and economists. In addition, we have a huge legacy of previous misuse of the environment and its resources. The cost of cleaning up and repairing past environmental damage is beyond calculation because many of the previous mistakes have yet to emerge in full. The greatest of these unknowns is that of global warming. Our profligate use of fossil fuels has been made with little understanding of the dangers caused by the accumulation of excess greenhouse gases (mainly carbon dioxide) in the atmosphere. We now know that the atmosphere will be about 4.5°C *warmer* by the end of the present century than it was in 2000. Politicians have responded to this threat by agreeing, among other things, to a reduction in CO_2 output by about 12 per cent by 2020 compared with the year 2000 (Department of the Environment, 1998). The figure of 12 per cent is a compromise between what can be achieved at an affordable cost and what scientists thought was the level of cut-back necessary to curb global warming. New figures suggest that a 12 per cent reduction will be far too little and a figure of 60 per cent reduction is now being spoken of among atmospheric scientists. The repercussions to society of such massive, but essential, cut-backs in greenhouse gases will pose an incredible strain on our politicians to construct an internationally agreed legislative structure that is affordable, sustainable and can result in a 60 per cent reduction in greenhouse gases. Fortunately, politicians will be able to call on radically new science and technology to help meet the new targets.

1.5 An environment suitable for life

All living organisms are compelled to spend their lives in a thin zone that surrounds the surface of the Earth where conditions suitable to support life can be found. This zone is called the biosphere. Based on current levels of knowledge, scientists think our planet is unique within our solar system in that it alone possesses a biosphere and, as such, is the only planet that can support life forms we would recognise. The biosphere, therefore, is of fundamental importance to our planet. Despite its importance, the biosphere does not possess a concrete shape or form. Borrowing terminology from the world of computer games, the biosphere is a 'virtual' object! If we look out of a window in our home, depending on the precise conditions, we will see sky, clouds, trees, grass, the ground, buildings and perhaps some other people and other animals. During daylight hours, our view will be visible because of the presence of sunlight. What we can

see when we look out of any window are individual components that, together, form a small fragment of the biosphere.

The biosphere comprises three main ingredients: part of the lower layer of the atmosphere (providing a suitable composition of gases necessary for respiration); the upper layer of the hydrosphere (mainly the oceans but also including fresh water sources such as rivers and lakes) and, third, a very shallow layer of solid rock surface, the lithosphere. The biosphere comprises an amalgamation of a very small portion of each of the great spheres (see Figure 1.5), each of which would be unable to support life on its own, but when united with the other parts provides all the necessary components for life on Earth.

Two quite different life zones exist – the aquatic life zone and the terrestrial life zone. Within each group the requirements for life are very broadly similar for most species. For an individual organism to complete its lifecycle, the biosphere must provide a safe living space from which it can obtain air to breathe, food and water and a place to dispose of its wastes. The biosphere is the space in which all species live, reproduce and eventually die. In respect of the basic demands we make of the biosphere, humankind is no different from other plants and animals. However, our species is

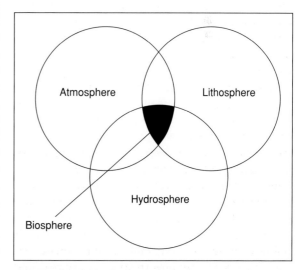

Figure 1.5 Location of the biosphere at the intersection of the atmosphere, lithosphere and hydrosphere

unique in that it uses resources in a materialistic way to accumulate an elaborate infrastructure of implements which are used to achieve further benefit for ourselves. It is true that other species, such as birds, sometimes build elaborate 'homes' but these are semi-permanent structures made from materials that can be collected from the biosphere. Humans are different in that they process raw materials into secondary products that are usually more durable in nature and, in the process, acquire a value. The use made of natural resources has varied over the millennia depending on the technological skills, the affluence level and the political maturity of society using the resources. Our potential 'pool' of natural resources has been confined to quite a narrow zone mainly on the solid crust of the Earth. We have literally been scraping the surface of the solid crust, restricted by technical inability to gain access to resources beneath about two kilometres of the surface. Even so, the mineral resources that are available to us have been sufficient to allow the industrial era of humankind to flourish and to build great cities (as shown in Colour Plate A). Until comparatively recently, we assumed that the resource base was inexhaustible and that we could continue to remove those materials necessary for supporting the unique human lifestyle. From the early 1950s the first signs began to emerge that natural resource exhaustion was taking place, indicating that our exploitation of the biosphere was not an infinite routeway to success. Vast iron ore reserves located to the west of Lake Superior in North America had been considered 'inexhaustible' in the early 1930s yet as a result of demands made during the Second World War had become exhausted 20 years later. In retrospect, the 1950s was an era of environmental ignorance. New deposits of iron ore in other countries were opened up and transported to the consumer nations of the northern hemisphere. Ideas of conservation of resources, reuse of scrap material or measures for improving the efficiency of the iron and steel industry were all in their infancy. Gradually, the concept emerged that almost all natural resources were 'finite'. For mineral resources, the total volume to which we have access at a point in time is more or less fixed. If the rate of consumption is greater than the rate of discovery at which new resources can be brought to a usable state then resource exhaustion will occur. Organic resources differ in that, in theory, they are infinitely renewable, but only if the quantity of resource harvested is less than or equal to the natural rate of renewal. An ever-increasing demand has meant that organic resources have also become depleted in a similar way to mineral resources.

It is tempting to draw a parallel between the ever-increasing use of resources by humans and the exponential curve of population growth shown in Figure 1.2 but that analogy is generally invalid because of three factors. First, prospecting for new resources has generally more than kept up with demand; second, improved efficiencies in the methods of processing and using resources has greatly increased the economic life of a resource, and third, new technology has resulted in uses being found for previously worthless materials. The development of alternative resources in place of traditional ones has opened up a choice of resource use, in some cases removing the threat of shortages. During the twentieth century, radical improvements were made to the way in which we used technology. It can be assumed that the twenty-first century will see this trend accelerate further especially as the opportunities brought about by space exploration for discovering new resources and for new conditions in which resources can be utilised increase.

Although new technology will inevitably make possible the exploitation of new reserves in the future, our immediate resource base remains fixed to that which has already been discovered and is accessible with current technology. The bulk of the resources needed to support our growing population, such as soil, water, minerals, atmospheric gases and non-human biological components (plants and animals), remain located in the three great physical resources reservoirs, namely the solid land surface, the waters comprising the oceans, fresh water and atmospheric water vapour and, third, the atmospheric gases.

1.5.1 The lithosphere

The lithosphere comprises the solid geology materials that form the Earth's crust, a layer on average about 16 kilometres thick. Scattered on the surface of the crust is a discontinuous layer of superficial deposits comprising mainly wind-blown sands, water-deposited sands and gravel and glacially deposited materials, for example boulder clay, outwash sands and gravel. The superficial deposits have been mainly involved in the formation of soils that, in turn, have allowed the development of natural vegetation and agriculture.

The solid geology, or the so-called hard rock deposits, contributes two separate resource types. The most basic resource requirement made of the crust is that it must provide a surface on which all land-based animals and plants reside. Rocks comprise the most significant of the 'substrata' on which terrestrial life can exist. It provides the location for a safe home area on which the life history of plants, animals and humans can take place. The main land-based substrata are soils and bare rock. Not all the geological materials provide suitable substrata, some types being resistant to the natural weathering processes, such as igneous materials. Other rock types may be hazardous to some life forms because of their chemical composition. The ancient pre-Cambrian rocks sometimes have an excess or deficiency of trace elements such as molybdenum or cadmium and prevent the healthy growth of plants and animals that feed on the plants. Such examples are, however, rarities. Most geological substrata can support life. For this purpose they must possess the necessary macro- and micronutrients shown in Table 1.1.

The value attributed by humans to the lithosphere has been less to do with the provision of a substratum and far more to its ability to provide mineral resources for economic use. By providing resources it has enabled successive generations of humans to cultivate the weathered surface layer (the soil) and to quarry and mine the solid geological deposits that lie below ground. The crust provides all fossil fuels (lignite, coal, oil and natural gas). It also provides all our minerals used in industrial processing, our building materials and radioactive materials (uranium) for use in civilian and military nuclear energy and weaponry.

1.5.2 The atmosphere

The uniqueness of our planet is due largely to the existence of the atmosphere, a collection of gases and water vapour that both protect the planet from harmful short wavelength solar energy and also provide a partial barrier to incoming space dust. Along with water, discussed in the next section, the atmosphere provides one of the two fundamental 'media' that enables life to exist on our planet. All organisms have evolved to respire (breathe) in a medium either of air or water. Once selected, it is virtually impossible for an organism to change its medium. As humans, we are born with lungs and an associated blood circulatory system that pre-designates us, as all other land animals, as 'air breathers'. In view of the fact that we have no alternative medium it is surprising, therefore, that we have mistreated the air we breathe by polluting it with particulate and gaseous material.

The atmosphere surrounding the Earth forms only a very thin layer. If it were possible to rise vertically in a hot air balloon through the atmosphere we would travel to the extremity of the atmosphere in about two hours, a distance of approximately 90 kilometres. Contained within the atmosphere are all the essential gases needed by plants and animals to breathe, a varying amount of water vapour (as low as 0.2 per cent over hot deserts to as high as 1.8 per cent in equatorial regions (Strahler and Strahler, 1974)), all the atmospheric weather systems that are so influential to life on the surface of the Earth, all the heat energy

---

Table 1.1 Essential macro- and micronutrients necessary for plant growth

Macronutrient	Micronutrient
Nitrogen	Iron
Phosphorus	Manganese
Potassium	Boron
Calcium	Chlorine
Magnesium	Copper
Sulphur	Zinc
	Molybdenum
	Vanadium*
	Sodium†

* relevance uncertain
† beneficial for some plants

transfer systems that allows the planet to regulate its temperature, as well as very specific chemical and ionic structures that enable modern telecommunications systems to work.

Because we consider the atmosphere, as well as the closely related oceans and hydrological cycle to exist in a state of communal ownership, with no apparent 'cost' involved in their use, we have treated these natural resources with little respect for their long-term security. It is probable that because of the vastness of the atmosphere, along with the oceans, we have treated them in a manner that is quite different from that of other natural resources. For most people, the atmosphere is not a natural 'resource' in the same way as fresh water, crude oil or tropical forests. Because of the combination of its immensity and the daily and seasonal variations in weather and climate, we have considered the atmosphere a limitless and self-perpetuating resource. In addition, because many of the processes at work in the atmosphere are invisible we remain genuinely ignorant of how and when we have been responsible for causing damage. Chapter 4 examines the major environmental problems associated with the misuse of the atmosphere and also discusses ways of solving the current problems.

1.5.3 The hydrosphere

About 71 per cent (361 million square kilometres) of the surface of our planet is covered by water and, when the great depth of the oceans is taken into account, the potential living space is some 96 times greater than that of the land surface. Science fiction writers often portray the oceans as the future home for humankind living in vast submarine cities. Such scenarios conveniently overlook the technical problems of living in the oceans. First, the salty oceans are highly corrosive to most of our building materials. The salinity of the seas would quickly corrode metal and cause concrete structures to decay. An even greater problem is that of coping with the great pressure that exists even at shallow depths. At the surface of the ocean we can imagine a column of atmosphere pressing down onto the surface. Calculations have shown that the 'average' pressure exerted by the atmo-

sphere on every square centimetre of surface area at sea level is equivalent to a column of mercury one square centimetre in cross-section and 760 millimetres high. This figure is said to equal one 'atmosphere' of pressure and it is this pressure that prevents all living things from floating off into space. Moving down into the hydrosphere leads to an increase in pressure by one atmosphere for every 10 metres descent so that, at only 100 metres depth, water pressure is equivalent to ten times that at the surface and is sufficient to compress all life forms with free air spaces (lungs, stomach cavities), as well as creating many technical problems for human-made structures. At the bottom of the Mariana Trench (10,500 metres below sea level), the pressure force is equivalent to about 1 tonne per square centimetre.

The problems of pressure and salinity present few difficulties for the life forms that have evolved to live in water. Indeed, the oceans were almost certainly the location of the earliest life forms on the planet and still play a central part in creating conditions that are suitable for all life on Earth. Of the solar energy that reaches the Earth's surface, some 90 per cent is absorbed by the oceans and enables the average temperature of our planet to fluctuate around an average of 14.5°C. The solar energy input is also responsible for causing evaporation to occur from the surface layer of the oceans and the resulting water vapour accumulates in the atmosphere, leading to the formation of water-laden clouds. This process forms an integral part of the hydrological cycle responsible for transferring water from the oceans to the atmosphere and eventually to the land as precipitation. In so doing the salinity is removed and atmospheric water and precipitation becomes fresh water suitable for human consumption.

Fresh water is an essential component for healthy life and is a constant requirement of all land-based plants and animals. Plants comprise 85 per cent of the body weight by water and animals about 75 per cent. Even though there is a superabundance of water on the planet we have encountered an increasingly serious water shortage at both continental and local levels. Chapter 5 examines the crisis of water availability in greater detail.

1.6 Biological components

Viewed from a human perspective it is easy for us to assume that we are *the* most important organism that inhabits the biosphere. It is because of our prodigious brain power and the associated physical dexterity of our appendages, especially our opposable finger and thumb, that we have been able to manipulate other species and make use of mechanical tools to modify our immediate surroundings. The considerable achievements of the modern industrial era, such as space travel, performing organ transplants and of building cities of immense size and complexity, have certainly set us apart from the far more modest capabilities of our nearest relatives, the apes. Despite these achievements, *Homo sapiens* remains very much a part of the biological kingdom. Our basic requirements as an animal species are virtually identical to those of more simplistic animals. All plants and animals require a set of basic requirements in order to live. These are:

- a food supply to satisfy an energy need to support cell maintenance and growth
- a clean atmosphere for respiration (breathing) purposes
- a means of dispersing body wastes (from breathing and from digestion)
- a means of reproducing in order to sustain the species
- a physical living space in which to perform these four functions
- a period of time in which to complete the life history.

As our population size has increased, so demand for the first five of the basic life requirements has increased. In addition, demand has been further modified by the increasing age to which many of us now live. In order to meet our increased requirements we have made use of technological skills to raise the yield level of resources from the biosphere. Some resources have been diverted from our competitors resulting in some species succumbing, as they are unable to gain access to sufficient resources. For the most part, the transfer of resources to assist the development of our own species has been a reflexive process, done for the most part with little deliberate attempt to disadvantage other species. The process has been in operation since our earliest ancestors began to diverge from other advanced apes (Leakey and Lewin, 1992).

In spite of what some ecologists believe may be a loss of thousands of species a year, the biosphere remains, at first sight, to be a healthy functioning entity. Despite massive extinction of species numbers by both natural and anthropogenic causes, the diversity of life remains so abundant and complex that it defies our full comprehension and this despite the extinction, by natural processes, of 99.9 per cent of all species that have ever lived (Thorne-Miller, 1999). Chapter 8 examines in more detail the reasons why it is necessary to retain maximum biodiversity. Precisely how many other species cohabit the biosphere at the present time is hard to tell. About two million species have currently been identified (56 per cent of them insects). Some biologists, for example Pimm *et al.* (1995), have claimed that we have identified and named only a small proportion of the total number and conservative estimates place the total number of species at between 10 and 13 million. Raven (1995) has calculated that as many as one-quarter of all current life forms may become extinct by 2025, due mainly to human competition. Based on the twentieth-century rate of extinctions as well as new pressures created by an ever-increasing human population, Raven has suggested that 5 per cent of all species will become extinct every ten years. In numerical terms, about 50,000 species *a year* will be lost of which only about 7000 will have been scientifically named and recorded. These figures are based mainly on land-based extinctions. Our poor knowledge of marine habitats (discussed more fully in section 6.1) implies that actual extinction rates may be far greater than those calculated by Raven.

Gradually, as we come to understand more about the diversity of life on Earth and the inter-relationships that exist between species, it becomes apparent that even the most insignificant have a role to play somewhere in the vast milieu of the environment. As a species, we make use of only a small number of the inhabitants of Earth but this does not imply that all the other species have no 'value'. All species play a specific role in maintaining the general well-being of the biosphere.

1.6.1 The kingdoms of life

A characteristic of our species has been our attempt to classify the complexity of the natural world in the belief that we will be better able to understand it. For example, chemists classified all the chemical properties of elements into a periodic table and physicists classified elements according to their atomic structure. In biology, Carolus Linnaeus (1707–1778) made the first successful classification of all known plant types and, later, a similar scheme was extended to animals. This classification system remains virtually intact to this day. Linnaeus's system classified organisms into a series of successively more specific categories, with each stage of the classification being given a specific name. This sequence is shown in Table 1.2. The Linnean classification provides precision in identifying and naming species. No two species can have the same name. The names are always Latin and the genus and species names are customarily printed in italics (or underlined if handwritten). The same Latin names are used throughout the world, thus allowing scientists to know exactly the species to which they are referring. Many species may also have a common name but these names are often used in a very local sense. Using a common name can lead to considerable con-fusion as shown by Keeton (1980) who states that the edible garden fruit plant known to inhabitants of Europe or North America as 'raspberry' can actually apply to more than 100 different plants in other regions of the world.

Linnaeus worked about a century before the work of Darwin revolutionised the way in which biologists recognised that plants and animals had evolved from pre-existing forms by means of natural selection. Linnaeus had based his classification on morphological details (i.e. external visual differences and similarities) and it was fortunate that morphological characteristics are products of evolution and can therefore be used as surrogates for more modern phylogenic relationships.

While Linnaeus was concerned with classifying living organisms into two main kingdoms, those of plants and animals, discoveries made possible by electron microscopy in the twentieth century allowed further kingdoms to be discovered. The new kingdoms were populated by microscopic organisms which often showed conflicting characteristics, some plant-like features being linked with animal characteristics and vice versa. In particular, the *Fungi* were neither plants nor animals and were allocated their own kingdom. Most modern-day classifications recognise four or five kingdoms. Uncertainty exists over the position of the so-called *Protista*, some classifications suggesting they are a unique kingdom, while others propose that they can be subdivided and placed within the three major kingdoms of *Plantae*, *Fungi* and *Animalia*. Viruses are also difficult to position in this sequence because they differ in so many ways from the other living organisms and yet have some similarities with the simplest of the kingdoms, that of the *Monera*. Because of the scientific uncertainty that surrounds the classification of viruses, they have been omitted from further consideration in this section.

1.6.2 Monera

The *Monera* form the most primitive group of organisms and comprise three divisions: the bacteria (responsible for disease in humans, domesticated animals and cultivated plants), the blue-green algae and the *Prochlorophyta* (Keeton, 1980). Contrary to popular belief, far more

---

Table 1.2 The classification of living things on the basis of phylogenic relationships. Each category (taxon) in this hierarchy is a collective unit containing one or more groups from the next lower level in the hierarchy. The example relates to humankind

Rank	Example
Kingdom	Animalia
Phylum or division	Chordata
Class	Mammalia
Order	Primates
Family	Hominoidae
Genus	Homo
Species	sapiens

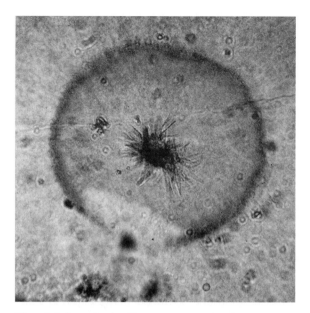

Plate 1.1 Example of a *Monera* organism. Fossilised blue-green algae from the Gunflint deposits, Ontario

very wide range of environments, some of which are exceedingly hostile to other forms of life. They can survive extremes of temperature and gaseous concentrations. They require an almost constant supply of moisture to ensure their survival.

1.6.3 *Protista*

A major question mark still hangs over where this group best fits within the modern classification of organisms. This revolves around whether a separate kingdom is required for organisms that show conflicting evidence of belonging to either, or both, plant or animal kingdoms. *Protista* display a lack of specialisation that suggests they may be a very primitive form of life. They have a similar appearance to the *Monera*, being single-cell structures and show little evidence of specialisation beyond the development of a cell nucleus (see Plate 1.2). Some *Protista* show animal-like characteristics such as the development of a *flagellum* (tail) that allows

beneficial bacteria exist than harmful ones. The group of bacteria known as 'decomposers' perform the essential task of breaking down the dead remains and waste products of other living organisms. Without their constant activity our living planet would soon grind to a halt, smothered beneath a growing layer of waste and deprived of the essential resources to support new growth. The blue-green algae (or *Cyanophyta*) contain chlorophyll and are able to photosynthesise, a highly complex chemical process and one that would not be expected to be associated with such simple and unspecialised organisms. Fossil evidence suggests that *Monera* were among the first life forms to evolve and today, have evolved into a very numerous and essential component of the biota. In body structure they are exclusively single-cell organisms, showing no differentiation into male and female types (see Plate 1.1). Each cell contains a nucleus and reproduction is achieved by asexual splitting of the nucleus and cell. Sometimes members of the *Monera* arrange themselves into colonies, but each cell remains an individual with no biological connection to its neighbouring cell. *Monera* display no powers of deliberate movement, being transported by water or air currents. This group inhabits a

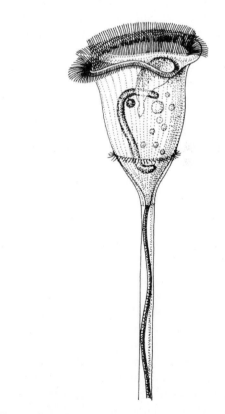

Plate 1.2 *Protista* organism. Unicellular *Vorticella*

movement to occur, although this is unlikely to be conscious movement and instead will be due to stimulus by chemical, temperature or illumination attraction. Others are more fungi like, some are plant like, while yet others display features of all three groups but at different stages of their life history. The problem of placing the *Protista* group into a specific category reveals the artificial nature of classification and suggests that the process is merely an aid to our understanding of a complex real world.

Protista that resemble animals are termed *Protozoans*, each of which can be considered the equivalent of a complete animal and not a single cell. There are five groups of *Protozoa* and while they lack differentiation into tissues and organs they can show considerable functional differentiation. Instead of organs they have functionally equivalent *organelles* that allow a highly complex anatomy to exist within a single cell. *Protozoa* are usually solitary organisms and can be found in most water-based environments. Although resembling animals, some *Protozoa* possess the plant-like feature of containing chlorophyll and are thus capable of photosynthesis.

A separate group of *Protista* are the fungi-like organisms and many of these have developed parasitic or saprophytic feeding habits on other algae, on plants or animals. Also in this group are the curious slime moulds that are animal like for part of their life history and plant like at other times. Slime moulds are usually found living in damp soil, rotting wood, leaf mould or other decaying material. As their name suggests, they look like masses of gelatinous slime, usually white in colour but sometimes red or yellow. Their structure comprises a large net of fine strands called a *plasmodium* that moves across decaying material, extracting nutrients and decomposing its food supply in the process.

The more plant-like *Protista* are also unicellular and possess chlorophyll but many also have animal-like flagella typified by the *Euglenoids*. They are highly mobile and some lack chlorophyll and are therefore *heterotrophs*, thus adding even more to the confusion of deciding whether they are plants or animals! Other common plant-like *Protista* are the *Dinoflagellates* (yellow-green algae and golden-brown algae) and the *diatoms*. Some species of Dinoflagellates are responsible for the formation of the 'red tides' common in the Adriatic Sea and off the Florida coast and which cause the death of millions of fish (Anderson, 1997).

Protista form the first step in aquatic food webs and those capable of photosynthesis can be classed as primary producers (or *autotrophs*, meaning 'self-feeders'). They are highly efficient in combining water and low-energy basic minerals into manufactured sugars via photosynthesis. Up to 80 per cent of the manufactured energy is available for transfer on to the next stage of the food chain. *Protista* require constantly damp or wet environments. In water, they concentrate in the top one metre of oceans, rivers and lakes where the high levels of sunlight stimulates photosynthesis. Few *Protista* are found below 10 metres depth in water due to a lack of sunlight. Despite the important evolutionary changes that allow photosynthesis to occur, *Protista* remain highly simple organisms and show little evidence of other development.

Wherever environmental conditions are suitable, *Protista* occur in prodigious quantities. Five litres of sea water can contain up to two million diatoms (Keeton, 1980), yet despite their abundance we make relatively little use of them as a human food source even though they comprise a rich source of iodine, potassium, mineral salts and vitamins. When diatoms die, their bodies fall to the bottom of the sea and in previous geological eras have been responsible for contributing to thick deposits of chalk and limestone rocks. Nowadays, the diatom remains are harvested as diatomaceous earth and used in the manufacture of detergents, polishes, deodorising gels and fertilisers. The Dinoflagellates are second in profusion to the diatoms in terms of primary production in food webs. They also have one unexplainable characteristic in that they act as vital symbiants to other invertebrate marine life. Without the presence of a specific Dinoflagellate species most corals would die.

1.6.4 The plant kingdom – *Plantea* or *Metaphyta*

This is a vast group and contains the so-called higher plants and includes all trees, grasses, agricultural crops and our ornamental garden species. This group also includes plants that non-biologists consider 'primitive', in that it includes plants that do not show differentiation into distinct leaves, stems, roots etc. This latter group is called the *Thallophyta* while the more recognisable higher plants are called the *Embryophyta*. All members of this kingdom contain chlorophyll and are therefore green in colour although some are shades of blue-green, brown, red or orange in colour.

The *Thallophyta* comprise three main groups: the green algae, brown algae and red algae, and first appear in the earliest geological period, the Cambrian, at about 550 million years before present. The first of these, the green algae, probably represent the group from which the first true land plants emerged. Most green algae are now found in fresh water although some live in damp areas on land. The *Thallophyta* show many major evolutionary advances over the *Monera* and *Protista*. Thallophytes are multicellular and the plants show differentiation into visibly distinct parts. Most significantly, sexual reproduction occurs although separate male and female organisms do not exist. Instead, each thallophyte plant can produce cells that become differentiated into a male structure that produces sperm and a female structure that contains an egg. Another significant development is that the lifestyle is no longer that of an organism that floats in water and in which transportation is dependent on the chance movements of water currents. Instead, at least part of the lifestyle involves a stationary (fixed) component in which the organism develops a 'holdfast' that anchors the plant to its substratum. The brown algae, which are almost exclusively marine species and comprise the seaweeds, grow to 45 metres in length and are permanently attached to rocks by means of massive holdfasts. The holdfast is not a root, merely a device to anchor the plant to a solid surface.

Members of the brown algae group today occupy inter-tidal habitats and may offer evidence to show how members of the *Thallophyta* gradually evolved to a life away from water. None of the present-day thallophytes can survive in truly terrestrial environments. To break away from an aquatic lifestyle required them to develop the following abilities:

- To hold sufficient water within the plant when the plant body itself was no longer surrounded by water.
- To develop a mechanism to transport water from the point of entry to the plant to the rest of the body and, in return, a process was needed to carry the products of photosynthesis to all parts of the plant body.
- To prevent desiccation of the plant through evaporation.
- To respire from the leaf surface to the surrounding medium of air instead of the more customary medium of water.
- To keep a large plant erect without the supporting buoyancy of water.
- To ensure reproduction could occur between separate male and female organisms that might be separated by many metres or even kilometres.
- For land plants to adapt to many different climates with extremes of temperature, moisture, wind and exposure being commonplace.
- To ensure land plants could obtain the necessary nutrients from soils that could show considerable variation between excessive and restrictive quantities of nutrients.

Taking account of the profusion of land plants we now find on the surface of the Earth, it is clear that the problem of moving from a life in the relative protection of water to one on the exposed land surface was not only achieved but was done in a most successful manner. Undoubtedly, there would have been many unsuccessful attempts by members of the *Protista* group and by the *Thallophyta* to colonise the marginal areas around water bodies. Evidence suggests that it was the green algae (the *Chlorophyta*) of the thallophyte group that eventually made the successful migration onto land. In their new environment, they evolved the necessary changes that allowed a tentative existence on land. A new group of plants evolved to take advantage of the opportunities and these new species formed the *Bryophyta*, comprising liverworts and mosses (see Plate 1.3). The bryophytes

Plate 1.3 Example of *Bryophyta* organisms. Moss-covered trees, temperate rainforest, New Zealand

retained many thallus-like characteristics; they were small and insignificant species living in damp areas alongside streams or on rotting wood. To combat extremes of climate the bryophytes were able to cease growing when conditions became unsuitable, that is they developed the important characteristic known as dormancy. Despite 500 million years during which adaptation to life on land could have been fully achieved, the bryophytes retain many of their ancestral characteristics and remain tied to the presence of water for their success.

The bryophytes cannot be considered true land plants and a further separate line of evolution was required to enable plants to complete the colonisation of the land surface. This was eventually achieved by the group of plants called the *Trachaeophytes* and took place towards the end of the Silurian period, circa 400 million years ago.

1.6.5 The true land plants – vascular plants

The bryophytes discussed in the previous section failed to become truly successful land plants because they lacked many essential adaptations. A new line of plant evolution led to the vascular plants or *Trachaeophytes* and in this group we finally find the necessary specialisation of the body structure into distinctive external parts such as roots, stems, leaves, flowers and seeds. Internally, the cell structure displays differentiation to suit specific tasks such as supportive tissue, transportation tissue and *meristem tissue* from which all other cells are produced. The life history of these plants is controlled by the production within the plant itself of complex chemical growth hormones known as auxins, gibberellins and cytokinin. These substances are produced in response to specific stimuli that occur both externally and internally to the plant. *Trachaeophyte* species are no longer totally dependent on the permanent supply of water as the outer 'skin' of the plant is covered in a protective cuticle that prevents excessive loss of moisture.

Five subdivisions of vascular plants exist, four of which comprise mosses, horsetails and ferns, each of which briefly came to dominate the terrestrial vegetation, only to be succeeded in turn by a later and more successful group. The fifth and final subgroup, called the *Spermopsida*, eventually emerged in the late Devonian and Carboniferous era some 350 million years ago and now dominate the world vegetation. This subgroup includes the true seed-bearing plants that have become so important a commercial resource for humans.

Although modern-day *Spermopsida* are highly adapted and efficient land plants the subgroup has undergone very considerable fine tuning over the millennia. There are two classes of *Spermopsida*,

the Gymnosperms, that are further divisible into five subclasses, and the Angiosperms. The Gymnosperms were the first to evolve (410–400 million years ago) and are less advanced than the Angiosperms especially in the degree of protection given to the seeds. At first, Gymnosperms resembled giant ferns and it was these that dominated the Carboniferous era (345–280 million years) decomposing slowly in conditions that gave rise to the coal measures. Another early member of the Gymnosperms were the cycads, which today have become the palms. The most abundant of the present-day Gymnosperms are conifers, or softwood trees, for example pines, firs and larches (see Plate 1.4). Conifers did not evolve out of the other Gymnosperm subclasses but appear to have evolved from an earlier, more primitive ancestor. They emerged quite late among the Gymnosperm fossils, appearing in the Carboniferous era and dominated the vegetation throughout the Mesozoic

Plate 1.4 Mature Coniferales. Corsican pines on the island of Corsica.

era (225–65 million years before present). They remain an important, but minor, component of the Earth's vegetation. Conifers have soft, resinous wood, leaves that are narrow and needle like and are covered in a thick, protective cuticle. The seeds are formed within very distinctive cones that provide only a modicum of protection for the developing seed.

The final class of land plants, the Angiosperms, appeared for the first time in the early Cretaceous period, about 135 million years before present, and since that time have expanded to dominate the world terrestrial vegetation. In contrast to the conifers, Angiosperms produce flowers that form to set seeds contained within a highly protective 'jacket'. The flowers have often become highly specialised to attract specific insects, ensuring that cross-pollination occurs. Angiosperms have diversified into tens of thousands of species and include woody trees and shrubs, flowering herbs and the grasses (see Plate 1.5). The wood of the Angiosperm is non-resinous and very resistant to decay. It is described as 'hard wood' and in tropical trees becomes highly coloured with a strong aroma, for example rose wood, teak, deal and mahogany.

The great diversity shown by this group has been made possible by its highly dynamic genetic structure. This feature has been capitalised on by modern-day plant geneticists who have bred specific genetic combinations to allow agricultural crops to be 'designed' for almost every possible environmental location on this planet (apart from permanent icefields and the hot, dry desert regions). Without the diversity of Angiosperms to provide a wide variety and abundance of food plants, it is probable that animal life would be less well developed. It is even possible to hypothesise that with a less dynamic evolution of animal types, our own species may never have developed.

1.6.6 *Fungi*

Fungi are quite unlike the other plant types. While they resemble other higher plants in that they possess multicelled bodies they do not possess conventional 'cells' with distinct walls. The structure

Plate 1.5 Example of *Angiosperm* organisms. Large hardwood trees forming part of the tropical rainforest, Ecuador

comprises not cellulose as in other plants, but a chitin-like substance akin to the body structure of insects. The body mass is arranged into a whitish branched network of filaments called *hyphae* and the entire mass is called a *mycelium*. The myclium is usually confined below ground or beneath the bark of trees or rotting vegetation. *Fungi* are parasitic or saprophytic and secrete enzymes onto their food and carry out extra-cellular digestion. Most *Fungi* live on or within plants, both living and dead (see Plate 1.6). They contribute greatly to the process of decomposition and to the recycling of materials. Unfortunately for the human economy, *Fungi* do not differentiate between natural vegetation and agricultural produce and, as a result, destroy millions of tonnes of agricultural crops

Plate 1.6 Example of *Fungi* organism. *Fungi* growing on a rotten tree stump

resulting in countless economic losses and destruction to much of the valuable harvest especially in developing countries where storage of food may be in non-refrigerated buildings. *Fungi* will feed on any organic substance, paper, leather or fabrics and cause immense damage to human property in the form of wet rot and dry rot. Other *Fungi* are parasitic on animals and humans. Many of the common human skin infections are fungal infections (for example, athlete's foot, ringworm and a number of lung infections). Many *Fungi* are pathogenic and responsible for immense loss of life among animals. There are, however, many beneficial *Fungi* and these are vital to the manufacture of bread, cheese and brewing products. Many pharmaceutical products rely on fungal cultures for their synthesis, notably penicillin. Finally, we eat mushrooms and some toadstools, the fruiting bodies of *Fungi*.

1.7 The animal kingdom – the *Animalia* or *Metazoa*

Members of the *Animalia* comprise multicellular organisms that obtain their food supply by consuming other plants or animals. Unlike the life forms discussed in the previous sections, animals are incapable of manufacturing their own food. Instead, animals search out their food, consume it and digest it internally. The fundamentally different feeding habits of animals have led to a series of complex modifications and specialisation of

Plate 1.7 Example of *Animalia* organisms. A Brahman bull and calf, Northern Transvaal, South Africa

body form compared to plants. Not least is the need for animals to move around in order to search out food supplies. This requires a means of movement based on limbs controlled by a cartilage, bone and muscle system and coordinated by a nervous system. A means of detecting food is required and this is achieved through the development of specific senses such as sight, smell, hearing and taste. Once a food source has been located, a means must exist to eat and digest the food, to extract nutrients from it and finally to expel the wastes from the animal. This process has been achieved through the development of a highly specialised digestive system. The body form of animals usually shows a distinct head and tail (front and rear) and clear bilateral symmetry (see Plate 1.7). The *Animalia* display highly developed specialisation of cells to form organs (brain, eyes, kidney, blood, etc.) and entire systems are developed for highly specialised functions (nervous system, respiratory system, circulation system, digestive system). Each area of specialisation contributes to the overall well-being of the individual and is under the control of a complex internal hormonal and nervous system which responds to external stimuli received from the environment.

In order that this diverse and complex set of components can be synchronised into a fully functional form, a powerful 'brain' has been developed and located in the head of the animal. The brain receives information from all the sensory perceptors as well as hormonal and endocrinal sources and, after processing the information, is able to control the animal in a safe and sustainable manner.

The evolutionary origin of animals is unclear, as is the reason why they evolved. A living world that comprises the four kingdoms of *Monera*, *Protista*, *Plantae* and *Fungi* would have been able to achieve a balance in terms of energy and material movements. The development of a fifth kingdom,

Animalia, adds another layer to an already complex biological world and, in the natural world, greater complexity provides greater stability. A more likely explanation as to why animals evolved can be found by examining the concept of energy flow through the natural system. The four non-animal kingdoms are highly efficient in the way in which the Sun's energy is captured and converted via photosynthesis to manufactured food energy. Losses of energy through movement are virtually nil as deliberate or conscious movement in plants is rare. The non-existence of a digestive system, the absence of a blood circulation system and consequently the need to keep warm are other areas of great energy saving. In terms of energy consumption, animals are the exact opposite of the other life kingdoms. Animals consume about 80 per cent of the energy contained in their food simply to fulfil their bodily needs. Consequently, animals constitute a vast energy drain on the remainder of the natural world. Without animals, however, there would be a vast energy surplus that would constitute a potentially dangerous, unstable energy source. The purpose of an animal kingdom may be nothing more than a means to neutralise surplus energy by consuming the energy and diverting it into an additional layer of complexity.

The animal kingdom has diversified into three subkingdoms, 28 phyla, six subphyla and a multitude of classes and subclasses and comprises about one million different species alive on the planet today. Although some animal species are the product of recent evolution, the very first animal fossils have been found in rocks of Cambrian age (550 million years) and thus are of an antiquity that is equal to that of the *Monera* and *Protista* kingdoms. It is reasonable to hypothesise that the origins of the *Animalia* occurred well before 600 million years ago, because a wide variety of fossils, representing most major animal groups (phyla), already appear in the oldest Cambrian rocks. It is not correct, therefore, to think of animals as evolving 'last' in an increasingly complex chain of evolution.

Our own species belongs to the phylum *Chordata* and include a diverse group of 42,000 vertebrates (animals with backbones) half of which are fish. Our designated class is the *Mammalia*, characterised by the possession of hair and mammary glands that secrete milk. This group is also warm blooded. The *Animalia* do not have the capability of manufacturing their own food supply. Instead, they are totally reliant on the primary producers (the *Monera*, *Protista* and *Metaphyta*) and are defined as consumer organisms or heterotrophs (feeding on others).

The diversification of types within this group has allowed all available space on land, in the air and in water to become colonised. It includes the microscopic insects to the largest ever species, the blue whale. The level of sophistication ranges from instinctive and involuntary behaviour as shown, for example, by an earthworm, to the highly intelligent and knowledge-based human societies. In the latter example, we appear to have moved away from the remainder of the *Animalia* in that we have resorted to the use of technology to alter our lifestyle. However, we are committed to remain a member of the *Animalia* in that we retain the same basic life requirements as all other animals.

In one special way our species differs from all the other *Animalia*. Our species has attained its pre-eminent position by exploiting all other life forms. We are unique in that we occupy the apex of the global food chain. We are the only species that does not suffer the insecurity of being preyed on by a higher order organism. The exploitation of the biosphere and its inhabitants by *Homo sapiens* has inadvertently resulted in the extinction of countless thousands (and possibly millions) of other species. In addition, we have destroyed habitats and changed the chemistry of the atmosphere and the oceans. The most substantial changes have occurred through our ignorance of the natural world. The paradox is that this has occurred mainly in the twentieth century when we have achieved our greatest scientific advances such as sending astronauts into space and surgically transplanting organs and tissues to extend our lives and when we can genetically modify other species to help our own survival. It is to be hoped that during the next 100 years of our development we will recognise the need to understand and manage our biosphere with far greater regard to a sustainable, long-term future. If we cannot achieve this then our very survival may well be in jeopardy.

Useful websites for this chapter

Kyoto Climate Agreement
http://unfccc.int/resource/convkp.html

NASA Landsat satellite data
http://geo.arc.nasa.gov/sge/landsat/landsat.html

United Nations, Rio + 10 Conference
http://www.un.org/events/wssd
http://www.un.org/esa/sustdev/index.htm

University of Minnesota, Department of Statistics
http://www.geom.umn.edu/education/calc-init/population/

Zero Population Growth
http://www.populationconnection.org

2 Changing perceptions of the environment

From a galactic perspective, Earth is not terribly impressive. The planet, one of the smaller ones of the solar system, is only 25,000 miles around the equator. . . . Earth is, to be frank, an inconceivably insignificant mote in the universe. But it's all we have.

Ehrlich and Ehrlich, 1987

This chapter examines the way in which our attitudes to the environment have developed, in particular from the 1970s onward. It contrasts the range of different outlooks from the committed 'deep green' viewpoint to the beliefs of the 'cornucopian technocrats'.

2.1 The earliest perception of the environment

Although the existence of hominoid species extends over at least seven million years (Leakey and Lewin, 1992; Aiello and Collard, 2001; Brunet *et al.*, 2002) this is a mere fragment of the time during which life has existed on Earth. Short though this time span may be relative to the 550–600 million years for which life has been present on Earth, it has been sufficient for our species to have made a very clear impression on the environment. For the majority of the seven million years, our species occupied a niche that was little different from that of other large mammals. Our ancestral species roamed the landscape, hunting prey, collecting plant food, fighting among themselves over access to food, a place to sleep and reproductive entitlement.

Even at the earliest stages of development our relationship with the environment was unlikely to have been a static one. Seasonal variations in climate would probably have ensured that early humans followed a nomadic lifestyle, tracking the migration of other animals and following the flowering and fruiting seasons of plants. However, our ancestors displayed one unique characteristic that other apes, from which we evolved, did not possess – inventiveness. This characteristic allowed our species to use the environment in ways that were totally different from all other species. Exactly how or when this unique skill was developed is not known. It may have happened gradually during the evolution of our species or, alternatively, there may have been cataclysmic periods of advance that punctuated hundreds of thousands of years when advancement did not occur. Quite suddenly, the archaeological record presents site evidence to show a highly developed human society with a well-established settled agriculture based on the cultivation of a limited range of crops and some evidence of domesticated animals. This site, near the present-day town of Jarmo in Iraq, has been dated to about 7000 years BC (Braidwood, 1971; Jones and Brown, 2000). Such was the sophistication of this system that it must certainly have been under development for thousands of years prior to 7000 BC. Clearly, the inhabitants of Jarmo had replaced the migratory lifestyle of the hunter–gatherer community with that of a sedentary settled lifestyle.

The implications of the new lifestyle found at Jarmo, both for humans and our planet, were immense. The transient pressures exerted on the environment by migratory hunter–gatherers were replaced by the continuous demands of a settled people. The environment in which they lived was required to provide a continuous supply of resources – food, fuel and clean water. The population was required to develop new social skills of living together, of sharing resources, and of not polluting or exploiting their neighbours' resources. To overcome these immense new challenges, our ancestors were required to develop considerable powers of ingenuity, invention and discovery, which gradually resulted in the development of more technical skills and scientific knowledge that allowed further advances to occur. By means of this interactive, iterative process, human development advanced in leaps and bounds to the present-day situation, which in itself represents a transient stepping-stone to some unknown future state of human development.

Somewhere along the pathway of human development a substantial proportion of the population appears to have become detached from the most fundamental and intimate of links that once existed between humans and their environment. The 'position' of the human animal within the general arrangement of living organisms with that of the physical environment changed from one in which humans were 'inclusive', as shown in Figure 2.1, to one in which they became separated from other animals and operated as an independent

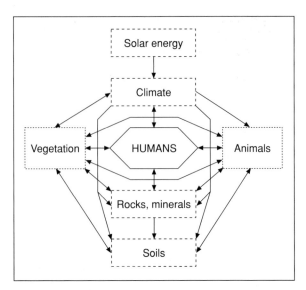

Figure 2.2 Main components and linkages in an ecosystem modified by humans

unit, as shown in Figure 2.2. The transformation had already occurred at the Jarmo settlement and has continued to the present day with an estimated 2.7 billion out of the total human population of approximately 6 billion people living in an urbanised environment (World Bank, 2000). After only a short time, urban dwellers begin to lose many of the ancient skills that enabled their ancestors to achieve self-survival in adverse conditions. Few people living in an urban environment know how to hunt and catch food in the wild, how to skin a rabbit or preserve food without access to a refrigerator. Our total reliance on modern medicinal cures for relatively simple ailments such as cuts and abrasions, or the common cold, means that knowledge of traditional medicinal cures has lapsed. Infusions made from plants often provide cures for a wide range of ailments. People still living in close contact with their natural environment retain their ability to treat their ailments by using plant extracts. Box 2.1 provides details of medicines used at the present time by the Capirona Indians, located on the Rio Puni River, tributary to the Rio Napo, in the Ecuadorian Amazon region.

Depending on the perspective of the observer, the loss of this so-called 'indigenous knowledge' may be seen as either virtue or vice. By relying on the advances of modern medicine, biochemistry and

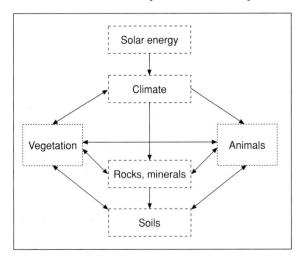

Figure 2.1 Main components and linkages in a natural ecosystem

BOX 2.1

Medicinal use of rainforest plants

The following list of forest plants represents a small proportion of species with medicinal or hunting uses and are currently used by the Capirona Indians in the Ecuadorian Amazon basin.

Chugriyuyu: for inflammation. Cook the leaves in water and apply directly to the part affected. Has antibiotic properties when applied in liquid form from the leaves.

Anguila panaga: used as a relief from rheumatism. Leaves are heated over the fire and applied directly to the body. Also inhale the vapour from leaves cooked in water.

Quibiyuyu: used to relieve sprains. Crush the leaves and spread on the affected area. Must be prepared and applied by a specialist (*pajuyo*).

Cajali: **POISON**. Both the leaves and seeds are crushed together and put into a basket and dropped in the river to kill fish. A much stronger poison than **Barbasco** but not as plentiful.

Cruscaspi: for regulating menstrual flow. Cook bark in water. Drink three cups per day as required.

Chuchuhuasu: cook bark in water to produce a red liquid. Drink to cure cramps, colic and rheumatism. Sometimes taken with alcohol which prolongs the effectiveness of the cure.

Agengibre: tuber is chewed by shaman (witch doctor) for blowing away sickness and by others for scaring away forest animals. Also used to cure influenza. Two drops of extracted liquid are placed in the nose. When mixed with garlic and onion can be used to cure yellow fever.

Huandu: **HALLUCINOGENIC**. Used by men to cure body aches, infections and scars. Also used to help find lost or stolen items. Liquid is extracted from crushed leaves and taken orally.

Vervena: liquid extracted from crushed leaves, mixed with salt and lemon and taken orally to lower fever and cure dysentery and diarrhoea.

Tabaco: used by the shaman to blow away evil spirits. Can also be used to reduce inflammation by heating leaves and applying directly to affected parts of the body.

Ayahuasca: **HALLUCINOGENIC**. Used by the shaman to contact spirits of the jungle. Leaves are boiled in water until the liquid becomes thick. Only one or two swallows are taken followed by a drink of tea.

Condizum: used to cure influenza. Crush leaves and place two drops of extracted fluid in nose. To lower a fever associated with 'flu, drink one glass of extracted liquid.

Huayabs: used to cure diarrhoea and colic. Cook the bark until water is brown. Drink one cup twice a day. Also used to cure influenza and yellow fever.

Yahuaticaspi: used for treatment of both diarrhoea and as a contraceptive. Cook bark in water. Drink one to two cups per day for three days after menstruation.

Curarina: used as an antidote for snake bites. Cook leaves in water. Drink one cupful.

Ismacaracha: cook leaves until soft then mash in the leaf of **Rumipanga** until very hot. Apply to the heads of children with boils or blisters.

Shunaunanaypanga: mash leaves and apply to abdomen for liver disorders. Alternatively, heat leaves to produce a liquid that is drunk. Prescribed by the *pajuyo*, a specialist who cures specific illnesses.

genetics, we are now on the verge of attaining control over our own destiny so that, in theory, we no longer need to rely on the unpredictable events associated with the natural world. The beginning of the present century witnessed the scientific breakthrough associated with the discovery of the entire human genetic code (International Human Genome Sequencing Consortium, 2001) and within the space of less than a year some research institutions were intent on creating the first 'designer baby' (Knight, 2001). The ethical and moral repercussions involved in creating such a baby places a huge new burden of responsibility on society. Religious groups argue that such an action would be a complete rejection of the power of God the Creator of Life while some biologists suggest that moving away from the process of natural selection, which involves a chance combination of DNA, would cause a serious oversimplification in the genetic viability of our species.

2.2 Changing perceptions of the environment

During almost the entire time over which the human species evolved, our position within the vast milieu of physical and biological components has been that of an insignificant and subservient organism. Our very existence lay at the mercy of natural events. However, a long time before the development of the way of life associated with the agricultural settlement discovered at Jarmo outlined in section 2.1 of this chapter, the inventiveness of *Homo sapiens* had directed our line of evolution away from that of the other large apes. Many millennia before the Jarmo civilisation, our species had taken the decision, albeit subconsciously, to take control of our destiny. At first, the level of 'control' was minimal and unpredictable. Climatic variability still dictated vegetation productivity and in turn, times of famine and abundance (Jones, 1979). Disease and pestilence were largely uncontrolled. Technological skills were such that little actual 'control' of the environment in which we lived was possible. At a pace so slow that it was hardly detectable, successive human generations gradually gained control of their circumstances.

Forests were replaced by agricultural land, isolated settlements grew into small towns (Getis *et al.*, 1996). Social organisation made possible trade between peoples of different skills. The dawn of the modern world remained a long way off and took effect with the Scientific Revolution of the seventeenth century and the Age of Enlightenment (also known as the 'Age of Reason') of the eighteenth century, a period of history that questioned almost all previously accepted ideas. For the first time in the history of humankind, deliberate, reasoned decisions were taken about the role of society as a whole and of individual men and women in particular. By this relatively late stage in the social evolution of humankind, our species had clearly taken the conscious decision to move out of the ecosystem component labelled 'animals' in Figure 2.1 and into the new, dominating unit labelled 'humans' in Figure 2.2.

By the standards of the twenty-first century the control exerted on the environment during the Age of Enlightenment was minimal but assessed in the context of the seventeenth and eighteenth centuries the newly found knowledge enabled a totally new perspective to be taken on the environmental circumstances in which humankind existed. From about 1750 until 1970 the population size of European countries and, later, North America, underwent a major increase. This increase was made possible by the generation of personal and national wealth brought about by industrialisation. However, it also marks the point in our development when a radical change occurred in the way in which humankind regarded its surroundings. Emphasis became almost entirely focused on expansionism founded on industrial wealth. Both inorganic resources (such as iron ore, coal and bauxite) and organic resources (such as timber, fisheries and wild herbivores) were utilised to further the economic wealth and the well-being of the industrialised nations. Agricultural production more than kept pace with the need to feed the growing population. Concern for what is now known as 'environmental issues' hardly existed. Within two centuries, the Age of Industrialisation brought about more species extinction, more pollution, more land degradation and more resource degradation than throughout the entire previous duration of human existence on this planet (Simmons, 1996). The species we call *Homo sapiens* had undergone a cataclysmic change and had emerged as the dominant species, not only over other plants and animals, but also of the environment that supported it. Viewed from the perspective of humankind, industrialisation brought with it employment, economic security, personal wealth, transportation, vast improvements in healthcare and in nutritional standards. An equivalent rise of education standards and of scientific achievement also took place.

Not all events were beneficial and major loss of life occurred in a succession of wars, culminating in the two world wars of the twentieth century. Countless other premature deaths occurred in peacetime as a result of pollution. In old industrial cities of England such as Leeds, serious air pollution problems were being reported in the first decade of the twentieth century (Cohen and Ruston, 1911). From 1950 onwards the number of authenticated cases of environmental damage increased in number, in severity and in distribution (see Box 2.2).

BOX 2.2

Identifying environmental damage to humans

Evidence of serious, local environmental damage was first observed in the old industrial areas of Great Britain, France, Germany and Belgium and from the New England region of the USA. The most commonly reported examples of environmental damage involved dangerously high air pollution levels that occurred during specific weather conditions and from chemical pollution of land and water resulting from indiscriminate disposal of toxic wastes. All the early occurrences were localised in geographical extent and therefore differ from the present-day conditions that involve a global deterioration of the environment.

One of the earliest 'modern' examples of environmental damage occurred at Ducktown, Tennessee, during the first decade of the twentieth century, when gases from a copper smelter destroyed vegetation across the interstate border in Georgia. In the 1920s another copper smelter at Trail, British Columbia, caused the first transnational pollution incident with the USA, while in 1935 the first known human deaths caused specifically by industrial pollution occurred following an air pollution incident in the Meuse Valley, Belgium. As industrial activity accelerated so did the occurrence and severity of environmental damage and in 1948, 20 deaths occurred in the industrial town of Donora, Pennsylvania, due to air pollution, while the worst recorded incident was the death of up to 4000 elderly and chronically sick people in London when, in mid-December 1952, a severe air pollution incident occurred. The pollutant that caused the deaths was never ascertained (Faith, 1972).

In all these examples, the impact of pollution was readily identified. Elsewhere, the damage caused by pollution was obscured by a considerable time span before visible damage became apparent. Such an example occurred at the small town of Woburn, Massachusetts. The leather industry had been established in Woburn in the eighteenth century but it was not until 1971 that a cluster of leukaemia incidents in children revealed that drinking water from two wells had been seriously contaminated by the indiscriminate dumping of heavy metals and organic compounds (Cunningham and Saigo, 2001).

Improved public health legislation combined with planning legislation has eliminated the majority of problems in developed countries. However, as recently as the 1980s severe environmental pollution and damage to humans still occurred in developing countries. In São Paulo State, Brazil, the town of Cubatao became known as the 'Valley of Death', holding the unwanted title of most polluted place in the world as a result of unregulated pollution output from a steel plant, a major oil refinery and fertiliser and chemical plants. Birth defects and respiratory diseases reached alarming levels and eventually a clean-up policy costing US$300 million resulted in ammonia levels falling by 97%, hydrocarbons by 86%, sulphur dioxide by 84% and particulates by 75%! With the necessary anti-pollution legislation in place it became possible to minimise pollution output and still retain an industrial base.

(See http://www.greenpeace.org/pressreleases/toxics/1999jan12.html and also http://habitat.aq.upm.es/bpal/onu/bp033.html – text in Spanish, but a translation option is available.)

Spurred on by the direct and indirect costs of paying compensation to communities and people who suffered as a result of environmental damage, governments and industry have worked hard to minimise further environmental damage. Until the 1970s management of the environment was largely directed through a policy of 'best practical means' (BPM) whereby 'practical' was taken to mean 'reasonably practical' having regard for the state of technology, the prevailing local conditions and to financial considerations. Such a policy often meant that different countries, or even parts of countries, operated under different levels of BPM leading to severe variation in environmental quality, for example at Cubatao (see Box 2.2). Such a policy not only allowed different management standards to be followed, it did not enforce industry to operate to the highest possible standard. Towards the end of the twentieth century a major change in environmental management policy occurred in which BPM was replaced by 'best available techniques not entailing excessive cost' (BATNEEC). This approach to environmental management involves three fundamental concepts:

1 Use of the 'best' to imply the use of the most effective method of preventing and minimising the release of prescribed substances or rendering harmless other polluting emissions.
2 Use of 'available' to imply that the method of pollution control must be accessible and procurable by operators of the prescribed process.
3 'Technique' to involve both technological and operational considerations.

The size of financial compensation payments was certainly an incentive for industry to agree to changes in the way in which the environment was managed. For example, in the USA alone it is estimated that there have been 36,000 serious cases of environmental damage and by 2001 some $38 billion had been allocated on clean-up and compensation (Cunningham and Saigo, 2001). Financial costs are not, however, the only reason for our changed perception of the environment. Of critical importance is the improved knowledge we now have about our planet. We now understand that the planet consists mainly of a complex set of finite resources – the basic natural resources of soil, water and air, and the economic resources such as minerals, fossil fuels and organic resources. In reality, the consequence of this knowledge has proved difficult to accept: we live on a planet that has a finite limit to what it can yield in terms of supporting the demands placed upon it by humans. The planet is, as the quotation at the beginning of this chapter states, 'all we have'. Apart from the addition of a small amount of space dust that penetrates the Earth's atmosphere and the loss of materials contained within spacecraft fired out beyond the gravitational pull of our planet, the mass of our planet has remained constant at 6×10^{21} tonnes ever since it was formed, approximately 4500 million years ago.

However, contrary to the finite dimension of resource availability, during the majority of the time human civilisation has existed, our planet appears to have been more than generous in supporting our material needs. Our predecessors grew up in the belief that our planet could provide a never-ending cornucopia of resources. Slowly, society has awoken to the fact that the combination of a growth in numbers of humans combined with an increased demand for mineral and biological resources has necessitated a reassessment of the ability of our planet to continue to yield a never-ending supply of resources at a constantly accelerating rate. Figure 2.3 provides an example of the increase in basic material resources in the most prolific of consumers, the USA, during the twentieth century.

2.3 Factors that influence our perception of the environment

Our perception of the term 'environment' is probably as unique to each person as are our fingerprint patterns. While it is possible to place people's attitudes towards the environment into general categories, when it comes to a detailed analysis of how each individual responds to the environment there are an infinite number of different views. In this respect our involvement with the 'environment' can be as active or passive as our views on a host of other key areas of modern life.

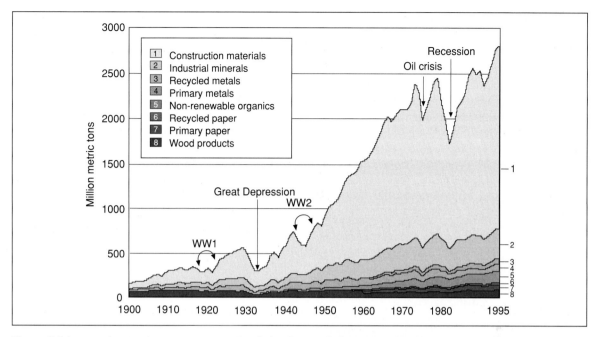

Figure 2.3 Increase in natural resource consumption during the twentieth century in the USA

Despite the importance of the environment to our well-being, relatively few people in the developed world have their own original thoughts on the environment. Instead, we have become influenced to a greater or lesser extent by the information and attention given by the media to environmental issues. Political party manifestos, government policy and economic factors also make an impact on our environmental attitudes and last but not least our involvement with the environment has been exploited by pressure groups and non-government organisations in an attempt to bring what they consider critical environmental issues to our attention (Huckle and Martin, 2001). To a large extent, the coercion of public attitude to environmental issues has been subsumed within the emotional arguments that surround the loss of 'wilderness' areas or the conservation of animals such as pandas, seal cubs and whales (Regan, 1993; Singer, 1993). However, the need to retain wilderness or conserve specific species forms only a small part of the more important requirement of caring for our environment.

As individuals, we assume that our own actions can make little impact on the environment,

preferring instead to concede responsibility for the environment to our elected politicians in the expectation that they will be able to introduce legislation that will ensure a 'safe' environment. However, environmental policy may not be uppermost in the minds of our political leaders. The economic state of the nation usually takes priority, but occasionally the priority may be defence or combatting terrorism. By tradition, the politician is concerned with legislating for a 'safe' environment in terms of public health. Only in the last quarter of the twentieth century have politicians in developed countries become concerned with quality of life issues, the reasons and implications of which are considered in Chapter 3.

Forfeiting responsibility for the environment to our elected leaders may be an accepted part of the contemporary democratic process but in terms of what may be best in the longer term for the environment, transferring responsibility away from the individual has distinct disadvantages. This is particularly so when the methods used by politicians in reaching environmental decisions is understood. Cutter and Renwick (1999) provide an account of the most common decision-making processes used by governments. Three general categories can be identified: *satisficing*, *incrementalism* and *crisis management*. Box 2.3 examines these three processes along with a fourth and potentially more suitable approach, that of *eco-management*.

As our knowledge and perception of the environment has increased so, too, has the intensity and level of the environmental debate. Until the 1970s, environmental arguments were essentially perceived as being relatively small scale and locally based. In retrospect, many of the problems were far more complex than were realised at the time, as shown by the problems caused by acid precipitation (see Box 2.4).

BOX 2.3

Strategies used in environmental management

Until the 1970s, politicians had little direct experience in debating and legislating for environmental management. Most experience had been gained from clean air legislation, for example the Clean Air Act (UK). Eagles (1984) has provided examples of environmental legislation introduced during the 1970s and 1980s in Canada and the UK. It was inevitable that during these early attempts at environmental legislation politicians and civil servants would rely on methodology that was already familiar to them in other areas of government.

For example, **satisficing** involves the sequential evaluation of two policy alternatives. Plus or minus points are allocated for each attribute within the different strategies. The policy that gains most points is taken forward to the next stage of the evaluation process where it can be compared with other policy strategies that may have been assessed by an alternative routeway. The main objective of satisficing is to establish a course of action that satisfies a minimum set of targets. This approach can be conducted quickly and cheaply; it involves a minimum of original research and will probably not evaluate the full range of alternative options. It can be made as a desktop study and has the advantage in that it is a well-tried technique in other management situations. Its main disadvantage when applied to the management of the environment is superficiality due to the consideration of only a limited range of options.

Incrementalism is best suited for situations in which strategies are not clearly definable or where conflicting attitudes, strategies or objectives exist. The management policy used in this approach is one of *making do as best one can*. A clear end point may not be known, but, provided the assessor knows what endpoint is *not* required, incrementation can be a useful approach to help with short-term, day-to-day management decisions. Disadvantages of incrementalism are that the approach can appear structureless to observers from outside the system while staff whose job it is to apply the policy decision may receive conflicting orders over a short timescale. This approach is especially unsuited for environmental management because of the medium- to long-term objectives that characterise most environmental problems. Despite its inherent disadvantage, in the light of the often poorly defined ambitions we set for conservation policies, this approach is sometimes the best that can be achieved.

The final and most widely used approach to problem solving is **stress (or crisis) management**. For governments that possess a slender majority it can become advantageous to practise *brinksmanship* in order to wrong foot the opposition! In this approach, a solution is delayed until the last moment – when a 'crisis' has arisen. There is no time for research into the problem and there may be little concern with finding a long-term solution. Instead, problem solving is based on short-term expediencies, in the knowledge that future refinement will be necessary. The nature of most environmental problems requires research and careful consideration of a wide number of options before a possible solution can be reached. Consequently, this category of management provides the worst option for conservation and environmental problems.

Attitudes and perceptions of our elected representatives towards the environment changed rapidly in the last quarter of the twentieth century. Assisted by environmental scientists, environmental lawyers and civil servants, the political structure of all developed countries now firmly embraces the need for strong

---

BOX 2.3 CONTINUED

environmental management. In particular, politicians have recognised the need to adopt radically different management systems to cope with environmental issues. The development of **eco-management** principles, in which the biosphere has been recognised as consisting of many complex and interrelated probablistic systems operating over long timescales has been accepted as one of the best environmental management practices. Lengthy and complex research is often necessary to establish a database on which eco-management decisions can be made, a good example of which is the Millennium Ecosystem Assessment (MA), launched by UN Secretary-General, Kofi Annan, on 5 June 2001. The MA consists of a global scientific assessment of ecosystems intended dramatically to increase the ability to manage ecosystems. While few would denegrate the ambitions of the MA many activist NGOs argue that enough studies already exist to show where and how ecosystem deterioration has occurred and what is needed is the political will to improve the current state of affairs.

(For details on the Millennium Ecosystem Assessment see the following websites: http://www.millenniumassessment.org/en/index.aspx and http://www.wri.org/wr2000/mea.html.)

BOX 2.4

Acid precipitation and damage to conifer trees

Acidification of the atmosphere by sulphuric and nitric acid is a major threat to the environment of the northern hemisphere. Major urban and industrialised areas release an estimated 90 million tonnes of sulphur dioxide (SO_2) pollution per annum directly into the atmosphere. Fossil-fuelled power stations and industrial sources are major producers of SO_2, while vehicle exhausts provide the main source of nitrogen oxides (NO_x group). The gases are transported by the prevailing winds for many hundreds of kilometres until the pollutants are washed out of the atmosphere (or 'precipitated') in rain, snow or fog. The acidity of 'pure' rain water is about pH 6.5 whereas acid rain usually has a pH of below 4.5 with an extreme low of pH 2.5 (Park, 2001: 251–5). Southern Sweden, Norway, parts of central Europe and the eastern seaboard of North America have shown the highest levels of acid rain. The acid fallout is believed to be directly responsible for the death of fresh water fish and the progressive decline, and ultimately the death, of trees.

Controversy exists as to the real danger of acid deposition on the environment. This has been partly due to conflicting field and laboratory evidence. It would appear that the impact of acid deposition is very site specific and is highly dependent on the 'buffering capacity' of the soil. Some soils, notably limestone or sandstone, appear to be able to neutralise the effect of acid deposition whereas thin glacial soils or thick granite bedrock (as in Scandinavia and Canada) fail to provide an adequate buffer against the enhanced acidity levels. The problem of acid deposition can be attributed to lenient BPM attitudes (see page 34) that allowed heavy industry to emit unsafe levels of particulates (ash, dust, grit and smoke) and gaseous pollutants (especially sulphur dioxide, hydrogen sulphide and oxides of nitrogen).

Action to reduce the amount of acid deposition shows great variation and depends on the extent of damage already caused by this hazard. In West Germany, for example, financial losses to the timber industry from acid deposition have been calculated at $800 million per year, while agricultural losses from acidification of the soil amount to $600 million. It has been estimated that 50% of the acid deposition in West Germany originates from outside that country (Mannion, 1992c; Press and Siever, 2001: 567).

The first signs of acid deposition damage to vegetation emerged in the early 1960s. Conifer trees growing in the vicinity of large industrial plant such as iron and steel mills in the USA and in Europe began showing signs of premature loss of foliage. Conifers are 'evergreen' species and the needles remain on

BOX 2.4 CONTINUED

healthy trees for between two and four years. Unhealthy trees growing near steel mills shed their needles after only two years and have needle surfaces pitted with yellow or brown spots. Attempts were made to explain this phenomenon by implicating soil nutrient deficiencies, adverse climatic factors, insect-borne or viral disease. Research showed that the visual damage that occurred on the surface of the needles was the penultimate stage in a process in which the internal water balance of the cells comprising each needle was severely altered. Initially, the cells within the leaf (the mesophyll layer) become dehydrated, followed by collapse, resulting in death of the leaf surface which changes colour from green to yellow. Finally, individual needles are shed from the branch. This problem became known as 'post-emergent chronic tip burn' (PECT) from the most severely affected sites where newly produced needles quickly developed damage symptoms (Berry and Hepting, 1964).

The science that surrounded the identification of acid precipitation took almost a century to develop to a point where the problem was understood (Kowolak, 1993). Once the problem had been scientifically understood, the mechanisms necessary to control gaseous pollution of the atmosphere were gradually put into place by governments following the First World Climate Conference in 1979 and advisory group meetings on greenhouse gases held in Villach (Switzerland) and Bellagio (Italy) in 1987. At Kyoto (Japan) in December 1997, 113 countries thrashed out the principles of an *international* agreement that was considered vital to improve air quality worldwide and to counter the threat of global warming (Ott, 1998). In 2001 the Kyoto Protocol was thrown into confusion when US President Bush overturned the earlier US agreement made by former President Clinton, clearly indicating how attitudes to the environment can change when a new government comes to power.

2.4 The main environmental viewpoints

Modern-day society is characterised by its highly dynamic nature and rapid response to political, market and environmental forces. It is unusual for any one particular set of views to remain in force for very long due to the speed with which events unfold at local, national and international levels and these ensure that we constantly reconsider and re-evaluate relationships, including those with the environment. Preceding sections of this chapter have shown that between 1750 and 1970, the predominant attitude of the old industrialised northern hemisphere countries has been one of almost total indifference and ignorance of the environment and the way in which humans interact with it. Until the 1970s the environment was considered a resource base from which materials were extracted on an 'as-required-basis', and at minimal cost, with little or no attention given to the impact a particular action might have upon the environment (Morse and Stocking, 1995).

In post-1970 years human attitudes to the environment have changed, and have generally become far more diverse in their outlook. At least ten different ways of 'thinking about' the environment can be identified, as can be seen from the list that follows. These different perspectives can be described as environmental 'viewpoints', some of which are *not* mutually exclusive:

- **Radical** – all living things have rights.
- **Emotional** – the need to take care of (protect) plants and animals.
- **Religious** – considers the environment as part of an individual's existence.
- **Feminist** – environmental exploitation has been due to male-centred dominance.
- **Political** – local, national or international perspectives predominate.
- **Scientific** – interest in environment is to see how things 'work'.
- **Vested interest** – economic assets determine the outlook on the environment.
- **Utilitarian** – considers the environment as a resource asset.

- **Humanist** – in which the environment has been established for human benefit.
- **'Blue sky technologists'** – no limits to our capabilities of discovering new resources and developing new technology to solve problems.

Most people will find difficulty in associating exclusively with only one of the viewpoints in this list, preferring instead to select parts from several viewpoints. Furthermore, the extent to which an individual concurs with a specific viewpoint will vary, depending on a wide range of circumstances that operate at any point in time.

The ten viewpoints in the list have been used to categorise people into distinct attitude groups. Broadly speaking, Ehrlich *et al.* (1973) divide environmental attitudes into two groups: those who believed that human inventiveness ensured there were no limits in our capabilities to discover and develop new resources and products were classified 'cheermongers' and the opposite group who believed we should exercise caution and not be optimistic about discovering new resources, as 'doomsters'. These terms have largely been replaced by 'technocentrists' and 'ecocentrists' (see Table 2.1) (O'Riordan, 1981; Pepper, 1984; Jordan and O'Riordan, 2000). In the first of these, technological solutions are seen as the key that enables society to advance. By means of technological skills, it is argued, the efficiency with which society operates can be improved. By so doing, we gain an ever-greater understanding and control of the environment. Belief in the technocentric model has dominated our approach to the use of the environment throughout most of the industrial era (Wagaar, 1970). The alternative attitude, ecocentrism, views the environment and its resources as finite and proposes that the only permissible development is *sustainable development*, defined as 'development that meets the need of the present without compromising the ability of future generations to meet their own needs' (World Commission on Environment and Development, 1987).

In its most extreme form, technocentrism has been called the *cornucopian* or *egocentric* approach, and is based on a belief that human development is best served by an unfettered pursuit of growth and maximum use of resources. It is argued that human ingenuity is so great that it will find new reserves of existing resources or alternatives for resources that become depleted. Cornucopianism is usually associated with the patterns of mass consumption that had been heralded as early as 1910 in the US with Henry T. Ford's production line philosophy of manufacture applied to the automobile. The term 'Fordism' has been applied to the mass consumption of manufactured products (Bocock, 1993). Fordist attitudes dominated the immediate post-war years and spawned marketing and political propaganda slogans such as 'you've never had it so good' (Huckle and Martin, 2001: 90–91, 121–3). The egocentric approach is based mainly on self-interest that allows entrepreneurs and governments to use natural resources for profit while claiming that such use also ensures wider public benefit and economic prosperity.

The Fordist belief that the economic growth model could be sustained indefinitely held sway until the 1970s but after the 1973 oil crisis (see Pickering and Owen, 1994) returns on capital investment began to diminish and a new development model was required to inject new vigour into the old industrialised nations. The rise of information technology, robotisation of the production line and biotechnology and, in particular, the search for lower labour costs provided a new strategy for industrial development. Post-Fordist development is now centred on more sophisticated consumer knowledge and special 'lifestyle' marketing that continues to evolve at breakneck speed. At its most extreme, the speed of computer processors doubles every 18 months while some 'high street' clothing for women has a 'shelf life' of only 14 days. The cornucopian or Fordist attitude still prevails although it has been forced to adopt cleaner production technology, now legally required by environmental protection standards and also to satisfy the expectations of an environmentally aware public.

Some companies have moved further towards achieving a minimal impact on the environment by adopting rigorous recycling of wastes and the use, where appropriate, of renewable products. Although still firmly embedded within the technocentric group, these companies are known as the 'accommodators' (see Table 2.1). This group argue that development is possible without an

Table 2.1 A summary of the different environmental viewpoints

Ecocentrists		Technocentrists	
Deep environmentalists	Soft technologists	Accommodators	Cornucopians
Lack of faith in modern, large-scale technology and its need for elitist expertise, central authority and inherently undemocratic institutions		Believe that economic growth and resource exploitation can continue indefinitely given (a) a suitable price structure (possibly involving taxes and fees); (b) the legal right to a minimum of environmental quality; and (c) compensation for those who experience adverse environmental or social consequences	Believe that humans can always find a way out of difficulty, through politics, science or technology
Believe that materialism for its own sake is wrong and that economic growth can be geared to provide for the basic needs of those below subsistence levels		Accept new project appraisal techniques and decision review arrangements to allow for wider discussion and a genuine search for consensus among affected parties	Believe that scientific and technological expertise is essential on matters of economic growth and public health and safety
Recognise the intrinsic importance of nature to being human	Emphasise small-scale (and hence community) identity in settlement, work and leisure	Support effective environmental management agencies at the national and local level	Accept growth as the legitimate goal of project appraisal and policy formulation
Believe that ecological (and other natural) laws determine morality	Attempt to integrate work and leisure through a process of personal and communal improvement		Are suspicious of attempts to widen participation in project appraisal and policy review
Accept the right of endangered species or unique landscapes to remain unmolested	Stress participation in community affairs and the rights of minorities		Believe that any impediments can be overcome given the will, ingenuity and sufficient resources (which arise from wealth)

(*Source:* after Jordan and O'Riordan, 2000)

overexploitation of natural resources. Instead, by using the best available technology, minimising waste, and maximising the reuse of previously used resources the accommodators have moved some way towards accepting the need for sustainable development although considerable scepticism exists among supporters of green ideology for the real motives behind the change (Rees, 1991: 303).

Whereas cornucopian and accommodator viewpoints rely to varying degrees on the use of natural resources and technology to enable development to occur, the ecocentric argument challenges the belief that winning of new resources is an essential component of the development process. Jordan and O'Riordan (2000) have shown that at least two variants of the ecocentric argument can be identified. In the first, the so-called 'soft technologist', consumption of new resources is seen only as a 'topping-up' option, permitted when all economic sources of recycling have been used up, and when industry has shown itself to be using the most energy-efficient processes. All new developments should be subject to independent assessment to ensure that they will provide real and necessary benefits for society. Of equal importance, new development must not create a dis-benefit for the environment. The latter includes the impact of pollutants on the environment, the exhaustion of finite resources, and the destruction or impairment of natural ecosystems and species. The full ecological, as well as economic, costs of development must be calculated and the total cost is that which the market must pay for development. In this approach, all new development must pass strict environmental, economic and social criteria. The criteria can be internal criteria relevant to a region, to a country, a trading bloc or may comprise internationally agreed criteria. Such an approach requires greater commitment from both industry and consumer groups, and can probably only be achieved through tougher regulatory control (Ekins et al., 1992). Soft technologists accept the need for essential development and are not inherently anti-technology. The approach requires the existence of a clear set of conditions under which development is allowed to occur.

The final viewpoint, 'deep environmentalism', is located at the opposite extreme from that of the cornucopians, and argues that any development that is human centred is intrinsically wrong, as it would inevitably bring about a deterioration of conditions for the rest of the environment and all non-human inhabitants (Naess, 1993). In practice, it is difficult to see how this approach can be a realistic alternative to the others that have already been considered. If pursued to its extreme, deep environmental policies would inevitably lead to a period of great social readjustment resulting from de-industrialisation and a move away from an intensive agriculture. Social wealth would decline and agricultural output be substantially reduced. A period of massive human population readjustment could be expected to follow. It is difficult to avoid the conclusion that no matter how ethically correct the argument of deep environmentalism might be, its application would result in a severe disruption to society.

2.5 Post-1975 – an environmental awakening

Although the early 1970s marks the point when modern environmental awareness can be said to have begun, it is necessary to look beyond that time to find the roots of the modern concern with the environment. For more than 100 years prior to the 1970s, European nations had been embroiled in colonial expansionism that led to rampant nationalism, social unrest and a succession of wars starting with the Franco-Prussian War of 1870–71 and ending with the Second World War, 1939–45. European expansion had reached its zenith by 1914 with an estimated 84 per cent of the world's land area under its influence (Huckle and Martin, 2001: 96–7). One particularly unfortunate development associated with this period was the rise of *colonialism*, an especially exploitative form of imperialism whereby cheap labour was used to extract natural resources for minimal reward and used by the 'motherland' for its economic expansion. While military and political domination was undoubtedly at the heart of imperial expansionism, there is little doubt that an associated benefit was seen to be access to vast supplies of additional raw materials to bolster industrial growth in Europe. Crosby (1986) has called this process *ecological imperialism*, in which the winning of natural resources resulted in a breakdown of the traditional local economy, degrading local peoples and their culture and resulted in *environmental racism*. Developing countries were left with a severely polluted and degraded natural environment, bereft of many of the basic natural resources necessary to support human life. The birth of the 'Third World' nations took place largely as a result of ecological imperialism (Haggett, 2001: 559–600).

At the end of the Second World War, the European protagonists emerged with an industrial infrastructure that was at best outdated and, at worst, almost totally destroyed. The 30 years from

1950 to 1980 witnessed a remarkable rebuilding of industry and a growth of urbanism. The renewal was based on innovative technology, especially telecommunications, electronics and computers, mass communication using road and air transport, and massive increases in agricultural output. Within the 'white heat of the technological revolution' (Harold Wilson, British Prime Minister, 1964–1970, 1974–1976) there appeared to be little time for concern over the state of the natural environment (Slater, 1997).

The changes in environmental attitudes at all levels of society that occurred between the mid-1970s and the early 1990s have been little short of amazing! The process of change involving government, management and the general public are discussed more fully in Chapter 3. In this section it is necessary to examine some of the main reasons that underlie those changes.

A number of events have been responsible for changing our attitudes towards the environment and its resources. These can be arranged into several groups:

- a belief that natural resources would run out
- that the increase in human population was out of control
- that extinction of plant and animals would lead to a collapse of the world ecosystem
- that rapid global change involving climate and the great natural biogeochemical cycles would lead to such massive environmental change that society would not be able to adapt.

A belief that a lack of natural resources would inevitably lead to a stifling of economic growth has been a recurrent worry throughout human history. In the 1970s the work of Meadows *et al.* (1972) led governments and industrialists to believe that shortages of key resources would limit growth at some point in the not too distant future. Wealthy nations began stockpiling commodities in an attempt to safeguard their economic growth and the belief that resources (especially minerals) would one day run out. This viewpoint dominated central economic and political thinking for a decade or more (Soussan, 1992). The fears that the supply of petroleum products would dwindle prompted the Organization of Petroleum Exporting States (OPEC) to raise the price of oil on the world market by a factor of 200–300 per cent in two phases during 1973–74 and 1979–80. Discovery of new oil reserves outside the areas of traditional supply were accompanied by considerable technological improvements allowing oil to be extracted from smaller and more difficult sites. Improvements in the efficient use of oil have helped offset the fear of an oil shortage and the price of oil on the world market has, apart from a few occasions when regional instability threatened oil production, returned to a price level that, when adjusted for inflation, is at an all-time low. As a result, attention was redirected not on resource shortages per se but on the impact on the environment of extracting and using these resources.

The second reason for an increased concern with the environment was the belief that the human population was increasing at an exponential rate. In 1830 the world population was about one billion. It took 100 years to double in size but thereafter reached four billion in 1960 and five billion in 1990. By 2000, world population was six billion and by 2050 is predicted to rise to approximately nine billion (UN Population Division, 1998). The problem is not so much the actual increase in human numbers but the way in which the population makes use of increasing amounts of natural resources. Between 1980 and 1997, the annual value of global economy almost tripled to US$29 trillion and yet the world population increased by 'only' 35 per cent over this period (World Bank, 1999: 194). Furthermore, the majority of the increase in consumption occurred in developed world countries where population increase is minimal. An outright increase in human population numbers is not, therefore, cause for alarm, but the consequent way a small proportion of that population uses resources is very worrying.

The third concern was that of the rapidly increasing rate of species extinction, especially in the last remaining undeveloped areas of the world. Chapter 8 examines these issues in detail. It is relevant to note that the loss of species such as the Giant Panda in China or the Condor from the southwest cordillera in the USA or the Spotted Owl from the redwood forests of the Pacific northwest have provoked great public concern – but some doubts exist that the concern has more to do with emotive values surrounding the protection of animals and less with a genuine concern that conservation measures must be applied to *all* species.

The fourth reason for the change in the public perception of environmental issues has been due to the message emanating from scientific research about the scale and magnitude of environmental degradation. Public attitudes gradually became receptive to the idea that the quality of life was dependent on the condition of environment. Its ability to absorb pollution of many different types and render it harmless was just as important as its ability to yield resources. The rise of alternative environmental viewpoints such as deep ecology and eco-feminism (see section 9.9.1) has done much to raise an awareness that different viewpoints regarding the ways in which we interact with the environment not only exist, but may offer solutions to seemingly intractable environmental problems. Many of the alternative viewpoints have existed since the mid-1960s when 'green' viewpoints first began to be heard, mainly in the context of a protest against the technocentric views that dominated social and political thinking. The way in which 'green' ideology has moved from the fringe of acceptability to that of a more central position in which it helps shape modern political thought and social behaviour has been one of the most radical changes of the late twentieth century (see section 3.8).

Acceptance of environmental as well as economic arguments when deciding on the future course of modern society has been assisted by the apparent increasing frequency and severity of environmental crises and even the occasional environmental disaster. Our interaction with the environment has been conditioned by events such as the first major breakup of an oil-carrying super tanker – the *Torrey Canyon* in 1967 – or the explosion of the number 4 reactor in the Chernobyl nuclear power station in April 1986 (McKinney and Schoch, 1998: 210–14) and the eventual acceptance that human activities were causing an acceleration of global warming that led to the Kyoto Agreement (Ott, 1998).

The phenomenal growth in scientific and technological advancement that has occurred since the mid-1960s has inexorably led society to an ever-greater reliance on technology as the means of solving environmental problems. Many people consider this method of solving environmental problems to be the same as running down a one-way street that is becoming increasingly narrow, which prevents us from making a U-turn. An attempt to counteract this trend took place first in North America during the 1960s and was soon followed by Europe, Australia and New Zealand and nowadays by most developed and aspiring nations of the world. A wave of new, unofficial organisations concerned with the environment sprang up. Some of these organisations eventually became the so-called pressure group organisations while others developed into non-government organisations (NGOs). All relied for their support on the young, well-educated and often disaffected cohort of the population. The broadening of the education base of the free world to include all sectors of the population allowed the environmental groups to draw membership from a wide range of social groupings. In this respect they differed from the strongly middle-class organisations such as the Sierra Club (USA) or Royal Society for Protection of Birds (UK) that had been established a generation or so earlier. Many members of the new organisations had expressed dissatisfaction with the establishment and it was inevitable that the ideological base of the environmental groups should lean strongly towards the left (see Huckle and Martin, 2001: 161–2). Because of their quite different social and political background the new environmental organisations were recognised as part of the *new social movements* (NSMs).

The disaffection that characterised many of the NSMs was due to many underlying and often conflicting reasons. Not least, it was the intense disillusionment over the long running Vietnam War (1959–1975) that characterised a generation of many young Americans. It was also the result of a new wave of adventure travel by affluent, well-educated people, able to travel to remote parts of the world and to experience at first hand alternative religions and lifestyles, and also to see the destruction of natural habitats. Environmental writers and journalists were responsible for bringing issues such as the deforestation of Amazonia, transnational pollution incidents, and the overcrowding and soil erosion in American national parks to the attention of people in their own homes.

Today, it is not unusual to hear children as young as five or six years of age discussing complex issues involving the environment, while in surveys of what worries teenagers about the future, environmental issues often appear at or near the top of their list. The concern of young people towards the environment is due to a large extent to the major improvements in environmental education in primary and secondary schools that have occurred over the last 30 years (SOED, 1991). Many high schools, colleges and universities offer 'elective' classes in environmental studies allowing students majoring in other disciplines to continue to upgrade their awareness of environmental matters while specialist degree-level courses both at undergraduate and postgraduate levels variously entitled 'environmental science' or 'environmental studies' are provided in many universities.

Information about the environment is now widely available on the TV, radio and in the press while more recently, the World Wide Web has provided an excellent means of providing world coverage on local environmental issues (http://www.iclei.org/). In short, members of the public are now better provided with accurate information on the main environmental issues than at any previous time.

Public attitudes are still unduly influenced by environmental disasters or particularly eventful natural phenomena such as a major hurricane or earthquake, even though many of these events should be considered 'normal' in the long-term incidence of natural disasters. Much greater concern should be levied at the changes that have been and are continuing to be caused by human activities such as the tropical deforestation, release of greenhouse gases into the atmosphere and desertification caused by intensification of farming.

Although we are better informed about the environment, the way in which it operates and the way it has been changed by humans, one intriguing question remains to be answered. Is our recent concern over environmental issues actually changing the way in which we behave towards the environment and to other non-human species (Eder, 1996)? It is relevant to ask this question because if society has understood the principles by which the environment works – such as through self-balance (homeostasis), recycling and control mechanisms that restrict the occurrence of extreme events – then we should also be able to introduce similar control mechanisms within human society. For example, greater use could be made of mineral resources recycling or more sophisticated use of control apparatus in buildings to provide a more stable temperature regime combined with a lower consumption of energy use. Where strong economic advantages can be gained, society can be persuaded to adopt many of the natural control methods that operate in the environment.

2.6 Back to the future!

This chapter has examined some of the ways in which our understanding of and relationship towards the environment has changed over time. It is safe to conclude that future historians will identify the late twentieth century as a time of environmental reawakening when the prevailing attitude of most developed nations became cognisant of the need to prevent pollution and restrict the consumption of virgin resources. Whether events that have occurred since the early 1970s will be recognised as one of the periods in which humankind has made one of the sporadic conceptual or technological advances that have punctuated our evolutionary development remains to be seen. During the time this book was being written one of the remarkable sequences of events was the emergence of ever-more spectacular scientific breakthroughs especially in terms of human and animal genetics. The successful cloning of animals and the completion of the mapping of the human genome were major milestones in our scientific and technological capabilities. As a result of these breakthroughs, society has been required to assimilate scientific advances that are capable of changing the future path of evolution. In reality, these achievements *have* made major news stories on our television screens and as newspaper headlines but the *significance* of these achievements for society as a whole has scarcely been examined. The implication of the recent scientific breakthroughs bring with them immense ethical and moral questions that both society and individuals should consider, as they will inevitably have a direct bearing on the next generation. One of the greatest challenges for society will be how to incorporate these very significant technological changes. What was considered 'future science' only ten years ago has become the present!

Goudie (2000: 433), in the concluding statement of his book *The Human Impact on the Natural Environment*, asks whether 'national governments and international institutions have begun to ponder whether the world is entering a spasm of unparalleled humanly induced modification'. The answer, undoubtedly, is 'yes'. When judged by the growing number of international conferences and the drawing up of treaties and agreements that deal with ethical issues such as cloning of humans and with environmental issues connected with the possible escape of genetically modified organisms from trial sites to the wild, it is evident that political awareness of the need to achieve greater control over events has increased significantly since the 1970s. However, earlier sections of this chapter have asked questions about the origin of this interest, a theme taken forward in Chapter 3. The question to be answered is: 'Are politicians *really* concerned about the ethical issues facing our species that have emerged as a consequence of our new scientific capabilities? Or are they merely seen as another set of variables that take their place in the rank order alongside all the others that must be managed in an attempt to exercise control of society?'

Phillips and Mighall (2000: 369) ask a series of questions relevant to this theme. For example, 'how should nature and society be interrelated?' and 'is it possible to manage this relationship to overcome or alleviate the problems it is creating?' Until relatively recently, nature and humans were studied separately as 'them' and 'us'. Humans were separate from the rest of nature and could do very much as they wanted in terms of using nature in order to support the growth and development of society. In more recent times, and certainly since

the mid-1970s, the dynamics of different social groupings have been seen more as a group of inter-relationships that react with the environment that surrounds them. That relationship is a dynamic one, but also is usually exploitive in that we use resources obtained from the environment for our own benefit. In that capacity, the relationship is an uneven one in that the 'environment' cannot benefit from us making use of its resources. But perhaps the environment can exert a price for the uses we place on it.

The price of using resources comes in the form of resource shortage and consequent rise in price, environmental deterioration and environmental change. Examples are legion, but in response people have developed a range of methods with which to manage nature. As one commodity becomes scarce we use an alternative, if pollution becomes a problem we filter out the harmful substances. Phillips and Mighall (2000) claim that we have become good at responding to environmental problems but that we have failed to *prevent* problems from occurring in the first place. In other words we manage the environment in ways that are *reactive* and not *proactive*. Societies are all too ready to embrace new technology with little thought of the environmental consequences. Perhaps this criticism is too harsh, for how can we know in advance what impact a new technology will bring to the environment. It is generally the case that sound scientific practice insists that a new product or technique is thoroughly tested in the laboratory or in field trials before released for general use. But it only takes a small loophole in legislation designed to control a process or a government prepared to bow to pressure from a multinational company and sanction the premature use of a product, to allow new technology to 'escape' from the research laboratory. For example, faced with a combination of economic necessity, intense pressure from society and the desire to be seen to make progress in advance of an election, politicians often take decisions that are fraught with uncertainty and indecision, as shown in Box 2.3.

Already, new and challenging technologies are emerging that will test the ability of scientists, politicians and environmental managers to cope with a new dimension of change. Table 2.2 lists some of

Table 2.2 Major technological and scientific changes that can be expected to make an impact on the biosphere in the twenty-first century

Event	Opportunities provided by the event	Impact of event on the biosphere	Positive or negative reaction for humans
Genetically modified organisms (GMOs)	Designing specific plants and animals to suit particular environmental conditions. Boosting food production	Unknown effect of GM plants on the native flora and fauna	Present opinion negative in Europe, positive in USA. Longer term situation may change
Mapping the human genome	Understanding the causes of disorders, malfunctions and diseases both before and after birth	Increase in the survivorship level of humans. More old people survive in a better state of health	Positive – but note the potentially negative impact on structure of society
Cloning of animals	Replication of domesticated animals ensuring an exact copy of the single parent Extension of cloning to replicating wild animals facing extinction through dwindling population numbers	Replication of 'successful' species at the expense of less useful species. Simplification of the genetic diversity of species	Increase in economically useful species is positive. Reduction in biodiversity is negative
Intelligent computers	Removal of mundane tasks from human control; possibility of running the built environment to greatly reduced tolerance limits	Fear that robotic intelligence will stultify human ingenuity to complex problem solving	Positive in the short and medium term. Negative (possibly) in the longer term
Intelligent buildings	Controlled internal environments of large buildings will provide better living and working conditions resulting in improved levels of health and efficiency	Fewer large buildings will be more efficient than many small buildings, resulting in reduction in demand for land, reduced inner city transport, reduced pollution	Positive
Recycling	New recycling technology will allow the creation of new employment and wealth-generating opportunities. Savings on scarce and/or expensive resources	Reduction in mining for virgin resources will help reduce the impact on remaining undeveloped areas. New attitudes towards the reuse of materials assists the development of a conservation ethic	Positive
Fusion power	Cheap, abundant, clean electricity allows new phase of high energy-demanding industries. Economics of electric transport improved	Reduction in nuclear wastes from conventional fission power plants. Replacement of oil-based transportation and reduction in oil-derived air pollutants	Positive
Learning by thought transfer	Some 'futurologists' suggest that traditional methods of learning will be replaced by radical new methodology involving the transfer of digital information directly to our brains	None presently known	

the more obvious technologies that will soon become commonplace all of which have the potential for creating environmental change that dwarfs all that has happened previously. While it is the role of the scientist and engineer to provide the new technology, once perfected, it becomes the task of the social scientist, the politician and religious leaders to help advise the general population of the moral and ethical problems associated with new technology. In this respect it is reassuring to learn that the US National Genome Research Institute spends 5 per cent of its budget on an Ethical, Legal and Social Implications Program, while in Britain the Human Genetics Advisory Commission makes recommendations on the social significance of the new findings on genetic research (*New Scientist*, 1999: 5).

The way we think about the environment and its resources has undergone immense change for the better during the last quarter of the twentieth century. However, the predominant attitude remains one that leans towards the 'technocentric accommodator' as defined by Jordan and O'Riordan (2000) (see Table 2.1) and as such remain a long way from an ecocentric political theory which allows a 'maximisation of the freedom of all entities to unfold or develop in their own ways' (Eckersley, 1992). It is unlikely that either an extreme technocentric approach or an ecocentric

viewpoint will best serve society in the future. *Homo sapiens* has evolved to its present position because of the ability to operate as a *strategic opportunist*. Abandoning these skills in the face of the immense technological advances that are now appearing would result in disaster for our species. We need a range of approaches but, above all, we must ensure that human concern for the well-being and respect for other non-human life forms is given far greater priority in the future than it has been given in the past.

Useful websites for this chapter

Greenpeace Memorial site for Contaminated Workers in Cubatao
http://www.greenpeace.org/pressreleases/toxics/1999jan12.html

International Council for Local Environmental Initiatives
http://www.iclei.org/

Millennium Ecosystem Assessment
http://www.millenniumassessment.org/en/index.aspx

Natural disasters, Cubatao (Brazil)
http://habitat.aq.upm.es/bpal/onu/bp033.html

World Resources Institute, Millennium Ecosystem Assessment
http://www.wri.org/wr2000/mea.html

3 Politics and the management of the environment

We are consummately adaptable, able to switch from one resource base . . . to another as each is exploited or used up. Like other successful species we have learned to adapt ourselves to new environments. But, unlike other animals, we made a jump from being successful to being a runaway success.

Tickell, 1993

The first two chapters of this book have shown that the environment has always presented challenges for the survival of humankind. This chapter examines the ways in which science and technology has allowed our species to make the transition from 'Stone Age' to twenty-first century *Homo colossus*. It considers the dilemma that has arisen between those groups (the technocentrists) that support the maximum use of technology with which to enhance the quality of life for humans, and the opponents of technology (the ecocentrists) who argue that it has been overreliance on technology that has caused the current environmental predicament.

3.1 Introduction

Compared with the environmental problems faced by our ancestors, the problems of the twenty-first century differ mainly in magnitude and the speed with which they arise. The nature of the problems remain remarkably similar: they involve the attainment of living space from which we can grow crops and extract natural and mineral resources. The way in which we satisfy the basic requirements is at issue, and what Tickell describes in the quotation opening this chapter, as the 'runaway success' of

human beings, is the cause of the present concern with our environment. We do not need to look far to see examples of the impact of this runaway success. As this chapter was being prepared in February 2002, there were 83 river flood warnings in force in the UK, each one a testament to a climate that appears to becoming wetter (due to natural and/or human causes) and to land use changes (caused by humans) that encourage more rapid runoff from the land. These environmental problems were not 'new' problems, but their magnitude, and the speed with which they have made an impact on modern society has taken us by surprise. Pahl-Wostl (1995) suggested that the emergence of such a scenario, typified by unexpected river basin flooding, would trigger an alarm signal among the population living in areas of risk. In reality, many of the environmental 'problems' are entirely 'natural'. The geographical term 'river flood plain' means exactly what it says – an area of land that periodically will become inundated. If society chooses to ignore the meaning of the word, perhaps because flooding occurs so infrequently that it is forgotten about, or because a flood protection scheme provides a high level of security, then this reflects a failure on behalf of society to recognise the potential hazard of an area or a mistaken belief that society can control the environment. Pahl-Wostl (1995) suggests that in these circumstances, it is necessary for society to revise its priorities and may require major changes in individual and corporate attitudes towards the environment. Society should also recognise that changes that occur within our environment and that cause disruption for society (for example, by flooding) should not be viewed as restrictions or limit our development, but present opportunities and challenges for managers of the environment, for politicians and for society to evolve in new ways.

Environmental 'problems' operate at many different scales, challenging the stability of our individual lives and of entire regions and countries. For many people, the constant effort of combating these problems has caused a form of 'battle fatigue' to set in. Many people consider themselves not to be responsible for contributing to the environmental predicament. They ask, 'What can we do as individuals to combat environmental degradation? We are not cutting down rainforest, or pumping vast amounts of pollution into rivers or into the atmosphere. We are living our lives in the only way we know how.' Such a response conveniently overlooks the fact that the environment in which each individual lives is part of the vast natural biosphere which comprises all the great interrelated natural cycles of energy and material transfer. It is the *combined* impact made by every individual human being that affects the environment and while it may be difficult to measure savings made at the individual level, every individual action can contribute to the solving of an environmental problem. Without the presence of humans, environmental problems would not exist as there would be no one to experience the impact of the problem. The need to manage the environment has been caused by the success of humankind in increasing our numbers, and of diverting natural resources for our own advantage. By so doing we have created problems of overuse and misuse of certain parts of the environment. Persuading society to take responsibility for these actions has proved difficult, as it involves curbing our traditional behaviour and may also require expenditure on environmental management. Box 3.1 examines some of the ways in which humans have caused environmental problems and also ways of managing them.

At the beginning of the twenty-first century, environmental issues occupy a central position in the manifestos of all western political parties and hold a position alongside employment, health, education and transport issues. Indeed, it is recognised that a good-quality environment is necessary for most other aspects of modern life to function properly. It is hard to believe that this situation has come about only in the last 20 to 40 years. Before 1960, the environment was a non-event in political ideology. Throughout the 1980s the political persuasion of the USA and Britain under the leadership of President Reagan (1981–89) and Prime Minister Thatcher (1979–90) respectively were resolutely opposed to taking on board any of the emerging environmental arguments of the day. Those few people who were concerned with environmental issues were seen to be 'diffuse, incoherent, a hotch potch' (Pepper, 1996), with ideas that had been randomly drawn from across the entire political spectrum. People who suggested that it was necessary to avoid further use of technology to 'control' the environment were described as practicing 'green politics'. The 'greens' argued that major mistakes had been made in, for example, the ways technology had been utilised (Commoner,

---

BOX 3.1

Denying responsibility for environmental problems

Humans have an amazing ability to deny responsibility for causing environmental problems! As a species we have been causing changes to the environment and its organic and inorganic components for at least 10,000 years. Jacobson and Adams (1958) recorded evidence of salinisation in Mesopotamian agriculture circa 2400 BC. More recently, Worthington (1977) recorded that 50% of irrigated land in Iraq suffered from build-up of salts due to poor irrigation practice but it is doubtful if any farmer would admit to being responsible for causing salinisation. The excuse given would likely be 'we were only following instructions given by the ministry of land – how could we disobey what our government told us to do?'

BOX 3.1 CONTINUED

Does responsibility for environmental problems rest on the shoulders of policy makers or the individual? Is a political solution the only answer to environmental problems? Providing a straightforward answer to these types of questions is impossible. Behaving with a 'social awareness' and yet permitting development to take place within specific guidelines is probably the best solution. Aldo Leopold (1966) argued strongly that society should develop a 'land ethic' that would allow society to recognise its own responsibility for the impact it made on the environment. In this way, Leopold considered it possible that the individual could contribute to the monitoring of the local environment and possibly help prevent environmental degradation.

An example of Tickell's 'runaway success' can be shown by the growth of recreation and tourism numbers. Organised recreation began in 1841 when Thomas Cook operated the first railway excursion to carry passengers from Leicester to Loughborough. Since that time, the 'runaway success' of tourism has seen visitor numbers to Mediterranean holiday destinations rise to 135 million visitors by 1990. By 2050 this figure is predicted to reach 235–350 million tourists per annum (http://www.tourismconcern.org.uk/media/press_release_balearic.ecotax.htm). The Spanish Balearic Islands (Mallorca, Ibiza, Menorca and Formentara) capitalised on their perfect location for providing sun, sea and sand and mortgaged all their natural resources in pursuit of providing the desired holiday facilities that northern Europeans want. Fresh water resources were overdrawn, pine forests cleared, hotels, apartments, roads and entertainment facilities built with little regard for the local environment. By 2000 there were clear signs that the natural infrastructure was at breaking point. Disposal of refuse and treatment of sewage posed major environmental health problems. The rising consumption of electricity caused supply problems while the level of air pollution caused by incessant tourist traffic made the attainment of safe air quality standards set by the EU unattainable.

The solution to the problem came in a form that a decade earlier would have been thought to be the act of a madman! The socialist–green coalition government elected in 1999 promised to reverse the damage caused to the islands by three decades of mass tourism. A tourist 'eco-tax' amounting to one euro (US$1.4) per adult per night (two euros for guests in five-star hotels) has been levied on every tourist visiting the holiday islands since 1 May 2002 (http://traveltax.msu.edu/news/stories/observer.htm). Eleven million tourists visit the Balearics each year and the eco-tax revenue will be used to build new sewage and recycling plant, to restore heritage sites, create national parks and nature reserves and even to demolish hotels and apartments built in inappropriate locations.

The impact of the eco-tax is calculated to reduce the number of tourists by 5%. Loss in income per head of island population (about 800,000) will be about 320 euros (US$450) per annum but there will be an unquantified improvement in the quality of life for the islanders.

But what do tourists think of an eco-tax bill that adds between 50 and 100 euros to a family holiday of two weeks? Immediate responses from holiday makers reported in the *Daily Telegraph* newspaper on 23 February 2002 were 'the tax is unfair and an insult to . . . tourists', and 'tourist's aren't to blame for water shortages, traffic congestion, refuse tipping and decay. . . . The Mallorcan authorities are'. So much for Aldo Leopold's pious hope that we develop a land ethic! Clearly, the tourist's comments attempt to absolve themselves from any environmental responsibility and show little concern for the so-called ethical responsibilities of tourism (Prosser, 1992).

In this case study, the islands' governments are to be praised for a bold action which could lead to a vastly improved quality of environment. Only time will tell if the eco-tax revenue will be used to repair the damage brought about during previous years. Much also depends on the islanders themselves – will they be prepared to wait to see the results of the eco-tax, or will they resent the loss in tourism income and demand a return to the previous free market conditions? See postscript on eco-tax on pages 76.

1971), the inability of agriculture to feed the population (Anderson, 1972) and failure to curb our own population growth (Ehrlich *et al.*, 1973). These mistakes combined to paint a bleak future for humankind. Green ideology became associated with criticism of the 'existing society and conventional values ... together with beliefs about what future society should be like if it is to be sustainable and environmentally benign' (Pepper, 1996). A future based on green values would comprise a population that did not rise exponentially, of pollution output controlled at source and of recycling materials at the end of their lifecycle so requiring a minimal use of new resources. The greens also sought a new index by which the success of society could be measured. It was argued that traditional economic indicators of growth, such as GDP or GNP should be replaced by an index of sustainability (see page 248). The nearer this index approached zero the closer society had reached a steady state.

What some people see as environmental problems others see as 'opportunities' to demonstrate the capability of human ingenuity to overcome adversity. Modern-day scientists and technologists accept these opportunities of finding ways of solving environmental problems. In reality, what is often required is not *more* science and technology, but an increased awareness of the social responsibilities that rest on one species, *Homo sapiens*. Environmental problems have often been caused by inappropriate or excessive human activity. Somehow, from within the *milieu* that exists between society and nature it is our task to manage these problems. The environmental problems facing modern society take many forms and are caused in a wide variety of ways. The next section identifies some of the main types of environmental problem for which we need to find solutions.

3.2 What are environmental problems and how do they occur?

Environmental problems can be defined according to many different criteria. For example, an atmospheric physicist might consider an environmental problem to arise when a particularly severe sun spot eruption resulted in a surge of energy to be picked up at the outer edge of Earth's atmosphere, disrupting the interchange of incoming short-wave length energy with that of the outgoing long-wave energy. An environmental chemist may find an environmental problem associated with a particularly violent volcanic discharge that caused changes in the chemistry of precipitation following the volcanic explosion. However, these types of environmental problem do not form the central issue of this chapter.

Instead, this chapter focuses on environmental problems that have resulted from the interplay of natural processes and human activity. Glacken (1967) was among the first to examine this question when interest in environmental matters emerged in the second half of the twentieth century when he asked, 'in his long tenure of the earth, in what manner has man changed it from its hypothetical pristine condition?' Eighteenth-century philosophers had posed questions of a similar nature, but answers had been sidestepped in the concern to examine the impact that 'environment' had on humankind. Goudie (2000: 1–11) has traced the subsequent development of these two approaches. The first approach, based on the work of physical and biological scientists, became geocentric in its focus (i.e. a study of the physical, chemical and biological processes that regulate the Earth). An alternative human-centred focus also emerged in which change was seen to encompass a full range of global issues concerning natural and human-induced changes (Munn, 1996).

Coming to terms with environmental problems in the twenty-first century poses a massive conceptual problem for society. On the one hand, we live in an age when technical achievements seem to have no limits – apart from ensuring sufficient money is available to carry out the necessary research. On the other, we know enough about the environment to understand that many of our actions have been responsible for bringing about degradation from the 'pristine condition' contained in the previously seen quotation from Glacken. As shown in Box 3.1, taking individual responsibility for our collective actions on the environment is a difficult concept for modern society to accept. Indeed, Huckle and Martin (2001: 2) suggest that we have devised a 'language and discourse . . . constructed to enable, describe or conceal what is happening'. Society has chosen to ignore many of the environmental problems that have arisen in the twentieth century, preferring instead to believe that technology will solve all problems. One extreme example is shown by the public's attitude toward the safe disposal of radioactive waste from civilian and military sources (see Box 3.2).

The magnitude and global extent of environmental problems has made the traditional strategies of migration and postponement of action no longer applicable to society. Globalisation of environmental problems now means that, for example, deforestation of tropical rainforest in Indonesia is as much of a problem for the citizens living in northern hemisphere cities as it is for the local inhabitants that lose their rainforest habitat. The contribution that each of us makes towards using the environment adds to the collective total and as such, any reduction that we can make as an individual contributes to a reduction for the whole. The situation has become complicated in that we now realise that many of the actions we take, both at individual and collective levels, contribute to a range of impacts, some of which result in benefits to the planet while others are detrimental. These impacts become evident on a variety of different

---

BOX 3.2

Radioactive waste disposal and the environment

Radioactive waste is produced from a variety of sources: by nuclear power stations, from military sources and low-level wastes from medical use. Unlike all other waste materials produced by human activity, there is no known method for the safe disposal of nuclear waste. High-level nuclear waste (see below) such as uranium-238 and plutonium-239 from nuclear reactors is particularly dangerous even after thousands of years. Nuclear waste is currently stored in slurry or liquid form in surface tanks protected by concrete and lead shields or buried in impermeable strata in deep boreholes as vitrified solids. Dumping of radioactive material deep on the ocean bed is now banned by international treaty.

Nuclear waste is classified into three categories of radioactivity:

1   high-level waste, with a radioactivity greater than 3.7×10^{-10} becquerels (Bq) per gallon of slurry
2   intermediate waste, with a radioactivity of between 3.7×10^{-4} and 3.7×10^{-10} Bq per gallon of slurry
3   low-level waste, with a radioactivity of less than 3.7×10^{-4} Bq per gallon of slurry.

All radioactive materials have half-life values that indicate the amount of time necessary for radioactivity levels to subside to non-dangerous values. Radioactive tracers used in medicine have very short half-life values, for example, iodine-131 has an 8-day half-life and after 50 days its radioactivity is but 10% of its fresh value. Plutonium-239, a product from all nuclear reactors, has a half-life of 240,000 years and retains a lethal radioactivity level 500,000 years into the future. Phillips and Mighall (2000: 228–30) state that in the USA alone, some 420,000 metric tonnes of high-level waste and spent fuel rods from nuclear power stations are in storage. The decommissioning of the nuclear arsenal following the end of the Cold War has added many thousands of tonnes of high-grade military wastes, some of which can be reprocessed for use in civilian power stations. The economic cost of dealing with radioactive waste has been substantially underestimated over the years. Hollister and Nadis (1998) suggest that the full cost to the US government of cleaning up four nuclear weapons sites could be over $375 billion. Failure to find the best possible way of rendering safe the growing amounts of nuclear waste could have immense environmental costs as well as causing major health hazards for humans.

The World Nuclear Association website provides an excellent source of information: http://www.world-nuclear.org/education/wast.htm.)

timescales and their strength may vary from region to region. For example, the development of settled agriculture was accompanied by a gradual development of social maturity within the first agricultural communities, yet the very act of becoming 'settled' placed a massive burden on the soil in which settlement occurred. Butzler (1982) has shown that deforestation and the deliberate use of fire associated with land clearing are the most important reasons for causing an acceleration of soil erosion. As an agent of erosion, humans are at least on a par with natural forces and Judson (1968) has calculated that since settled agriculture began, the total sediment load transported by rivers has increased by 300 per cent! As human civilisation has advanced over the ages, it has created an ever-greater impact on the environment and all other non-human species. The perpetrators may be unaware that environmental degradation has occurred but that does not detract from the fact damage to or simplification of the environment and habitats has occurred.

3.3 The uniqueness of human actions on the environment

To understand the special role played by humans in the biosphere we must first place our species into its correct location within the environment. Humans are members of the animal kingdom and, as for all animals, we obtain our food from a variety of secondary sources, originally from within a natural food web, but today from an agricultural food chain. We differ in that unlike other animals that are merely *components* of the food web, we have taken *control* of the ways energy and matter pass through our food chain. We have achieved this through the management of our different agricultural systems. Second, we differ in that we rarely find ourselves preyed on by another animal and consequently have been relieved of the need to constantly guard our food supply or indeed our own survival. We have removed ourselves from the natural food web and have taken up a position as 'controller' of an agricultural food chain.

The earliest interactions between humans and their environment would probably have involved access to food (Nicholson, 1970). For example, the first hunter–gatherers ate seeds, berries, shoots and nuts collected from the natural vegetation. Birds and other mammals (notably monkeys and apes) competed for these same food sources but, provided the population size of our ancestors remained small and nomadic, direct conflict with other species searching for food probably remained slight. The situation changed when our ancestors became sedentary farmers. Other wild animals would then have attempted to take food from the protective stockades in which animals were herded. Farmers would have prevented this by scaring off the intruders by means of shouts or throwing sticks and stones. From this point onwards, the relationship between humans and the natural environment changed, with humans successfully reducing or removing all forms of direct competition.

Soon after adopting a settled lifestyle, a new series of environmental problems emerged. Unlike a nomadic existence, a settled community must ensure that it does not create pollution that itself becomes a hazard to the community. Gradually, the expansion of human influence over the natural world expanded the territorial influence of humankind. Tickell (1993) stated that such is the demand for organic resources by humans that we now make use of 40 per cent of all the plant growth (as agriculture, forest and industrial crops). Initially, lack of technology meant settlements were confined to the most easily managed sites, where the greatest yield of crops could be obtained for the minimum effort. As the skills of farmers increased, so the farmers moved onto sites that required greater technical capabilities until today, agriculture has extended onto all the potentially cultivable sites – and a good many unsuitable sites as well. By 2000, Park (2001: 618) has stated that up to half of the land surface had been substantially altered by agriculture or urban and industrial land uses.

As soon as settled agriculture was seen to possess advantages over nomadism, our ancestors set about maximising the advantages of a sedentary lifestyle. Hardin (1968) has explained the sequence of events in his now classic paper,

Tragedy of the Commons, in which, one by one, each family unit would maximise the agricultural potential of the patch of land, allowing more food to be produced. Section 7.6 explains this process in more detail and also the consequences for the environment.

The practical effect of the events described by Hardin were to initiate overgrazing of the land by domesticated animals, followed by the first signs of soil erosion and to environmental degradation. Middleton (1999) has likened these events to a game of chance, calling the process the 'global casino' because the element of chance plays a great role in the precise format of how environmental degradation might occur. As we study the environmental problems of the present day, the 'global casino' analogy becomes more relevant because of the huge financial motivation that now usually accompanies the use of natural resources. Middleton suggests that the casino metaphor can also be applied at a socio-economic level, with some sectors of society having greater opportunity to engage in contests with the environment.

3.4 Application of science and technology to solve environmental problems

In Chapter 2, it was shown that as the use of technology by humankind increased in extent and in sophistication it placed a greater burden on natural habitats and non-human species. It is now necessary to consider whether a contributory cause to environmental decline – the introduction of powerful, new technology, the repercussions of which were unknown – can itself be harnessed to solve the problems it has helped create. In the past, we have developed new technology to overcome problems of resource shortage so it is not inappropriate to suggest that new technical 'skills' might also be capable of counteracting environmental degradation. White (1967) was one of the earliest writers to draw attention to the way in which technological advances allowed our civilisation to develop. Advances in technology enabled humankind to outcompete nature in almost all respects

until today, the situation has been reached when technology allows society to make almost unlimited choice as to how we use natural resources. Lomborg (2001) has argued cogently that one of the greatest human success stories is the way in which we have successfully used new technology to support growth and expansionism within society. Supporters of the ecocentric argument however, suggest that expansionism cannot continue indefinitely. Drawing parallels with exponential growth in non-human species they suggest that shortages of key natural resources, combined with undesirable feedback from overcrowding, disease, and stress, will ultimately lead to sub-optimum conditions. Continuing the ecocentric argument, it is suggested, for example by King and Schneider (1991) and von Weizsäcker *et al.* (1998), that such is our technical prowess that it should now be possible to use technology in a quite different way from in the past. New technology should enable the quality of life to be improved, but at the same time exert a much lower impact on the environment.

We remain a long way from the science fiction image of a future technology that allows us to live in a perfect world in which the world population is stable and we recycle all our waste and generate zero pollution. In reality, the use of technology has paid scant regard for the environmental consequences. In the past, this has been partly due to ignorance of the ways in which the biosphere operates. However, this excuse is becoming invalid as scientific knowledge of the biosphere and its components becomes more complete. Subsequent chapters of this book provide case studies of the ways in which the physical and biological components of the environment have been damaged through inappropriate use of technology. Fortunately, there are also some success stories, such as river basin management and designation of conservation zones in which species biodiversity takes precedence. A new emphasis on the development of *appropriate* technology holds promise of an increase in resource use but without an associated deterioration of the environment.

Raising the use level of a resource or part of a natural system depends on understanding the safe level of use that can be placed on the system.

Calculating that level involves a detailed knowledge of the system and includes:

- the species of plants and animals comprising the system
- the morphology of the components (i.e. how the parts fit together)
- the internal linkages between the components involving the flow of energy and the movement of materials
- the means of recycling materials through the system
- the natural variability within the system from season to season and year to year
- the interconnection between the system and the adjacent systems
- an indication of how extracting resources for human use affects the system and the surrounding systems.

Providing answers to this list of points is a major task and often the answers we can provide are only partial. However, there are other ways with which to calculate the safe level of use of an ecosystem or natural resource. The exact method depends on whether the resource is abiotic (non-living) or biotic (living).

3.4.1 Determining the safe use of an abiotic resource

Abiotic resources comprise the non-living physical components of the biosphere. These include: the atmosphere consisting of a mixture of gases (examined in more detail in Chapter 4), fresh water (see Chapter 5), salt water (see Chapter 6) and soil, land and mineral resources (considered in Chapter 7). By tradition, the atmosphere, the oceans and fresh water have been considered as 'free resources'. They are owned by no one and are freely accessible by all. There have been many occasions in the past when this group of resources has been overused, resulting in shortages, or misused, resulting in pollution problems. Management of some of these resources has therefore become necessary to ensure a guaranteed supply of essential components in order that society can continue to function. Two of the abiotic resources have largely escaped management, at least until very

recent times. The oceans and the atmosphere have been exempt from management, partly because of their vast size, partly because of their constant turbulence and partly because of the sheer problem of obtaining international agreement on how to manage these resources.

The other abiotic resources, soil, land and mineral resources located beneath ground, being small in size and occupying a discreet space, have been more easily managed. In particular, their management has been made possible because a person, an organisation, or a government usually owns them. Ownership usually entitles the user to consume or utilise the resource, either by mining, abstraction or farming the resource. Farming the land involves care and maintenance of the land, for example, ploughing, draining, fertilising, fencing or terracing. Although the soil is an abiotic resource it is unique in that it provides a base in which vegetation can grow and thus forms a vital bridge between the non-living and living worlds. Used carefully, soil can be cultivated for centuries without loss of fertility or loss of physical structure. In other situations where inappropriate management or steeply sloping ground has been cultivated, soil that has taken thousands of years to form can be lost in a decade (see Colour Plate C).

Mineral resources are almost always carefully managed due to their potential economic value. Metal ores, coal, oil and phosphate-rich rock are typical examples of abiotic resources that are mined and sold for profit. Market demand will determine the value of the reserve, as will ease of access to the resource, distance from market and the refining costs of the product. The quantity of the reserve can usually be determined by geologic prospecting and the quality assessed by chemical analysis. Once these factors are known, economic factors take over to determine the rate at which the reserve will be consumed. At a certain cut-off point, the resource will become uneconomic unless new technology becomes available to revitalise the economic value of the resource.

The consumption of an abiotic resource to its point of physical exhaustion takes place because the resource has a renewal cycle that is measured on a geological timescale, that is in hundreds of thousands, or even millions of years. For this reason, abiotic resources are termed 'finite'.

3.4.2 Determining the safe use of a biotic resource

Biotic resources also provide vital supplies for the support and sustenance of humans. They differ in one important respect from abiotic resources. For the most part, they are capable of renewing themselves on a timescale that is coincident with the human lifespan. Apart from the growth of some hardwood tree species that take more than 100 years to reach maturity other organic resources take between six months (agricultural crops) to about 30 years (trees destined for conversion to pulpwood) to reach the end of the *economic* cycle. We make use of only a fraction of the total number of species. For example, of the 250,000 or so flowering plants, Tudge (1988) states that humans have exploited only about 3000 species and of these a mere 150 species have been commercially exploited. At the present time, our food crops are based on an incredibly small group of only 20 different species. A similar situation exists with the commercial use of fish stocks, while approximately 80 per cent of soft wood timber production in the UK originates from one tree species, the Sitka spruce.

It might be assumed therefore, that all the non-commercial species exist in complete safety away from human attention. This is not so. In an ecosystem, species co-habit in bewildering profusion. Order is provided by means of species arranged into *trophic* (or feeding) layers. Trophic layers comprise units in the food web, each organism occupying a specific trophic layer and form the prey of the adjacent feeding layer. In a survey of species diversity made on a forest in southern England, different tree species were shown to provide habitats for different types and numbers of species (see Table 3.1). If a tree species of economic importance – such as the oak – were cut down and removed from the forest, the existence of up to 284 dependent species would be jeopardised. Chapter 8 considers the importance of species diversity in more detail.

Management of a biotic resource becomes a difficult compromise between establishing a *sustainable* harvesting level of a resource with economic value, and the role that species plays in supporting the entire ecosystem, of which it is a part. Overharvesting not only depletes the valuable biotic resource but also threatens the survival of other species that are essential in providing the biodiversity of the ecosystem. Establishing a safe level of harvesting therefore becomes a critical management decision in the long-term survival of the ecosystem.

One way in which it is possible to establish a safe level of consumption is to establish the natural *carrying capacity* of an ecosystem. All ecosystems will possess a theoretical optimum number of species and individuals but this number is unlikely to be attained except in exceptional conditions. Instead, the population will constantly fluctuate depending, for example, on the climatic conditions that prevail over a period of time. On some occasions, the carrying capacity may be exceeded, while at others, spare capacity may exist. The fluctuating capacity of an ecosystem to support plants and animals is the antithesis of what good management practice hopes to achieve. For example, a modern farming system aims to control most of the inputs to the agricultural system, thereby maximising the growth rate of a crop, or of animals for the maximum duration. Clearly, a mismatch exists between the management objectives of an agricultural system and the natural fluctuations that occur within an ecosystem. By applying technology to the managed system it has been possible to raise the output of the agricultural system. This has chiefly been achieved through the input of additional energy sources, mainly the replacement of human and animal labour by machinery powered by oil or electricity. It is still necessary, however, to be aware of the carrying capacity of the environment in which the farmer is operating. As the output level of the farm is raised then so, in theory, is the risk of causing environmental damage, typically shown as soil erosion or desertification. An understanding of the concept of carrying capacity thus becomes increasingly important (see Box 3.3).

Table 3.1 Number of animal species supported by different tree species in southern Britain

Common oak	*Quercus robur*	284	Long-term native
Sessile oak	*Quercus petraea*		
Willow species	*Salix* spp.	266	Long-term native
Birch species	*Betula* spp.	229	Long-term native
Hawthorn	*Cretagus monogyna*	149	Long-term native
Blackthorn	*Prunus spinosa*	109	Native
Popular species, including Aspen	*Populus* spp.	97	Native
Crabapple	*Malus sylvestris*	93	Native
Scot's pine	*Pinus sylvestris*	91	Long-term native
Alder	*Alnus glutinosa*	90	Long-term native
Elm	*Ulmus glabra*	82	Long-term native
Hazel	*Corylus avellana*	73	Long-term native
Beech	*Fagus sylvetica*	64	Native but recent arrival
Ash	*Fraxinus excelsior*	41	Native
Spruce	*Picea* spp.	37	Exotic
Lime	*Tilia europea*	31	Native but recent arrival
Hornbeam	*Carpinus betulus*	28	Probably introduced by Romans
Rowan	*Sorbus aucuparia*	28	Native
Maple	*Acer campestre*	26	Introduced
Juniper	*Juniperus communis*	20	Long-term native
Larch	*Larix* spp.	17	Introduced in seventeenth century
Fir	*Abies* spp.	16	Introduced in seventeenth century
Sycamore	*Acer pseudoplatinus*	15	Native but recent arrival
Holly	*Ilex aquifolium*	7	Native
Sweet chestnut	*Castanea sativa*	5	Probably introduced by Romans
Horse chestnut	*Aesculus hippocastanum*	4	Introduced
Yew	*Taxus baccata*	4	Long-term native
Walnut	*Juglans regia*	4	Introduced
Holm oak	*Quercus ilex*	2	Introduced in sixteenth century
Plane	*Platanus acerifolia*	1	Introduced

(*Source*: data from Nature Conservancy)

BOX 3.3

The concept of the 'carrying capacity'

Carrying capacity is a well-established model used by ecologists and expresses the maximum population size that can be sustained by a habitat over a period of time. It is a concept that applies mainly to natural animal populations but it also has relevance to the size of the total human population.

A population of animals (the *heterotrophs*) will be limited in number by the ability of the vegetation (the so-called *primary producers* or *autotrophs*) to provide them with enough food. A population can exceed the carrying capacity of the environment for a short period of time, but the price for *overshooting* the carrying capacity is that some animals will be underfed and will ultimately die from starvation or will be forced to migrate to locations that have a population below the carrying capacity. Initially, when organisms arrive in a new area they will undergo rapid population increase (often following an exponential increase – doubling over each unit of time). This produces an increasingly steep population growth (see Figure 1.3), but the

BOX 3.3 CONTINUED

steepness of the curves flattens off as the maximum population size is reached and then fluctuates slightly over successive time periods. The reason for the stabilisation of the population size is due to an *environmental resistance* (represented by the K value). K comprises a complex set of values, including food supply, availability of territorial space, competition both within and between species, out- and in-migration and disease. The strength of K will determine the fertility of the population, the mortality rate and the life span (see Figure 1.4).

In theory, all species can be classified as adopting either predominantly 'r' or 'K' life strategies. Species that follow an 'r' strategy are likely to have populations that follow an exponential curve and are the result of density-independent growth patterns. On the other hand, 'K' strategists follow a logistic growth curve and have a population that responds to density-dependent factors. In theory, we would expect human population growth patterns to follow a 'K' strategy, with a life history that is adapted to relatively scarce resources and reproduction that can be controlled when conditions are harsh and competition becomes intense. Through the application of new technology, *Homo sapiens* has made available greater quantities of existing resources as well as discovering totally new resources. These have allowed our species to continually extend the carrying capacity to new limits and we have responded by increasing our numbers more in line with a 'r'-type species. The curve of world population numbers appears more like that of an exponential curve than that of a logistic curve. However, for individual countries (European countries in particular), it is often the case that population numbers have become relatively stable. Changes in the detailed shape of the population curve are mainly due to periods of population loss during warfare (World Wars One and Two) and to phases of international migration.

How can the concept of the carrying capacity be applied to human populations and can it help calculate the maximum sustainable population size? Demographers use the term *overpopulation* to describe the situation when the human population exceeds the carrying capacity of a region. In animal populations, overpopulation is accompanied by habitat degradation and, ultimately, a population crash. Some environmentalists (such as Ehrlich, 1977) believe humans have already caused habitat degradation (e.g. soil erosion, pollution, malnutrition), while others (notably Simon, 1996) believe that technological advances will allow the carrying capacity to be raised substantially beyond its present figure of 6.2 billion.

In order to understand how the carrying capacity operates we need to know a little more about the way in which populations increase in size. The population size for any species will either *increase* over time (positive growth rate) or *decrease* over time (negative growth rate). These changes can be expressed as an equation:

$$N_t + 1 = N_t + (B - D) - (I - E)$$

where N = number; t = base time; t + 1 = one unit of time after the base time; B = number of births in the unit of time being considered; D = number of deaths in the unit of time being considered; I = number of in-migrants in the unit of time being considered; E = number of out-migrants in the unit of time being considered.

In a situation where K = 0, the environment would exert no constraint on population growth rate (for example, unlimited living space, no climatic restrictions such as drought or frost, a constant and limitless supply of food, and an absence of competitor species or losses through disease). Under these conditions, a constant increase in population would occur, limited only by the reproductive limits of the species. This theoretical maximum rate of increase is termed the *biotic potential*. Each species will have a different biotic potential depending on the gestation period, for example 18–25 days in small rodents to 22 months for an elephant.

Figure 1.3 shows a population growth curve typical for a species that follows the biotic potential. There are relatively few instances when biotic potential is able to operate uninterruptedly for more than one or two

BOX 3.3 CONTINUED

successive time periods. *Environmental resistance* will quickly begin to operate and limit the population size. The initial J-shaped population curve is replaced by a *sigmoid curve* (or S-shaped curve), as shown in Figure 1.4 in which the growth rate is lowered due to the operation of a varying environmental resistance. The population responds to the environment rather than being controlled by it.

The number of individuals that a region can support over time will vary constantly. In mid-latitudes, the number is usually lower in winter than in summer due to winter cold reducing the rate of food production. Conversely, in Mediterranean regions the hot dry summers cause a greater reduction in numbers than the wetter, mild winters. Species respond to these differences by migrating at unfavourable times of the year or by timing their reproductive cycle to coincide with the most favourable living conditions.

(A fuller explanation of the principles of population growth and carrying capacity can be found in many textbooks on ecology, for example Jackson and Jackson, 2000 and Jarvis 2000.)

3.5 Technocentric versus ecocentric development: the dilemma

Many dedicated environmentalists consider that the very act of involving technologists and scientists to solve environmental problems to be the worst possible option as they claim it has been science and technology that has been responsible for getting us to the current environmental predicament. To an extent, this assumption is correct, but the argument fails because criticism is usually founded on the use of old technology and science in which an understanding of environmental science was almost non-existent. Inappropriate technology has undoubtedly been the cause of many environmental predicaments but there appears to be no alternative for humankind other than continuing to rely on the best modern technology. We have manoeuvred our societies into a cul-de-sac from which it is impossible to turn around and find another pathway into the future. Any change, no matter how slight, from our reliance on technology would lead to food shortages, power blackouts, a decline in medical care, a truncation of personal mobility – in short, a substantial retrograde step from current standard of living. Ecocentrists counter by asking what have we to lose? The longer we rely on technology, the further we move towards inevitable environmental disaster, so why not make the decision and abandon total reliance on technology? They argue that the losses incurred by turning away from traditional technology could be regained by reorienting society to work within a truly sustainable level of carrying capacity. In addition, further gains would be achieved in the form of an assured long-term, higher quality of life in which ethical, moral and aesthetic values superseded materialistic gains. Can we imagine a time in the future when environmental concerns outweigh all others? Technocentrists would categorically reply 'no' as technology will *always* be able to solve our problems. Ecocentrists would provide a more complex reply and suggest that there is no assurance that technology will always be successful. For that reason society should keep their options open, allowing environmental factors to rank alongside issues of economic well-being, of public health and of political and security issues.

Although the ecocentric argument may appeal to those who yearn for a lifestyle that is less reliant on greater use of technology, would ecocentrists alive today be prepared to take action that might restrict the benefits still to be gained from science and technology in the future? Can we afford to forgo the prospects of still-to-be-developed medical knowledge that would cure most illness, replace damaged organs and limbs by new ones grown in a laboratory or extend the lifespan to an unimaginable length? Less controversial, do we want to throw away the opportunities of a household that is managed by the microchip: a house that cleans itself when dirty, restocks the freezer when empty or controls the heating and ventilation according to our personal requirements? Could we give up the choice of products available at affordable prices found in hypermarkets and shopping malls? The answer to such a series of questions is almost inevitably 'no'. Any change that results in reduction of the well-being of the individual and for society would be strongly opposed. It would be catastrophic to propose that society turns its back on science and technology as the means of bringing new advances to society.

The challenge that faces the inhabitants of the twenty-first century is how to adapt science and technology so that it benefits not only the human component, but also all the non-human species and the environment in which we all live. O'Riordan (2000) has summarised the current concern expressed by some sectors of society that see the modern power of science and technology directed against the sensitive and complex real world. Science, so some conservationists argue, is concerned with the selection of a small number of environmental and biological variables, the manipulation of which makes it possible to achieve a better management of the environment for the benefit of society, but not always for the benefit of other non-human organisms (Wynne and Meyer, 1993).

The paradox of the situation is that without the discoveries and achievements of science our species would remain at the level of a Neanderthal society. We have relied on the achievements of science to raise our standard and quality of life. What is now at issue is the fear that by relying on science we may have created a belief that we possess a freedom to do what we wish with the Earth. Never before in the history of humankind has there been such a concern that we need to restrain the success of the human animal and to impose limits on the ways in which we use the planet. O'Riordan (2000) claims: 'Until now those qualities [reliance on science] have been the very essence of progress and material security. To challenge them requires boldness and a cast-iron justification.'

In spite of all the evidence that shows how technology has allowed modern society to attain its current position, there remains a belief among some sections of society that the application of modern science and technology is moving society into greater conflict with the environment. Of particular concern is the increasing way that powerful multinational companies are sponsoring research institutes to undertake scientific research on their behalf. The controversy over research to support the use of genetically modified (GM) crops provides an example of the classic conflict between multinational companies such as Monsanto, governments, environmental NGOs and individual members of society. Despite a major information campaign by industrial companies pioneering the use of GM crops, public opinion against GM foods, especially in Europe, shows clear signs of increasing resistance, although in America about one million hectares of land had been planted with GM crops by 2000 (Huckle and Martin, 2001). Worries over the effects of GM crops are threefold: effect on wild species, effect on human health and economic implications. The first of these concerns the inadvertent transfer of pollen from GM crops to wild species, resulting in a potential modification of the wild stock. Loss of the pure genetic structure of wild plants would be of concern to geneticists in that they have traditionally relied on wild stock to cross-pollinate domesticated species in what is called 'back-crossing' to reinvigorate agricultural crops after several generations of cultivation (Ehrlich and Ehrlich, 1982: 85). Once into wild stock, trans-genetic material could move through the food chain in totally unpredictable directions, unintentionally affecting many other species. The impact of foodstuffs containing genetically modified material on human health is probably infinitesimal. Monsanto publicity claims that: 'Research shows GM foods to be as close to 100 per cent safe for humans as science can ever be.' But the long-term effects of GM products on humans has not been certified as the first GM crops were only produced as recently as 1996. By 2002 GM crops were being grown in 15 countries, (predominantly in the US) covering 125 million hectares worldwide (Goddard, 2002). Soybeans and corn are the two crops most likely to incorporate modified genes (see Figure 3.1). Genetically modified crops have already allowed significant improvements in crop yields as resistance to

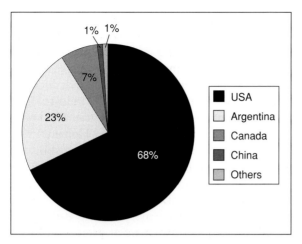

Figure 3.1 Percentage of genetically modified crops grown in different countries throughout the world, 2002.

diseases, pests, infertile soils and climatic restrictions (drought, frost) are incorporated into the modified crops. The benefits of GM crops in terms of increasing food output, especially in developing nations, has undoubted benefits. In these circumstances, how relevant are the attempts of environmental activists, who enjoy the comfort of a developed world existence, to block the use of GM crops?

Government regulations do not permit the use of new scientific and technologic processes outside the research laboratory without compulsory exhaustive testing. New products are carefully assessed for all *known* implications before a product or a technique is given a licence for use. However, such is the level of public concern over the safety of genetically altered crops that neither the licensing procedures nor the publicity machine of industry and commerce are sufficient to allay fears of possible damage to the environment. Despite the use of sophisticated marketing publicity some sectors of the public remain unconvinced that certain aspects of new technology, especially those that involve manipulation of genetic material in the production of our food, are totally safe. For many people, it is not the level of scientific assessment that is questioned, but fears that insufficient time has elapsed for side-effects to emerge or that too many vested interests exist between government and industry. In an attempt to understand

the mood of the protesting groups, companies that make most use of new technology in their business have been required to learn new ways of promoting the safety of their products. Whereas the public appears willing to accept radically new technology in the field of electronics (as seen in home computing, digital cameras, hi-fi equipment and in motor vehicle design), the same is not true for food production or medical research. In an attempt to help industrialists assess the mood of the general public, managers are being trained in the skills of corporate social responsibility (CSR) in which the possibilities offered by technological advance are matched to the mood of society (see www.ft.com/csr for a series of four articles on CSR). One of the objectives of CSR is to examine the motives of informal and temporary organisations known as 'pressure groups' or lobby groups that are formed to oppose technology that *might* have unknown effects on the environment or on minority interest groups unable to protest for themselves.

3.6 Environmental pressure groups

Hardly a day passes without our TV, radio and newspapers reporting the activities of a conference, a campaign or a report that tries to focus our attention on a critical 'current environmental issue'. The issues at stake cover almost every imaginable aspect of modern life. Unfortunately, the success of each case depends not only on the merits of the cause but also on the amount of 'air time', and hence sponsorship, of the issue.

Environmental pressure groups have existed from the 1870s, although for about the first 100 years they were not labelled as 'pressure' groups. Indeed, they were as far removed from the image of a protest group as it was possible to be. They took the form of local groupings of interested members of the public who set up local history clubs or ornithological societies, meeting in church halls or the homes of individual group members. Progress was greatest in the New World and the passing of the Conservation Act by the US Congress in 1872 was the first to allow land to be set aside for the exclusive benefit and enjoyment of the public.

In Europe, activities of the nineteenth-century environmentalists took quite a different pathway, concentrating their efforts on the study of local flora and fauna, recording local meteorological conditions, the keeping of diaries recording information such as arrival dates of the first migrant birds in springtime and departure dates in autumn. The first efforts of these small groups were to protect the existence of threatened species of plants and especially of animals (Sheail, 1998).

The first modern environmental pressure groups appeared at the beginning of the 1970s (Grant, 1995). Although the range of issues covered by environmental pressure groups has been remarkably diverse, they have the common aim of successfully influencing public attitudes towards environmental issues at a local, national or international scale.

Although pressure groups are now widely recognised and accepted as part of the democratic system, their purpose within the structure of society must be questioned. We can legitimately ask why we need pressure groups in a society governed by highly transparent, democratically elected governments. Do pressure groups help the processes of democracy or do they attempt to destabilise it? Are pressure groups intent on bringing about a fundamental reshaping of society or are they mechanisms used by politicians for concentrating power within the existing political system?

Many different categories of environmental pressure group now exist, ranging from the small local group that tries to prevent local development on a greenfield site, to the global operations of organisations such as Friends of the Earth and Greenpeace. The raison d'être of pressure groups has broadened considerably since the 1970s. At first, they were seen as dissent groups that had grown from the left-wing student groups in west coast university campuses of the United States. The establishment identified these groups as dope-smoking, long-haired, flower power dissidents who wanted to escape the US draft laws and avoid being sent to fight in Vietnam. That these people wanted a different lifestyle from that offered by the establishment was only poorly understood, as was the concept that the ways in which human

development was thought to lead to a greater confrontation with 'nature'. The newly emerging discipline of environmental studies, based on the much older academic study of ecology, had already drawn attention to a serious mismatch between the rate of resource consumption by industrialised societies. In an argument reminiscent of that raised by Thomas Malthus (see Jones and Hollier, 1997: 131–4), questions were asked about the ability of the environment to continue to yield resources and to absorb the growing quantity of wastes from industrial and domestic sources.

In the 1970s, most environmental issues were of local or regional concern. For example, the Los Angeles smog problem was specific to downtown areas of that city. One of the first major success stories for the environmental lobby was to ensure that Los Angeles Health Department monitored the air quality of the city for specific pollutants – even though the damage caused by the pollutants was not fully understood. On a wider scale, the diurnal movement of the infamous Los Angeles smog caused severe damage to citrus fruits grown throughout the state and particularly in Orange County and the San Joaquin Valley. Had this problem existed today, protests might have been more vociferous. Instead, the farmers preferred to sell their high-value farm land to real estate developers and to re-establish their farms away from the smog. Some farmers repeated this sequence several times as the scale of the smog problem became greater and today it is significant to note that a smog problem is found in many other areas of the world. In Los Angeles, a decline in urban air quality was due to the mixing of motor vehicle exhaust gases with fragments of rubber from tyres and particles of road surface material into a photochemical compound that became toxic in the presence of solar energy. Because of the synergistic actions of sunlight on the air pollutants ingredients (the peroxyacyl acids or PANs), the toxic mixture became known as photochemical smog. PANs were also the cause of respiratory disease in humans, especially those such as traffic policemen and taxi drivers, who spent most of their working hours breathing the downtown Los Angeles air. What had started as a local issue quickly became relevant to much of California and, later still, PANs were shown to occur in the atmosphere above most of the other large industrial cities of the world. Significantly, however, it was the development of new technology that solved the problem of urban smog. The compulsory use of unleaded fuel, the fitting of all petroleum-fuelled vehicles with catalytic converters to clean exhaust gases of nitrogen oxides and the engine management systems developed around fuel injection as opposed to the earlier carburettor has resulted in a significant improvement in urban air quality despite a massive rise in car use.

Pressure group organisers quickly learnt that to obtain maximum success, publicity campaigns should be directed towards two quite separate targets. First, key staff in public administration, for example the local mayor, and elected representatives, should be made aware of the political consequences of not taking action. Second, the concern of the general public should be activated through a publicity programme showing, where appropriate, the impact of the environmental problem on public health, the loss in revenue through reduced crop harvests or loss in quality of life. Some of the most successful protest groups have been centred around the emotional protection of endangered species – seal cubs around the Hudson Bay, the Spotted Owl in the Californian redwood forests, Greater Crested Newts along the route of a new bypass road in southern England. Having the support of a film star or sporting hero was also beneficial to the cause! Pressure groups gained considerable ability to influence public opinion and in such a situation, counter-pressure groups were set up to oppose the original debate. Such an example occurred in November 1990, when the Californian electorate faced a radical environmental programme, dubbed Big Green (*Nature*, 1990). The electorate were faced with a multitude of complex and often contradictory issues. Early opinion polls suggested a high degree of public acceptance for the new political initiative to cope with environmental problems. However, a counter-proposal was launched by opponents to Big Green and was publicised by a massive TV and press advertising campaign. Green campaigners countered by claiming the advertising campaign was paid for by anti-environment business interests. At the last moment, the Californian electorate suffered mis-givings, resulting in the defeat of the majority of the environmental issues.

Environmental pressure groups are not against all development. They are often seen as a thorn in the side of many multinational companies, because they ask questions that many individuals would like to ask but can't get access to the appropriate person. Asking appropriate questions of business and commerce requires detailed knowledge, involves the expenditure of time and money and, increasingly, requires a detailed knowledge of the planning and legal systems. Few private individuals can operate at the level required to challenge the actions of a powerful company. We can consider the work of the pressure groups as representing the conscience of society. Some pressure groups have taken on a quasi-official status at which point they change status from that of a pressure group and become a non-government organisation (NGO). NGOs comprise non-party political pressure groups, advisory agencies, aid agencies and consultancies that list among their aims the protection of the environment. Some NGOs, such as the Worldwatch Institute and the International Commission on Climate Change (ITCC) are part-funded by governments specifically to carry out independent research concerning international problems. While the acceptance of government funds by NGOs might be considered to lead to a loss of independence this has generally not happened. Instead, the sponsoring of NGOs to investigate specific problems enables governments to obtain an independent, and often alternative, viewpoint on environmental problems that would otherwise be lacking.

Some governments sponsor NGOs to act as 'think tanks' on controversial issues. For example, in June 2000, the Royal Commission on Environmental Pollution reported that Britain must achieve a 60 per cent cut in CO_2 output from fossil fuels if the nation was to meet its contribution to global warming – and to allow poorer countries room for their development (RCEP, 2000). The Commission found that unless significant efficiency gains from fossil fuel power plants could be achieved by 2050, a 60 per cent reduction in CO_2 will require the equivalent of 50 new nuclear stations the size of Sizewell B (1.2 gigawatts). In addition, it would need major investment in off-shore wind farms, wave power generation and 15 per cent of farmland devoted to energy crops and a tidal barrage across the River Severn to replace the output from fossil fuel power stations. It is clear that a 60 per cent reduction in CO_2 output would require a radical change in the structure and behaviour of society. When the Commission published its findings in 2000, its recommendations were sidelined in the belief that society would not be able to adapt to a change of such magnitude. However, one year later, in February 2001, the newly formed Department of Environment, Food and Rural Affairs (DEFRA) was openly proposing the need to reduce global CO_2 level by 60 per cent with 'many benefits for the UK from taking early action to cut emissions' (DEFRA, 2001).

In this example, we find the British government using a respected organisation to act as a pseudo-official pressure group to prepare society for an inevitable and controversial change of government policy. Lukes (1974) had predicted such use of a royal commission to study contentious issues stating that this would be an example of 'second-dimension' political power in which government divested itself of dealing with difficult decision making to a reputable organisation comprising leading scientists in the field. Such a policy will only be successful if the public holds science in a sufficiently high regard to comply with the recommendations. The EU monitoring organisation, Eurobarometer, has shown that while Europeans have high expectations of science and technology, recent crises such as BSE in cattle, foot and mouth disease and uncertainty over GM crops have resulted in science and technology no longer being held as the panacea for solving serious problems. Eurobarometer Report 55.2 (2001), provides the results of the survey (http://europa.eu.int/comm/public_opinion/standard_en.htm).

3.7 Changing public and political attitudes

By the end of the 1980s the work of environmental pressure groups, of green political parties and NGOs had exerted sufficient influence on members of the public that many environmental issues had become understood by a large proportion of the population, especially by professional and middle-income earners and by young people. School 5–14 curriculum guidelines and secondary (high school) syllabuses incorporated environmental studies and led to a far greater awareness of environmental issues. At university level, specialist courses on the environment became widely available while TV, press and publishing offered a wide range of detailed documentaries on environmental topics. In the space of 20 years, the amount of attention given to the environment had changed from that of a minority topic of specialist interest to one of wide acceptance by the general public.

But what *details* of the green argument had been given to the general public? Why did concern with environmental issues rise to prominence in the 1980s, only to subside in the 1990s? As the next section of this chapter will show, green politics in a European context reached a zenith in 1989 only to become incorporated into the *milieu* of the established political parties in the ensuing decade. Two reasons can be given: first, the established political parties defused the arguments of pressure groups by adopting many of the ideas and branding them as their own. Second, the more extreme green policies, commonly known as *dark* or *deep* green political thought, proved too radical for the middle-ground electorate. The true depth of the original green arguments can be gauged from the following passage taken from *The Coming of the Greens* (Porritt and Winner, 1988):

> The most radical [of the green aims] seeks nothing less than a non-violent revolution to overthrow our whole polluting, plundering and materialistic society and, in its place, to create a new economic and social order which will allow human beings to live in harmony with the planet. In those terms, the Green Movement lays claim to being the most radical and important political and cultural force since the birth of socialism.
>
> Porritt and Winner, 1988: 9

This concept of green ideology was probably in the mind of British Prime Minister, Margaret Thatcher, when in 1988 she described green politicians as representing 'the enemy within'. Green ideology was seen to be disruptive, of disturbing the status quo of society, in short, the antithesis of what established politicians wanted to happen. The dilemma facing the politicians who held power was how to introduce a sufficiency of green ideas into their established manifestos to enable those members of the electorate who felt a conscience about the misuse of the environment to continue to vote for their traditional party, yet at the same time not alienate other sectors of society including business, industry and commerce.

In many ways, events that occurred during the 1990s played into the hands of the establishment. The break-up of communist eastern and central Europe, followed by that of the Soviet Union has been credited with allowing western politicians more time to deal with environmental issues in place of time spent on more traditional 'Cold War' problems. Science also came to the aid of the establishment in the 1990s providing both a confirmation and an explanation of several long-standing environmental problems. Notably among these was an explanation of the causes of the El Niño climatic events that periodically disrupted southern hemisphere countries (Couper-Johnston, 2000). Politicians skilfully linked environmental issues with policy changes to give the impression that established governments were capable of mitigating environmental problems without resort to a radical change in society as advocated by the greens. Much attention was paid by the establishment to the benefit of using *appropriate* or *sustainable* technology without too much concern being given to explaining what these types of technology actually involved. For example, the Ford Motor Corporation claimed in 1998 that its latest medium-sized world car, the Focus, was some 250 times 'cleaner' when assessed on its lifetime pollution output compared to the Anglia model produced 30 years earlier. What was not said was that considerably more new cars were sold in the 1990s than in the 1960s and when traffic congestion and resource consumption was included in the equation an image of modern motoring creating 250 times less impact on the environment than in the past was incorrect. In a similar way, power-generation companies claimed that installation of new generating equipment made the plant more

efficient and less polluting, overlooking the fact that so great was the *demand* for electricity that any overall improvement achieved by the new technology had been offset by increased consumption.

3.8 Politics and the environment: the rise of 'green' politics

Green values are largely based on *ecocentric* and not *technocentric* criteria (Jones and Hollier, 1997) in which long-term environmental and non-economic factors are taken into consideration as well as human values (see section 2.6). When the first attempts to introduce a green agenda into a political context were made in the 1970s the general public was almost entirely unaware of what are now known as 'environmental issues'. Apart from some local environmental issues such as urban and industrial air pollution in the old industrial areas, water quality in the Mediterranean Basin and soil erosion in arid regions, members of the general public were blissfully unaware of the impending explosion of environmental catastrophes that awaited them in the final 30 years of the twentieth century.

Although the modern era of green politics emerged during the 1970s, the environment had figured in political decision making as early as 1832. In that year, Hot Springs in Arkansas, USA, was 'proclaimed by Presidential decree' as a 'reservation for the enjoyment of the public' (Rettie, 1995). As far as is known, local townspeople requested the designation, on the basis of what would now be called *green* reasons, that the area had an intrinsic value as well as an economic value and that the amenity provided by the site was to be formally recognised in the town charter. It was to be a further 40 years before the concept of formal, legal status would be given to the world's first national park (Yellowstone) made possible by the passage of the Conservation Act of 1872 (Allen and Leonard, 1966). (See also the Evolution of Conservation website, http://memory.loc.gov/ammem/amrvhtml/conshome.html.)

Green values first gained prominence in the United States of America at a time (mid-nineteenth century) when pressure on land and resources was minimal, yet there were sufficient people with a vision to recognise that the level of human impact on the land would change. Apart from the duration of President Reagan's time in office (1981–89) and that of the present incumbent, President George W. Bush (2001–present), the USA has retained its pre-eminence as a champion of green issues. For a period during the 1970s and 1980s Europe took the lead in establishing a 'green' political movement. Green political ideology made its first major impact in 1972 when a minority journal, *The Ecologist*, published in Britain, devoted one issue to a controversial series of papers entitled *A Blueprint for Survival* (Goldsmith *et al.*, 1972). The central argument contained in the *Blueprint* was that society was overusing and misusing natural resources and that if this trend continued, serious disruption of biospheric processes would occur. Readers were encouraged to lobby their MPs and consider standing for election in local and national elections. The contributors to *A Blueprint for Survival* were mainly academics and members of the planning profession concerned about the mismanagement of the environment. The proposal that green issues should be formally represented by a political movement dedicated to green issues bore fruit when the Ecology Party was formed in the UK in 1973. Success for this political party at national level was never achieved, although at local county council and town council level some success was obtained. Meanwhile, a highly distinctive radical green political movement was soon established in mainland Europe – especially in West Germany (the party known as *die Grünen*, the Greens) and with similar political groupings in Denmark (*De Grønne*) and the Netherlands (*Groenen Taxaties en Makelaardij o.g.*). Bahro (1986) has analysed the structure of the European green parties in the early 1980s. He showed that the idealism that powered the early green movement combined with the confrontationary nature of many of the issues made it inevitable that the green manifesto should become radical rather than reformist in nature.

In the USA, environmental activists were inspired by the success of the German *die Grünen* Party and in 1984 formed the Green Committees of Correspondence (GCoCs). Local political success was quickly achieved in the New England states through campaigns on issues such as stopping construction of incineration plants, and on watershed management. In 1989, a draft national green programme was prepared and the renamed US Green Party now has representation in 46 of the 50 US states (http://www.greenparty.org). The green lobby attained its maximum voice through the work of Al Gore who, in 1984, was elected to the US Senate. Although not a member of the Green Party, Gore, as a member of Congress, earned a reputation as an authority on environmental issues and pioneered efforts to clean up hazardous waste dumps and brought political attention to the depletion of the earth's ozone layer. Subsequently, he served as Vice-President to Bill Clinton (1993–2001), the most powerful position so far gained by any politician with a strong concern for the environment.

Within Europe during the 1980s, the political manoeuvrings of green politics became increasingly radicalised and chaotic. Many of the publicity-seeking approaches used by the European green parties during the late 1970s and early 1980s failed to make an impression on the general public and, even in countries with electoral systems based on proportional representation, failed to gain more than 10 per cent of the electoral vote. It was to be a decade later before public attitudes towards green politics began to change. Assisted by the first scientific concern over global environmental issues such as the destruction of the ozone layer by chloroflurocarbons (CFCs) and the first signs of global warming, public acceptance of the green argument began to increase. A high point was reached during the elections for European MPs in 1989. In the UK, the Green Party won 15 per cent of the votes but due to the first past the post electoral system the 2.5 million votes did not gain a single seat in the European Parliament. However, this success served to accelerate changes that had already begun among the established political parties.

Events in the rest of Europe differed from those in Britain. For example, in West Germany and Denmark, the green parties presented a direct challenge to the established parties by taking seats in the national parliament. In response, national governments incorporated an increasing element of green policy and thereby eroded the support of the green parties. The considerable success in Europe by green parties during the late 1980s was followed by the failure of the West German Green Party to retain its 28 seats in the national parliament in 1990 followed by disaster for the Swedish Greens (the *Miljöpartiet*) in the 1991 national elections (Rudig, 1992).

The economic situation in Britain during the 1980s was marked by recession and mass unemployment. The Tory government had little sympathy with controls proposed by advisory bodies that obstructed recovery and, consequently, proposals to impose, for example, limits on pollution output or dumping of wastes at sea were not implemented. This, in turn, led to an increasing frustration on behalf of the advisory bodies set up during the 1970s to advise on environmental reform. The early years of the Thatcher government saw the disbanding of many of these advisory bodies, including the Clean Air Council and Commission on Energy and the Environment. Many of these bodies were small and consisted of specialist groups that provided a point of access to government for various special interest groups. For the most part, these were not extremist pressure groups but comprised members of the academic and scientific community.

After the electoral successes in Europe in 1989 by members of the Green Party, the traditional political parties realised that green issues could win a substantial proportion of the votes (Flynn and Lowe, 1992). If left unchecked, the new parties could eventually challenge the supremacy of the established parties. As a consequence, the British Conservative Party, which had been resolutely anti-green, underwent a substantial change of view in the late 1980s (see Box 3.4).

As might be expected, the rapid changes that marked public attitudes towards the environment has meant that politicians, too, have found it necessary to constantly reassess their response to environmental issues. Green parties especially in Europe also found it necessary to reorganise and rethink their political strategy in the light of scientific knowledge and also to take account of changes in the responses from the established political parties. In response to the global dimension of environmental problems and solutions, eight European green organisations (Benelux, UK, France, Germany, Sweden and Switzerland) joined forces in 1984 to form the European Coordination of Green Parties. In 1993 a much enlarged European Federation of Green Parties comprising 28 national parties was created. The role of the Federation has been to promote the development of a sustainable and socially just Europe, which it claims can be achieved through eco-development. In this respect, the aim of the Federation does not differ from the original aim of the Ecology Party set up in Britain in 1973.

The concept of eco-development is founded on the principles of sustainable use of resources. Accordingly, the aim of the European Green Party is to achieve eco-development through ecological

---

BOX 3.4

Why the British Conservative Party turned 'green'

The 1980s was the decade when the upper echelons of British government policy makers realised that environmental issues were relevant to the electorate and could alter established voting patterns. The Conservative Party held uninterrupted power under Mrs Thatcher from 1979 until 1990. Until mid-1988, there was little evidence to suggest that she considered environmental issues to be anywhere but on the fringe of political thinking. Previously, the Prime Minister had referred to environmentalists as *the enemy within* while the Thatcherite think tank, the Centre for Policy Studies, had indicated that 'to be Green is to be wet' (Sullivan, 1985)!

Flynn and Lowe (1992) demonstrate that throughout the 1970s and 1980s the mainstream political parties had shown a periodic concern with environmental issues but at election times, this interest

---

BOX 3.4 CONTINUED

disappeared. Cynics have suggested this flirting with green issues to be a 'testing of the waters' but in 1988, a distinct change occurred in the agenda of all the main parties. This was the result of events that had started in an earlier Labour government when Prime Minister Harold Wilson had set up (in 1970) advisory bodies in an attempt to include moderate environmental opinion into policy making. The radical greens wanted no part in such inclusion and set up the Ecology Party in 1973. This party would later be renamed the Green Party.

A visit to West Germany by Mrs Thatcher in the summer of 1984 had provided an opportunity for her to assess the political significance of environmental issues but there was little evidence that these lessons were translated to the British context. However, in a remarkable speech to the Royal Society on 27 September 1988, Mrs Thatcher admitted that the protection of the environment and the balance of nature are *one of the great challenges of the late twentieth century*. What had made her make such a remarkable volte face?

The cause of this shift is not hard to find: 1988 had been designated the International Year of the Environment by the United Nations. By late spring, the topic had seized public attention in a quite new way. There was mounting popular anxiety about the so-called greenhouse effect, with the general warming of the earth's atmosphere (Mintzer, 1992) and the hole in the ozone layer above the Antarctic (Crutzen and Golitsyn, 1992). During the summer of 1988 several environmental incidents occurred: high mortality of seals in the North Sea (allegedly due to water pollution but in reality, a naturally occurring viral infection), and the public concern over the vessel *Karen B* and its cargo of toxic waste forced to ply the high seas looking for a country willing to accept the cargo and process it into harmless residues.

It is noteworthy that Mrs Thatcher's speech was made not to a political gathering but to a scientific audience. It had not been given advanced publicity by the Tory Press Office. The significant element of the speech occupied less than one-quarter of its length and was placed towards the end. Could it have been that the speech was yet again 'testing the environmental temperature'? Or might it have been intended to pre-empt the first annual conference of the new Liberal Democrat Party in which environmental issues played a large part?

It is unlikely that such a seasoned politician as Mrs Thatcher would have missed the wider political context that environmental issues would soon assume. Also, it must not be overlooked that Mrs Thatcher had been trained as a research chemist and the environmental topics that she would be most confident to speak about to the Royal Society were specifically concerned with the chemistry of the atmosphere: greenhouse gases, the depletion of the ozone layer and acid deposition. Several months earlier in June 1988 she had addressed the G6 summit of western nations in Toronto and had endorsed the Brundtland Report of the World Commission on Environment and Development and its concept of sustainable development (UN, 1987).

In spite of Mrs Thatcher's apparent acceptance of some of the main environmental concerns, there was evidence that events in Europe and at a world level were pushing the British government faster than it wanted to move! The British government had gained a reputation for employing stalling tactics over environmental issues as witnessed by the response over ways of reducing acid deposition in Europe. The government had dragged its feet on the EC directives on vehicle exhaust fumes and power station emissions designed to combat acid rain, reaching an agreement only under the most intense pressure from other European countries. The government was pitifully slow to recognise the threat from chlorofluorocarbons (CFCs). The US government banned them in 1978. Eighteen months later, the British government was still lobbying other European governments to settle for the lowest possible reductions in CFC use.

In the short term Mrs Thatcher had gained time by stating that 'we must ensure that what we do is founded on good science'. In the longer term, she had ensured that scientific argument would come to influence the decision making of politicians.

sustainability, equality and social justice and to achieve its ends requires a 'new citizenship – where human rights and rights of minorities, the civil rights of immigrants and the individual's rights to asylum are fundamental' (www.europeangreens.org). In advocating these rights, green politics in the twenty-first century differ little from the radical social changes that were justified by Bahro (1986) as setting green politics apart from the establishment during the 1980s.

3.9 Politics and the environment: the role of the European Parliament

A new unifying influence on the environmental policies of EU Member States was that exerted by the European Parliament during the 1980s. Initially, the approach used by the EU Parliament to raise the profile of green issues copied those of earlier pressure group tactics. Progress occurred slowly behind the public gaze but the setting up of a separate Directorate-General with special responsibility for the environment marked a major step forward (see http://www.europa.eu.int/pol/env/index_en.htm for the EU environment mission statement and details of the latest environment action programme). O'Riordan and Jordan (2000: 492) suggest that the slow progress may have been due to 'the unwillingness of member states to surrender power to a supranational institution' and 'unlike a federal state like the USA, the EU lacks . . . an . . . integrated political . . . culture that operates at a high degree of homogeneity.' In contrast, the EU claims that its environmental policy has been among the most successful of all EU achievements. The Sixth EU Environment Action Programme 2001–2010 headed by Margot Wallstrom, Commissioner for the Environment, has planned a wide-ranging approach to environmental challenges that are anticipated to be of relevance in the first decade of the new century. The commission recognises that a wider constituency must be addressed than in the past and that a move away from existing supporters of green policies is essential. Business and industry interests have been integrated into EU environment proposals to a far greater extent than previous programmes. In particular, four priority areas of concern have been identified:

- climate change
- nature and biodiversity
- environment and health
- natural resources and waste.

Unlike earlier initiatives that relied on legislation to impose environmental standards, the sixth programme relies on a broader participatory approach linked to a policy strategy based on sustainable development. Emphasis will be placed on enlarging the circle of participants, of active involvement and accountability both of members of the public, of business, industry and transport.

3.10 The state of the environment at the beginning of the twenty-first century

Historians urge that in order to understand the present we must study the past. In the recent past, we paid scant regard for the well-being of the en-vironment and, as a result, we have been left with a legacy of environmental damage caused both by ignorance of how the environment works and from human greed in extracting natural resources in ways that took little account of how the environment would readjust to the demands placed on it.

Our knowledge of the environment today is such that ignorance should no longer play a part in contributing to environmental degradation. That is not to say that we now understand everything about how the environment works. It often seems that as we learn more of the ways in which environmental components fit together we discover just how little we really understand! The backlog of environmental problems means that we still have a long way to go before we are capable of managing the environment and its resources in a sustainable, non-degradable manner.

Despite knowing that we should not dump our wastes into landfill sites, burn vast quantities of carbon-based fossil fuels or destroy the remaining areas of tropical rainforest we still show only limited signs of curbing such actions, as well as countless others we know to be environmentally unsound. We justify our lack of action on the grounds that it is impossible for industry to make sudden, major changes without incurring massive economic costs to society, that would lead to destabilisation of the economy. We also still cling to the hope that future technological achievements will be able to repair the damage that we cause today in the same way as existing technology can clean up the problems created by our predecessors. It might appear that all the efforts of the green political lobby have been in vain and the technocentric cornucopian attitude still prevails today much as it did in pre-green political days in 1972. However, such a conclusion would be incorrect.

The time from 1972 to the present, less than half a human lifespan, has witnessed a major advance and acceptance of environmental issues. The main difference between the 1970s and today has been the way in which the relative importance of environmental issues has risen through the list of political issues. In the 1970s concern with the environment represented a marginal viewpoint, the 'diffuse and incoherent views' identified by Pepper (1996). By the start of the new century many of the 'new citizenship' views have been or are in the process of being adopted throughout the developed world. Environmentalism has expanded from the local to a global scale of concern. How to ensure that our economies embrace ideas of eco-development and achieve sustainability has proved more difficult to achieve. These issues taxed the minds of delegates attending the UNCED Conference in Rio de Janeiro in 1992. No nation has yet developed a system that permits a transition from development based on economic growth to that of sustainability based on eco-development (Quarrie, 1992).

Perhaps the clearest indicator of the success of green political ideas has been the extent to which environmental arguments have been accepted into mainstream political ideology. Even though all European governments now actively incorporate environmental management into their manifestos, the success of green parties in many European countries occupies a higher position today than at any time in the past. Despite setbacks in the early 1990s, at the time of writing *die Grünen* hold 49 seats and occupy third position in the *Bundestag*, the lower house in the German parliament, while the Swedish *Miljöpartiet* recorded 15 per cent in the 1995 European elections – the highest figure yet recorded for a national green party. In both Finland and in Italy, green politicians hold the post of minister for environment. In Belgium, the Green Party, *Ecolo*, provides the Deputy Prime Minister while in London, the Green Party polled 11.1 per cent of the London Members' vote, sufficient to give it three seats on the new London Assembly elected in May 2000.

3.11 Matching development to human needs

Without access to suitable technology it is probable that our societies would be stuck in the pre-industrial era. Not content with using a minimum of technology to 'get by', we have *maximised* the use of technology and, in turn, *maximised* the direct benefits for humankind. By so doing we have changed the balance between ourselves and all the other biosphere components – biotic and abiotic – and have repositioned our species as an exploiter of biosphere resources. As a species we are not content with mere 'survival' and in the name of progress we have consumed increasing amounts of resources to bolster our growth and development (Goudie, 1992). We have made increasing use of scientific knowledge and technological skills to enhance the capability of our species. The two are now so intricately bound together that we overlook the fact that while scientific knowledge of the natural world has been of major assistance in allowing us to *understand* how the biosphere and all its components operate, by contrast, technology has provided the means of *exploiting* the biosphere.

It is difficult to foresee that we will ever attain a complete understanding of the environment because we continue to discover totally new mechanisms at work in the environment of which we know very little. Gaining new scientific knowledge is usually made step by step, often over a lengthy time period, whereas technological skills have tended to increase exponentially. An example of the slow progress of scientific knowledge can be seen from our understanding of the El Niño climate phenomenon. Many of the extreme climatic events that we have experienced since the early 1990s are blamed (often wrongly) on the El Niño phenomenon suggesting that El Niño is a recent discovery. The occurrence of extreme wet and dry phases in the climate of South America was well known as early as 1500 by the Inca civilisation (Middleton, 1999). Much later, in 1897, changes in barometric pressure between the east and west regions of the southern Pacific Ocean were identified but were not understood until the mid-1930s when the British scientist, Sir Gilbert Walker, Director-General of Observatories in the Indian Meteorological Service, studied and named the fluctuating air pressure as the 'Southern Oscillation'. Not until the 1980s was it realised that the El Niño and the Southern Oscillation were part of the same phenomenon, now called the ENSO phenomenon. Following the research of Sir Gilbert Walker, climatologists and physicists have been trying to understand the atmospheric science that causes the ENSO event. It has only been since the advent of new technology in the form of satellites specifically designed to monitor sea temperature and pressure and buoys tethered in the Pacific Ocean that record ocean temperatures both at the surface and at depth, that the mechanisms of ENSO have finally been revealed. In this example, science and technology have worked together to solve a particular problem that is responsible for massive agricultural losses, for forest fires and for the temporary disappearance of prolific ocean life.

One of the features of modern technology is its ability to deliver a reliability and standardisation of performance independent of the environment in which it works. Technology includes all the techniques, knowledge and organisation that we use in our day-to-day lives (Cunningham and Saigo, 2001: 47–51). For humankind, technology has provided the means to utilise the resources of the planet for the specific well-being of ourselves. Inevitably, there have been casualties as a result of technological achievements. The modern fishing vessel, equipped with echo sounders to locate fish shoals, radar and global positioning permits fishing

to continue in conditions that would have previously halted activity on grounds of safety. As a result, fish stocks have been placed under more intensive, continuous pressure. The short breaks in the fishing season due to inclement weather were often sufficient for fish stocks to feed, mature, breed or migrate. Under the old level of use, fish stocks could be maintained whereas modern fishing technology has so shifted the advantage towards the fishermen that severe fish stock depletion has occurred (see Chapter 6). Draining a swamp, eliminating pests and diseases through the use of chemical sprays, modifying the genetic structure of our food crops all have repercussions for non-human species – and ultimately for humans as well. Knowing what is *appropriate* technology is one of the great challenges facing our technocrats in the twenty-first century (Simmons, 1996). Technocrat cornucopians would claim that any technology that brings benefit to humankind is appropriate technology, whereas the environmental accommodators are prepared only to accept a level of technology that does not involve a cost to the non-human component of the environment.

A feature of the 1990s has been a move towards closer integration of science and technology, resulting in a blurring of where science ends and technology begins. Human organ transplants represents a classic example. Medical science is now totally dependent on technology to allow surgery to take place. Increasingly, we find that this trend has been fostered by 'sponsored' scientific research in which 'strategic science' is focused on those subject areas that will result in an increase in knowledge that, in turn, might assist the subsequent development of an as yet unidentified, marketable product that will result in benefits for humankind. The concern of many is that the multinational companies are funding science to further the technology that underpins the wealth and success of the company. If this is so, then sponsored science is no longer objective and dispassionate but becomes oriented to furthering the end of a specific sector of society – the multinational corporation.

Society is increasingly faced with deciding what combination of science and technology is *appropriate*, *ethical* and *beneficial* to humankind but at the same time is associated with a level of impact on the rest of the biosphere that we consider acceptable. This is, perhaps, the most challenging question we will be required to answer in the next 50 years and for which there is no absolute answer. For the deep green lobby, the requirement is for technology to exert *no* impact on environment whereas the technocrat accommodators would be willing to tolerate a level of impact that was positioned just below a point where the well-being of humans was being jeopardised by the use of the technology. The prevailing view has been that provided humans receive a net benefit from the use of technology then its use is justified.

Our judgment on the relevance of the way we have made use of technology will inevitably be assessed from our own standpoint. The benefits of relying solely on technological development has not been to our universal advantage. Doubts have now emerged as to whether we are justified to continue making maximum use of technology. Should we take account of technological developments on the other non-human members of the biota? And how far should we take account of the Gaia theory devised by Lovelock (1989), which insists that the science of the environment involves the entire dynamic system comprising all living matter and their interaction with the air, the soil and the water?

Despite the incredible contribution technology has made to our development it is possible to list many environmental disasters caused by inappropriate use of technology, for example, pollution of land, air and water originating from the mining, processing and manufacture of raw materials. Another disaster is the overcrowding of living conditions in many inner-city areas which have resulted in sub-optimum conditions for quality of human life. Technology has brought mechanisation to agriculture and released millions of people from the toil associated with agriculture especially in developing countries. Those people have migrated to shanty towns on the edge of cities where a total absence of an urban infrastructure has resulted in widespread destitution. Commoner (1971: 52–4) identified what he considered to be *fundamental faults in technology*, for example, inappropriate use of technology to prevent disposal of sewage water, reliance on inorganic fertilisers derived from finite resources as a means of sustaining modern agriculture and the use of very high-compression engines in modern motor vehicles. Reading Commoner's work some 30 years after it had been first published one finds that many of the problem areas he identified have been overcome, or side-stepped through the application of yet more technology. The use of technology to cure the problems caused by a previous technological generation typifies the progress of humankind throughout the ages.

The technocentrist would argue that slums can be eliminated if sufficient capital existed to allow urban planning and new housing to be built. The technology required is not 'state of the art'. The most urgently needed technologies include provision of piped water, sewage disposal, well-constructed housing and the provision of medical facilities. These requirements are already available to a large proportion of society. The ecocentric lobby argues that technology causes the problem in the first place. Replacing people power by machines driven by fossil fuel makes people unemployed and the machinery uses finite resources and causes pollution. The writings, for example, of Commoner (1971), Meadows *et al.* (1972), Nader (1973) and Ehrlich *et al.* (1973), all drew attention to the finite resource base on which modern, industrialised society was based. Ehrlich rekindled the concerns first expressed by Thomas Malthus in 1798 by drawing attention to the increasing gulf between the ability to grow sufficient food crops and feed the rapidly increasing world population while Schumacher (1973), in his book *Small is Beautiful*, argued for a move away from global development to that of small-scale, local and sustainable development. Not surprisingly, the attention given by these authors to the crises they considered was awaiting society resulted in them being branded as '*doomsters*'. Thirty years since these fears were expressed the world population has doubled in size and yet we do not appear to be closer to total environmental meltdown. What was wrong with the original doomster arguments? Put simply, the doomsters failed to realise the extent to which new technology would enable inhabitants of the developed world to enjoy a constant improvement in wealth generation, of material gain, of improved life expectancy and of quality of life, a theme investigated in depth by Lomborg (2001).

One of the main opponents to the doomster argument has been the academic business administrator, Professor Julian Simon. His research throughout the period from 1970 to the present has focused on the value of people in sustaining a vibrant economy. In his book, *The Ultimate Resource 2*, Simon (1996) claims that:

> The real issue is not whether one cares about nature, but whether one cares about people. . . . The central matters in dispute here are truth and liberty, versus the desire to impose one's aesthetic and moral tastes on others.
>
> Simon, 1996: p. xxxiv

In this short statement, Simon has encapsulated the essential differences between ecocentric and technocentric arguments. Contrast Simon's optimistic, technocrat quotation with the deep green viewpoint of Porritt and Winner (1988) given in section 3.7. Porritt and Winner argue for an ultimate environmental objective in which 'a new economic and social order which will allow human beings to live in harmony with the planet' is created, while Simon argues that the environmental lobby engenders personal guilt but fails to be explicit about our individual contribution to environmental problems. There are signs that environmentalists may have inherited the position once occupied by economists as purveyors of dismal science. It is the suggestion that society must move away from technology and accept an alternative, and unproven, system (based on sustainability and eco-development) that proves to be of such anathema to the technocentric lobby and that, in the final analysis, has proved unattractive to the general public.

3.12 Matching development with environmental sustainability

In Table 2.1, a range of the most common attitudes and approaches to the use of the environment was given. In practice, the extreme views of both the deep environmentalists and the extreme cornucopians have serious practical limitations and if implemented would lead to a substantial shift in the organisation of society. In a democratic society, conservative views held by the majority of the population ensures that in the long term, the extreme viewpoint does not prevail. However, over time, public attitudes can be reshaped and through a combination of deliberate and involuntary processes, public attitudes towards environmental issues can be changed. An example is provided by the gradual change in public attitude towards the required cut-back in CO_2 emissions. The EU target reduction of 12 per cent has been accompanied by senior political figures 'suggesting' that far greater cut-backs will be required if global warming is to be curbed. A figure of 60 per cent reduction in CO_2 output has been spoken of by the British prime minister as the future target.

Much of the stringent environmental legislation now being considered has become necessary because of the laissez faire attitude shown to the environment in the past. Previously, we have been guilty of not assessing the full impact of the current technology on the environment. In many cases this was due to ignorance of the true impact of the technology then being used. If it were possible to turn back the clock 100 years and redesign the twentieth century to take account of the current level of environmental knowledge, would society have chosen a different path? This is a difficult question to answer because the socio-economic conditions that existed in 1900 could not have supported a path that was fundamentally different from the one that was taken. The prevailing technology was firmly based on coal as the prime provider of energy. Knowing what we now know, it would have been possible to develop methods of trapping particulate pollution as it left industrial chimneys. Pollution of rivers with untreated wastes from coal mines could have been prevented and better nutrition for working-class people could have been provided. Had we known, we could have taken steps to reduce the problem of global warming we now face by making an earlier start on developing the technology required for renewable energy. But we have already utilised almost every possible option to generate hydroelectricity and the know-how and the materials needed to design large efficient wind turbines only became available as a result of research carried out for the aeronautics industry in the 1970s. Technology still has not provided a practical electric battery for use in the private car while the much-vaunted fuel cell as a means of propulsion is still some way from commercial viability. The inevitable conclusion appears that, for the most part, we have made the best application of the available technology at most stages in the past. Technology is driven by factors of commercial profitability; if the marketplace can afford the most recently developed technological breakthrough, whether it is the perfection of nylon yarn for use in clothing or the launching of the latest communications satellite, every attempt will be made to purchase it. This is the classic technocentric spiral in which progress is attained most rapidly by the individual ornation most able to afford it.

The ecocentric argument would question whether *all* technological development was necessary. It is because we have relentlessly pursued technological advance that consumption of energy has increased exponentially over the last 50 years and for the same reason, we now face severe warming of the atmosphere because of a build-up of greenhouse gases (see section 4.6). Global warming is one of the most threatening problems we currently face. Unless steps are taken to reduce the output of greenhouse gases the processes at work in the atmosphere will cause changes that would be inimical to ourselves. The implications if this occurred would be catastrophic. Global warming would change the way in which the hydrological cycle operates, it would bring about major changes in crop growth, it would cause changes in the pattern of human disease and cause a rise in sea level and substantial flooding of low-lying land (Hadley Centre, 2002; Tyndall Centre, 2002).

The speed and scale at which environmental change is now occurring is so great that solutions to problems require international action. Attempts have been made at approximately ten-year intervals to bring heads of state together to discuss the environmental problems and to achieve a political commitment to bring about change for the better. The first major international meeting to consider the environment took place in 1972, the same year that saw the publication of *A Blueprint for Survival*

(see page 66). The UN Conference on the Human Environment took place in Stockholm in 1972. Many heads of state refused to attend as the build-up to the conference was characterised by ill-feeling between developing and developed countries. The former were certain that an objective of the meeting was that the developed nations would attempt to impose a pattern of growth on the developing world. A positive outcome from the Stockholm Conference was the setting up of the UN Environment Programme (UNEP) (http://www.unep.org/about.asp) which did much to assist with the monitoring of the environment over the ensuing years.

In 1983 the UN organised the World Commission on Environment and Development, chaired by the former Norwegian Prime Minister, Mrs Brundtland. The Commission report, entitled *Our Common Future* (Brundtland, 1987), was widely acclaimed in the press but critics suggested that the central theme of the meeting, sustainable development, was so poorly understood by the participants that the outcomes from the meeting were insignificant (http://geneva-international.org/GVA/WelcomeKit/Environnement/chap_5.E.html).

In June 1992 the UN Conference on Environment and Development (UNCED) was held in Riode Janeiro (UN World Summit http://www.unep.org/wssd/Default.asp). Prior to the main meeting of world political leaders, four two-week preparatory committees (the so-called PrepComs) were held, at which scientists, members from NGOs and pressure groups met and prepared papers on a wide range of issues on which the politicians were required to take action. O'Riordan (2000: 39–43) has provided a comprehensive summary of the UNCED Conference. Six outcomes from the meeting can be summarised here:

1 Framework Convention on Climate Change – aimed to curb the anticipated rapid change in global climate.
2 Convention on Biological Diversity – aimed to conserve the dwindling stock of biotic resources.
3 Agenda 21 – the high-profile section of the conference. It attempted to identify how sustainable development should take place

and required signatory countries to prepare national and local plans to implement Agenda 21.

4 Rio Declaration – 27 guiding principles on environment and development.

5 Convention to Combat Desertification – an initiative focusing on low-latitude nations aimed to understand and limit the causes of desertification.

6 Forest Principles – an attempt to protect remaining areas of natural forests. Largely unsuccessful.

Was the UNCED meeting a success? In terms of lasting outcomes from the meeting or judged in terms of value for money, the answer is probably no. However, there were many intangible achievements, for example the bringing together of environmental activists, of networking and of the environmental solidarity between developed and developing nations that was achieved. The meeting raised the international profile of the environment to a level never previously achieved. On the deficit side, many fine words spoken by politicians at the conference were not translated into action. Policies remain highly fragmented; for every successful implementation of policy there exist many failures. The issue of Third World poverty has crippled the aspirations of many developing countries to implement UNCED agreements.

Ten years on (in 2002), another World Summit on Sustainable Development was held in Johannesburg, attended by 65,000 delegates (http://www.johannesburgsummit.org). The Summit focused on actions needed to achieve sustainable development. 'Inclusiveness' was the hallmark of the meeting, with the role of women, children and youth all playing a prominent role in the debate. In addition to representatives from individual EU Member States, a large delegation attended from the European Union Parliament. The European Commission was instrumental in setting many of the issues on the agenda for the Johannesburg meeting, believing that developed countries must take the lead in pursuing sustainable development. In the absence of political participation from the US government, the EU considers that it must take the leading position in the pursuit of global sustainability (http://www.europa.eu.int/comm/sustainable/pages/summit_en.htm).

3.13 Conclusion

How will historians of the future look back on the years from 1972 to the end of the twentieth century? Will it be seen as a time when society faced up to the hard fact that we live on a finite planet and accepted that limits to growth and development really do exist? Or will it be seen as an era of great scientific and technological breakthrough, when space exploration became possible, when human organ transplants became commonplace and when cloning of plants and animals became reality? There is plenty of evidence to answer yes to all these questions.

Perhaps of greater significance: will the period be considered a time when the world achieved environmental maturity? The answer to this question would be 'probably – but we could have done a lot better'! It was certainly the time when green political parties attained credibility and it was also the time when established political parties finally accepted that environmental issues could threaten long-term sustainability of society. It was also the time when warning messages from environmental scientists were finally recognised, if only because some of the messages were impossible to avoid.

If we wish to continue to enjoy an environment with clean air and water, with fertile soils that support natural vegetation and a diverse population of wild animals, with extensive natural forests, savannas and tundra then it is imperative we ensure that our industrial systems operate on clean, sustainable principles. This need not cause a restriction on personal freedom. We will still be able to own and drive cars, although their size, efficiency and method of propulsion may have to differ from those we are used to at present. Recycling used products will become the norm. Using energy generated from renewable resources will replace fossil fuels. Making better use of the environment will be possible as a result of improved understanding of the environment. Our political leaders and environmental scientists must recognise the need to work together to ensure we achieve the best possible outcomes for future societies. Politicians will be required to 'achieve the impossible' (Prins, 1993) in governing a technologically advanced society that is based on principles of sustainable development.

Although we have little idea of how, in the future, environment and society will choose to interact we must ensure that a political system exists that recognises our planet as a finite resource. We must accept that, as there are many unknown environment pitfalls ahead, we must remain vigilant and keep all our options open.

Postscript on the Balearic eco-tax

The eco-tax payable by all tourists to the Balearic Isles, outlined in Box 3.2, was scrapped in November 2003. The new conservative government elected in May 2003 bowed to pressure from the business community and tour operators to abolish the tax. Despite the eco-tax, tourist numbers had remained healthy and the tax had generated almost US$19 million (€17 million) since its introduction. Visitor numbers had dropped by 7% since introduction of the tax but there has been a world-wide decline in tourism since the attack on the World Trade Centre in New York on September 11th, 2001. The money raised had been invested in wetland conservation, tree planting using indigenous Mediterranean species such as almond, carob, apricot and olive, demolition of ugly beachfront hotels, and creation of museums and visitor centres. The new government has promised to invest an amount equal to that raised by the eco-tax, but fears have arisen that investment will be on revenue-earning ventures such as golf courses and marinas.

Despite being awarded first prize in 2002 by the British Guild of Travel Writers Globe Award, the Balearic tourist tax was strongly opposed by organisations such as the Association of British Travel Agents (ABTA). Other countries have watched the experiment with interest; The Gambia has introduced a US$8 (€7) a head tourist levy and in the Caribbean, cruise-ship passengers are charged US$20 (€18) a head to cover pollution costs. In Zanzibar, local dive operators charge US$1 per dive to compensate local fishermen for loss of earnings.

Useful websites for this chapter

Eurobarometer 55.2, Leading national trends in science and technology
http://europa.eu.int/comm/public_opinion/standard_en.htm

European Green Parties Portal
http://www.europeangreens.org

EU Environment Directorate General portal
http://www.europa.eu.int/pol/env/index_en.htm

Evolution of the Conservation Movement
http://memory.loc.gov/ammem/amrvhtml/conshome.html

Genetic engineering and its dangers
http://online.sfsu.edu/%7Erone/GEessays/gedanger.htm

Genetic Engineering News
http://www.genennews.com/

The International Centre for Genetic Engineering
http://www.icgeb.trieste.it/

The Observer
http://traveltax.msu.edu/news/stories/observer.htm

Tourism Concern
http://www.tourismconcern.org.uk/media/press_release_balearic.ecotax.htm

Monsanto and GMOs
http://news.ft.com/ft/gx.cgi/ftc?pagename=View&c=Article&cid=FT3AK7XVIYC&live=true

United States of America Green Party
http://www.greenparty.org

UN World Summit
http://www.unep.org/wssd/Default.asp

UN Conference on the Human Environment held in Stockholm (1972)
http://environment.harvard.edu/guides/intenvpol/indexes/treaties/STOCK.html#syn

UN Conference on Sustainable Development (2002)
http://www.johannesburgsummit.org

UN World Commission on Environment and Development (1983)
http://geneva-international.org/GVA/WelcomeKit/Environnement/chap_5.E.html

World Nuclear Association
http://www.world-nuclear.org/education/wast.htm

World Summit on Sustainable Development
http://www.europa.eu.int/comm/sustainable/pages/summit_en.htm

4 The atmosphere and the environment

The bias ... introduced by the relatively tiny size of the human frame, and its usually surface-based viewpoint, can produce very misleading impressions of the scale of the atmosphere at the bottom of which we spend almost all our lives.

McIlveen, 1992

The atmosphere is one of the three great unique components of our planet. It controls the heat balance of our planet ensuring that it is neither too hot nor too cold. It provides our climate, comprising distinct combinations of heat and moisture. It also provides a mixture of gases which allows us, and all other life forms, to breathe. Clearly, without a suitable atmosphere life would not be possible. This chapter looks first at the physical properties of the atmosphere and then considers some of the ways in which humans have been responsible for altering it.

4.1 Introduction

The atmosphere is inextricably linked to the uniqueness of the planet and it is inconceivable to imagine our planet with an atmosphere of any other composition than the one that exists at the present time. It has a fundamental role to play in our existence both in the present and also throughout the evolution of all life on the planet over a time span of about 600 million years. Despite major advances in our knowledge of the atmosphere during the twentieth century, there are still many aspects of the way in which the atmosphere works that we do not fully understand. The way in which the atmosphere affects our contemporary lifestyle is often underplayed, insulated as we are in artificially heated, cooled and illuminated buildings. Our knowledge of the mechanisms that drive the atmosphere is comparatively recent and there is still much to discover. We already know that the atmosphere undergoes constant change on a number of different timescales, for example, diurnal, seasonal and long term. Provided these changes are progressive and slow to occur, the biosphere has shown itself to be able to adapt to these changes. Occasionally, in the past, more rapid changes have taken place, for example at the onset of a cold, glacial period (Folland *et al.*, 1990). On these occasions considerable species mortality has occurred, but from which recovery has been possible. A major concern at the present time is evidence to suggest that the climate is changing faster than at any known time in the past. The consequences for our agricultural systems, for the supply of fresh water and the distribution of disease will be immense and will require major adaptations on behalf of society.

To fully understand how the atmosphere works would require a knowledge of physics, chemistry, meteorology, climatology, oceanography, industrial technology – and anthropology! We would also need to extend our study of the atmosphere back in time to some 3800 million years ago, when it is thought our proto-atmosphere began to form. Over the intervening millennia, the working of the atmosphere has become intertwined with that of the hydrosphere (the oceans) to create a set of conditions that are supportive of a *biosphere* in which life forms exist.

The atmosphere provides the correct combination and concentration of gases for humans and all other animals and plants in which to breathe. It also creates a barrier against the intense and harmful incoming short-wave energy from the Sun and which, if allowed to reach the surface of the planet in an unchecked form, would destroy organic life as we know it. It provides a means of regulating the temperature of the planet, working in conjunction with the oceans to keep the average surface temperature at about 14.5°C. Without the natural warming effect of the atmosphere (the so-called 'natural greenhouse' effect), the average surface temperature of the planet would be −18°C (Harvey, 2000). Finally, the weight of the atmosphere exerts a downward pressure that, at sea level, is said to equal one 'atmosphere' (see section 4.2.1). The atmospheric pressure is neither too great to crush our bodies, nor too little to allow us to float upwards into space (Ahrens, 1994).

In spite of its importance for us as human beings, we have largely ignored the atmosphere, believing it to be self-perpetuating and beyond the ability of humankind to alter it. Our main interaction with the atmosphere is through the daily weather patterns it delivers to us. Despite all the scientific and technological achievements of modern society, we have no control over the so-called macroclimate nor of the day-to-day weather patterns. Consequently, modern-day urban societies tend to treat the atmosphere as a 'backcloth' against which life is played out, and which only comes to prominence when it disrupts our normal lives due to the occurrence of climatic extremes. In the worst cases, the climatic extremes become climatic hazards and may result in severe disruption to our lives and cause substantial loss of life (Smith, 1975; Tobin and Montz, 1997). Although most people know (and care) little of how the atmosphere operates, we are totally dependent on its operation, not only for heat and moisture, but also as a vast depositary for our air pollutants. Ever since the beginning of the industrial era we have used it as a space in which we dump particulate pollution (comprising smoke, ash, grit and dust) and also a vast range of gases (most of which are the product of combustion). We assume that the atmosphere will disperse these pollutants,

reducing their concentration to safe levels, and by so doing, make it possible for life to continue. Only recently we have become aware that human activities on the surface of the Earth bring serious implications for the way in which the lower layers of the atmosphere behave (for example, see Leighton (1971) for an early account of the misuse of the atmosphere by humans). Although climatologists and meteorologists have made considerable advances in understanding the changes that have taken place in the lower atmosphere, it is the unpredictable and unquantifiable changes resulting, for example, from global warming, that cause such concern to early twenty-first century society. Seidman (1998) has suggested that while understanding the causes of atmospheric change is a necessary first stage, an essential requirement is that society can change its attitude so that it can 'address discontinuous, unpredictable and contradictory trends' that appear in the atmosphere.

4.2 Development of the atmosphere

The chemical constituents of the Earth's atmosphere were formed at the dawn of creation about 4.6 billion years ago (Folsome, 1979; Ahrens, 1994). The initial composition was most likely a mixture of hydrogen and helium, the two most abundant substances in the universe. Other gases were also present, for example ammonia and methane, both of which are compounds of hydrogen. Initially, these 'light' gases floated off into space, but gradually other denser gases emanating from volcanic eruptions originating deep within the molten Earth began to accumulate in a layer around the Earth to form a proto-atmosphere (see Table 4.1). Assuming the proportion of materials in the early volcanic outpourings were similar to those that occur in volcanic gases today, about 80 per cent of the outpourings comprised water vapour and a further 10 per cent was carbon dioxide (CO_2). During the period of Earth history between 4.6 and about 2.0 billion years ago volcanic activity was intensive because the interior of the planet was highly radioactive – it contained radiogenic heat that caused the interior composition of the planet to exist in molten and highly

Table 4.1 Composition of the early outgassing atmosphere of the primitive Earth

Major gases	Minor gases	Trace gases
Hydrogen	Carbon dioxide	Methane
Water vapour	Sulphur monoxide	Sulphurdioxide
Nitrogen		
Carbon monoxide		
Hydrogen sulphide		

unstable form (Strahler and Strahler, 1974). The large amount of water vapour contained in the atmosphere is not strictly a gas but its behaviour is so similar to that of a gas and so interconnected with the other atmospheric gases that it is appropriate to include it in this category. The water vapour emanating from the interior of the planet condensed as a thick cloud cover that totally enveloped the planet from which rain poured down in an almost continuous rainstorm. This early 'atmosphere' was deadly poisonous because of the high concentrations of methane and carbon monoxide gases and the planet would have been totally lifeless.

Apart from the internal radioactive energy there were three other sources of energy. Most important was the external source of intense radiant solar energy. With only a 'thin' atmosphere providing little protection, short-wavelength solar energy bombarded the surface of the Earth. The speed at which short-wave energy enters the Earth's outer atmosphere is 300,000km/s (Leroux, 1998) and, initially, much of this raw energy reached the Earth's surface. Thunderstorms were constantly raging and resulted in the second source of energy from the massive discharges of lightening. The final and minor source of energy came from meteoric bombardment and stellar dusts that rained down onto the planet.

About 3800 million years ago in the Earth's history a critical point was reached, setting future events on Earth apart from all other planets in the solar system. Until this time, oxygen, a very reactive element that readily combines with other elements, had been present only in small quantities as 'free'

oxygen. Most of the oxygen molecules were locked up in the rocks and volcanic gases. With so little available oxygen, the atmosphere was highly toxic and probably prevented any occurrence of life forms. From 3800 million years ago, additional oxygen was formed high in the proto-atmosphere when water vapour (released from volcanic explosions) interacted with the strong sunlight. By means of a process called photodissociation, oxygen molecules were split off from water molecules as shown in Figure 4.1 (Folsome, 1979). Photodissociation produced about 2×10^{12} grams of oxygen per year – very little in absolute terms. It was probably the only source of free oxygen between about 4000 and 2000 million years ago but towards the end of this period a new and very significant event occurred. From rocks called the Gunflint Cherts in Ontario and Minnesota, dated between 1700–1900 million years old, geologists have found evidence of increasing amounts of oxygen trapped in the crystal lattice of sedimentary rocks. Fine bands of iron (the so-called 'red beds') were deposited extensively after this time suggesting the oxidation of iron compounds had become common (Stebbins, 1971). The sudden increase in oxidised materials suggests that another means of generating oxygen in addition to photodissociation had become possible. This was most likely due to photosynthesis taking place in the green leaves of vegetation. Today, photosynthesis is responsible for producing the majority of the oxygen in existence on our planet, some 181×10^{20} grams of oxygen a year passing into the atmosphere (*Scientific American*, 1970: 57–68). The photosynthesis equation is shown in Figure 4.2.

Summarising the sequence of events leading to the present-day conditions of the atmosphere we find that from approximately 2000 million years ago the amount of free oxygen in the atmosphere

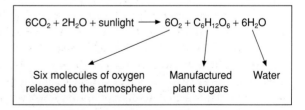

$$6CO_2 + 2H_2O + \text{sunlight} \longrightarrow 6O_2 + C_6H_{12}O_6 + 6H_2O$$

Six molecules of oxygen released to the atmosphere   Manufactured plant sugars   Water

Figure 4.1 The photodissociation process

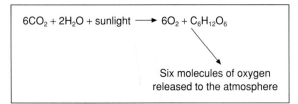

$$6CO_2 + 2H_2O + sunlight \longrightarrow 6O_2 + C_6H_{12}O_6$$

Six molecules of oxygen
released to the atmosphere

Figure 4.2 The photosynthesis equation

increased as a result of two separate processes – photodissociation and photosynthesis. The free oxygen accumulated in the proto-atmosphere where it absorbed the incoming ultra-high-energy, short-wavelength solar radiation (the alpha, beta, gamma and x-rays) that are known to damage living protoplasm. The accumulation of oxygen in the atmosphere triggered the formation of ozone (O_3) that, in turn, effectively diminished the entry of the short-wavelength rays. Once established, the protective effects of the ozone layer enabled the evolutionary process to begin, resulting in the highly diverse range of life forms we find on our planet. As the ozone layer became effective in filtering out the harmful solar rays about 2000 million years ago, increasing signs of life appear in the fossil record. Conditions that allowed life to form probably started many times over and in many different parts of the planet. Initially, these attempts at creating life were short lived as conditions suitable for life were unsustainable. Eventually, a successful and robust strand of life became established and one that has persisted to this day.

By about 600 million years ago, physical and chemical stability of the atmosphere had been achieved. Biological stability was probably still some way off but all the necessary life support systems had been established and the way was set for the emergence of the plant and animal kingdoms.

4.2.1 Structure of the atmosphere

The atmosphere comprises a layer of gases that extends upwards from the Earth's surface for some 10,000 kilometres before its density becomes so thin that 'space' is encountered. In reality, 99 per cent of the gases occur in a layer only 30 kilometres thick above the Earth's surface (Strahler and Strahler, 1974) and 50 per cent of the atmosphere occurs in the lower five kilometres (Barry and Chorley, 1998).

The atmosphere has been subdivided into layers based on the temperature and rates of temperature change. Figure 4.3 shows the main divisions of the lower atmosphere. Rising from the surface of the Earth, the temperature falls at a uniform rate known as the *normal environmental lapse rate*. This average lapse rate is 6.4°C per 1000 metres ascent. Numerous departures from this average rate occur, depending on latitude, season of year and on the humidity of the air. At a height of about 14 kilometres the air temperature has fallen to about −60°C but the lapse rate suddenly changes and air temperature begins to rise slowly. This marks the point of the *tropopause*, where the lowest layer of the atmosphere, the *troposphere*, gives way to the second zone, the *stratosphere*. Climbing further, we pass through the *mesosphere* and *thermosphere* layers, at the top of which the temperature reaches 1650°C. Such high temperatures have little significance because the air is so thin as to approximate a vacuum.

Our planet is kept warm by the atmosphere, especially the lower 30 kilometres or so, but somewhat surprisingly, it is not the incoming short-wave solar radiation that is primarily responsible for the warming process. Because the short-wave energy is exceedingly powerful it passes quickly through the atmosphere until it reaches the troposphere. Here, the existence of water vapour and greenhouse gases checks its speed of passage. As the short-wave radiant energy comes ever closer to the Earth's surface it is absorbed, scattered and reflected by clouds, water vapour and eventually, by the surface of the Earth. The incoming short-wave radiation undergoes a progressive transformation to long-wave energy and becomes reflected back into the lower layers of the atmosphere as heat energy. In effect, the principal means of heating the planet is the Earth's surface itself and is only indirectly dependent on the Sun (Leroux, 1998).

Human interest in the atmospheric layers is concentrated on events that take place in the troposphere. Almost all climatic and weather events occur in this layer. One of the most relevant

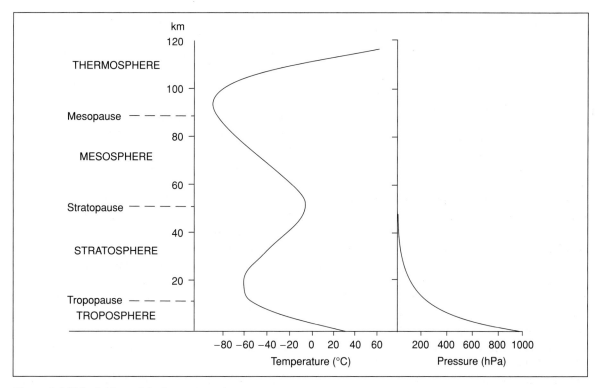

Figure 4.3 Main divisions of the lower atmosphere

components is the concentration of water vapour contained within the air, a value measured as the *humidity* content. Water molecules are evaporated from free water surfaces (oceans, lakes and rivers) and when subject to further heating, begin to rise through the troposphere. Eventually the heating process stops; the water molecules become cooler than their surroundings and condense back into liquid droplets to form clouds from which precipitation may occur. The troposphere also contains multitudes of dust particles, some of which are of natural origin (desert dust, salt nuclei from the oceans, volcanic dusts) but increasingly are of human origin (from industry and the burning of fossil fuels). Together, the dust and moisture droplets are termed aerosols and when present in large quantities can reflect incoming radiant energy and consequently change the heat balance of the planet (Goudie, 2000: 340–41).

The atmosphere provides one other critical factor for the Earth's environment – that of *air pressure*. At sea level, and under conditions described as *normal temperature and pressure* (NTP), the weight of a column of air exerts a pressure of approximately one kilogram per square centimetre of surface area. It is this pressure that keeps all life forms pressed on to the surface of the Earth and to which our bodies have become adapted. Air pressure is usually measured by means of a barometer and the unit of measurement is the millibar (mb) although the internationally recognised unit of pressure (the 'SI' unit, from the French *Système International d'Unitiés*) is the Pascal (Pa). A full explanation of all the units used in measuring environmental values is given in Bradbury *et al.* (2002). The average air pressure at sea level is 1013.2mb, equivalent to one 'atmosphere' unit. Moving away from sea level brings immediate and significant changes in air pressure and at 6000 metres in the troposphere the pressure has fallen by about 50 per cent, sufficient to make breathing difficult for many humans and often resulting in acute mountain sickness (AMS), more correctly termed high-altitude pulmonary oedema or high-altitude cerebral oedema, both of which are life threatening to humans. Very different pressure patterns occur beneath the surface of the Earth, particularly in the oceans (discussed in Chapter 6).

4.2.2 Chemical composition of the atmosphere

The chemical composition of the lower most 80 kilometres of the atmosphere is highly uniform. This layer is called the *homosphere* and is chemically distinct from the diverse layer, the *heterosphere*, that occurs beyond 80 kilometres (Strahler and Strahler, 1974).

The gases are held in place by gravitational attraction, all completely mixed so as to give pure dry air the properties of a single gas. One gas dominates our atmosphere – nitrogen – forming 78.084 per cent of the atmosphere by volume. Nitrogen is a relatively inert gas and does not easily form compounds with other gases. It is, however, one of the basic requirements for successful plant growth but is not available in its atmospheric state. A means must therefore be found of converting nitrogen to a more readily available form. This is achieved through the processes of ammonification and nitrification, the former achieved by specific microbial activity in the soil and the latter taking place during lightning activity (Van Loon and Duffy, 2000).

Oxygen comprises 20.946 per cent of the atmosphere and readily forms compounds with a wide variety of other substances. Pure oxygen is a highly volatile and explosive gas and an atmosphere comprising raw oxygen would be a very unsafe place. Section 4.2 has already explained how very little 'free' oxygen was initially present in the atmosphere, yet its accumulation was vital in eventually forming the protective ozone layer that was a precursor to the formation of a 'safe' life zone for plants and animals.

Two other atmospheric gases, argon (0.934 per cent) and carbon dioxide (CO_2) (0.033 per cent), exist in exceedingly small amounts. Carbon dioxide assumes a relevance way beyond its percentage presence. First, it has the ability to absorb outgoing long-wave energy and controls the temperature of the troposphere. In this respect it is the pre-eminent *greenhouse gas*. Second, CO_2 is of fundamental importance for the photosynthesis equation (described in section 4.2). Further consideration of this gas will be given in section 4.6.

The remaining gases are neon, helium, krypton, xenon, hydrogen, methane and nitrous oxide. Their concentration is minute, krypton, for example, forming one part in every million part of air. Tropospheric air is never 'pure'. It contains gases such as sulphur dioxide (SO_2), hydrogen sulphide (H_2S) and carbon monoxide (CO). These gases are the product of combustion processes – both natural (forest fires) and from human activities (central heating boilers and industrial sources). In addition, there are countless other gases released to the atmosphere from specialised industrial processes (paper making, ceramics, petro-chemical and industrial cleaning processes) (Graedel and Crutzen, 1993). It was estimated that by the early 1970s the quantity of industrial and urban gases released per capita in the USA exceeded one tonne per year (Stoker and Seager, 1972). The implications of these additional gases on the chemical and physical properties of the atmosphere have reached crisis proportions and will be considered in sections 4.4 onward.

4.3 Processes at work in the atmosphere

This section deals briefly with some of the major processes at work within the atmosphere. It is beyond the scope of this book to examine the detail of the different theories but it is necessary to know a little about the mechanisms that are responsible for creating the *weather* that affects people, plants and animals living on the surface of the Earth. Many of the processes remain only partly understood and one of the most exciting aspects of research into the way in which the atmosphere operates is that entirely new features are still being discovered. For example, although we know the reason for the annual wind reversal that causes the Asian monsoon, the detailed reasons for the different dates at which the 'burst' of the monsoon occurs is only partly understood. If we could fully understand and predict the date of the arrival and severity of the monsoon in different parts of Asia, the benefits would be enormous. Farmers could prepare their rice fields knowing precisely when and how much rain would fall. Preparation for water-borne diseases could reduce human suffering and hydroelectric dams could have their water levels adjusted to suit the forecast rainfall amounts. These, and many other detailed preparations, would have an immense implication on the lives of approximately one-third of the world's population, the majority of whom remain totally dependent on the monsoon cycle for the successful production of the staple food crop – rice (Douguédroit, 1997).

Most of our weather is formed within the troposphere, but some major processes occur in the lower reaches of the stratosphere and these have a significant impact on long-term *climate* conditions we experience in the troposphere. We need to understand the processes at work in the atmosphere in order to understand our weather and climate and also to help us understand why the events taking place within the troposphere are subject to unpredictable change. (See Box 4.1 for a fuller explanation of the different meaning of weather and climate.)

The air contained within the atmosphere is constantly in motion due to the heating action of the Sun's energy. There may be relatively short-lasting phases when parts of the atmosphere are less active than other areas but because of the constant flow of heat from the Sun into the atmosphere and the constant loss of heat from our planet back into space, our atmosphere is said to be in a constant state of flux. A simple experiment in a school physics laboratory will show that when heat is applied to a gas it warms, becomes less dense (lighter), expands in volume and occupies a greater space. Conversely, when a gas cools, it becomes denser (heavier), contracts in volume and sinks to the lowest position it can find. Identical processes

---

BOX 4.1

The difference between weather and climate

The *weather* affects us on a day-to-day basis and also varies from place to place. Weather comprises the condition of the atmosphere at a particular time and place and is a combination of the following:

- temperature of the air
- pressure of the air
- relative humidity of the air
- cloud cover
- precipitation (if any)
- visibility
- wind speed and direction.

By continuously recording these seven main weather variables over a period of 30 years it is possible to determine the 'average' weather or *climate* for a specific location. While weather conditions are often unique to a particular recording point, climate is usually said to occur over a large area. Climatologists speak of climatic regions covering thousands of square kilometres and over which a *macro-climate* occurs. Where rapid climate changes occur, as in mountainous areas, at the seashore or between urban and rural locations, then medium-scale *meso-climate* is said to occur. The scale of these changes may occur across narrow zones of a few kilometres. Finally, when small-scale climate change occurs within a building, beneath a forest canopy or contrasting conditions on the exposed and protected sides of a rock, *micro-climate* conditions are said to exist.

Long-term climate records of 100 years or more will inevitably include extremes of weather. These long-term records allow us to calculate the probability of the one in 100 event such as a severe storm, heavy snowfall or high summer temperatures.

are at work in our atmosphere, albeit at a very much larger and more complex scale.

The heating of the Earth's atmosphere by the Sun's energy is influenced by four main factors:

1 The orbit of our planet around the Sun is slightly elliptical resulting in a varying distance from the Sun during a 365¼ day year. Consequently, the amount of energy that reaches the Earth's atmosphere will vary, depending on the exact position of the orbital path.

2 The Earth rotates once every 24 hours on its axis that is tilted 23½° from the vertical. The combination of diurnal rotation and inclination, combined with the elliptical orbit results in distinct day and night and summer and winter variations for the northern and southern hemispheres. Seasonal differences are due to the considerable variations in energy inputs depending on which latitude in a specific hemisphere is nearest the Sun.

3 The layer of atmosphere through which the Sun's energy must pass varies, depending on the season. At the winter solstice in the northern hemisphere, the rays pass obliquely through the atmosphere and disperse their energy across a much greater volume of atmosphere than in the summer when the energy passes more directly through the atmosphere. The situation is reversed in the southern hemisphere. The transparency of the atmosphere also varies due to changes in humidity, cloud cover and aerosol content.

4 The interface between the base of the atmosphere and the Earth's surface varies greatly in texture, colour and substance (water, solid rock, soil or snow and ice). The amount of energy that is reflected back from the surface into the atmosphere is known as the *albedo* and varies from about 10 per cent over tropical forests, 25–30 per cent over deserts to a maximum of 90 per cent over snowfields (Dury, 1981). The albedo of the southern hemisphere is much lower than that of the northern hemisphere due to the predominance of the darkly coloured ocean surface.

If we consider a hypothetical situation in which the Earth was neither tilted on its axis nor rotated on its axis, then a point located on the equator would face the Sun. The position of the point would change as the Earth circled the Sun. This point (known as the point of solar constant) would be the location of maximum heating. The lower layers of the atmosphere would become warm, air pressure would become lower, and consequently, the air would become unstable and rise. As the air began to rise it would expand in volume because of the lowered pressure, the heat energy would be dissipated through the larger volume and the temperature of the air mass would cool. Ultimately, the cooling process restricts further upward movement of the air. However, the heating process at ground level continues to displace air upwards and drives the movement of air above. The vertical movement of air eventually gives way to lateral movement, pushing the air away from the equator towards the mid-latitudes. The process would continue with the air mass cooling further, becoming denser and descending towards the poles where it would form a stable high-pressure zone. A return flow from the poles towards the equator would occur at ground level so completing the cycle (see Figure 4.4). Two separate cycles would encircle

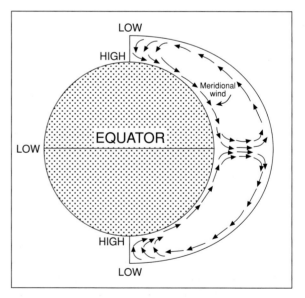

Figure 4.4 Theoretical movement of air through the atmosphere surrounding a stationary Earth

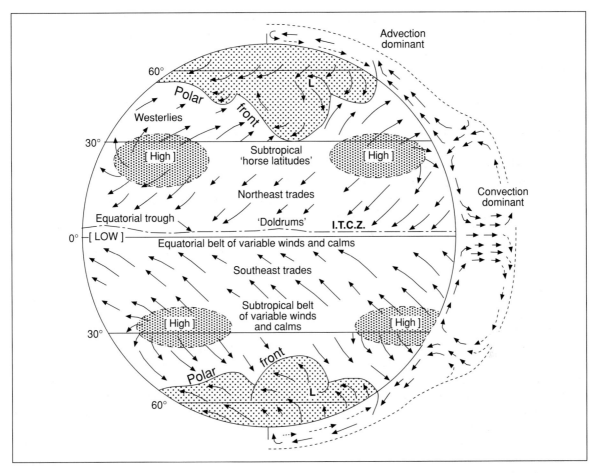

Figure 4.5 Cross-section showing the global atmospheric movement of air and the associated global pattern of winds

the planet, one in both hemispheres. This model was devised by the English physicist, Hadley, in 1735, who then refined his argument by theorising that the rising air at the equator would sink back to Earth at about latitude 30°N and S due to radiative cooling (see Barry and Chorley (1998: 124) for a full explanation of this process).

In reality, the combination of a rotation on its axis once in 24 hours, a tilt of 23½° and an elliptical orbit around the Sun of 365¼ days ensures that the point of solar constant swings between two extreme latitudinal positions of 23½°N and 23½°S of the equator (the Tropics of Cancer and Capricorn respectively). The circulation system mirrors the movement of the solar constant, with the zone of maximum heating moving north to 23½°N by 21 June and to 23½°S by 21 Decem-

ber. Figure 4.5 shows a simplified cross-section of the circulation system. Note that the theoretical single circulation cell shown in Figure 4.4 has now split into three main parts, the Hadley cell, the mid-latitude (or Ferrel cell) and a polar cell. The main driving force for the planetary circulation is thought to be the uprising air that occurs in association with the inter-tropical convergence zone (ICTZ) located within 20°N or S of the equator.

By 1948 the circulation model based solely on the vertical movement heat shown in Figure 4.5 underwent major change. The existence of the subtropical high-pressure systems in the vicinity of latitude 30°N and S associated with descending air of the Hadley cells are clearly key elements in the pressure pattern yet their origin remains something of a mystery. Leroux (1998: 46–7) argues that descending air should undergo compression and become warm and in the process become lighter and rise yet this does not appear to occur. In practice, the Hadley cells, once thought to be permanent features of the global pressure pattern, appear to undergo continuous breakdown and reform. While the existence of Hadley cells implies meridional air movement (north–south), Sir Gilbert Walker, working at the beginning of the twentieth century, identified zonal movements (east–west), especially in mid- and high latitudes and within these latitudes horizontal mixing and movement of heat energy appears to be of greater significance than meridional movement.

4.4 Impact of climate and weather on society

Such is our preoccupation with climate-related disasters as borne out by the plethora of television programmes on hurricanes, floods and droughts that it is easy to overlook that the climate of our planet is mainly benign to life. For most of the time humans have existed, we have had few ways of counteracting unfavourable climates. To avoid periods of heavy rain, cold or drought, our ancestors behaved in similar ways to the rest of the animal kingdom – they migrated with the seasonal movements of climate. The use of fire as a means of keeping warm was a relatively late development. At first, fire was used as a means of protection from wild animals and, accidentally, for cooking food. Only when buildings became relatively watertight and draught proof was it possible to use a fire to heat buildings. Clothing was a much more practical means of allowing our ancestors to survive winter cold. It was in terms of food production that our ancestors were most restricted by climate, being confined to those areas of the planet where a suitable combination of heat, moisture and soil type allowed crops to be grown. The proverbial years of famine and plenty were the result of climatic events. Cold spring times or wet summers decimated the amount of food that could be harvested whereas an absence of spring frosts combined with warm, moist summers provided ideal crop growing conditions and provided opportunities for bumper harvests.

Smith (1975) has discussed the problems inherent in interpreting statistical climate information in the context of the well-being of humans. People show infinite variation in their preferences for specific combinations of warmth, moisture, exposure (wind speed) and amounts of sunshine, whereas other climatic variables, such as number of days with thunder or amount of cloud cover are usually of less importance. Climatic preference will vary with the age, health and employment of an individual. People employed in the skiing industry will welcome heavy snowfalls, whereas anyone working with transportation will recognise snow and ice as the cause of severe work-related problems. Astronomers prefer working in cloud-free climates, while fruit farmers avoid areas prone to hailstorms as the hail pellets can destroy fruit flowers and berries.

For the proportion of the population that are urban dwellers, day-to-day weather has become far less significant than it was in the past. Centrally heated and air-conditioned offices, shopping malls and homes have insulated vast numbers of people against the extremes of the weather. But what makes a comfortable and healthy climate for people? Numerous attempts have been made to link and weight climatic variables in such a way as to reflect those climates that we consider favourable and those that we wish to avoid. A simple measure of temperature and precipitation is not satisfactory as people respond differently to individual values (Tobin and Montz, 1997). Work by Maunder (1962) considered 13 different climatic variables to identify the most and least preferred combinations in New Zealand. Terjung (1966, 1968) devised a *comfort index* first for the USA and later for the rest of the world. His work suggested that low-latitude highland regions and land adjacent to cool-current oceans offered ideal climatic conditions for humans. A combination of temperature and humidity is often used to indicate human 'comfort levels' (see Table 4.2). Smith (1975) has provided an extensive discussion of the merits of specific climatic types on human health and on the way in which climate has influenced building styles and, in turn, the culture of societies.

Table 4.2 Temperature–Humidity Comfort Index for humans

Relative humidity

Temp °C	10%	20%	30%	40%	50%	60%	70%	80%	90%	100%
19	16.5	17	17	17	18	18	18.5	18.5	19	19
20	17	18	18	18.5	18.5	19	19.5	19.5	20	20
21	18	18.5	18.5	19	19.5	19.5	20	20.5	20.5	21
22	18.5	19	19.5	19.5	20	20.5	21	21.5	21.5	22
23	19	19.5	20	20.5	21	21.5	21.5	22	22.5	23
24	19.5	19.5	20	20.5	21	21.5	22	23	23.5	24
25	20	20.5	21	21.5	22	23	23.5	24	24.5	25
26	20.5	21	21.5	22	23	23.5	24.5	25	25.5	26
27	21	21.5	22	23	24	24.5	25	25.5	26.5	27
28	21	22	23	23.5	24	25	25.5	26	27	28
29	21.5	23	23.5	24	25	25.5	26	27	28.5	29
30	22	23.5	24	25	25.5	26.5	27	28.5	29	30
31	23	24	24.5	25.5	26.5	27	28.5	29.5	30	31
32	23.5	24.5	25	26	27	28.5	29.5	30.5	31	32
33	24	24.5	25.5	26.5	28	29	29.5	30.5	31.5	33
34	24.5	25	26.5	27	28.5	29.5	30.5	31.5	33	34
35	25	26	27	28.5	29.5	30.5	31.5	33	34	35
36	25.5	26.5	28	29	30	31	33	34	35	
37	26	27	28.5	29.5	31	32	33.5	35		
38	26	28	29	30	31.5	33	34	35		
39	26.5	28.5	29.5	31	32	33.5	35			
40	27	29	30	31.5	33	34.5	35.5			
41	28	29.5	30.5	32	34	35.5				
42	28.5	30	31.5	33.5	35					
43	29	30.5	31.5	33.5	35.5					

Temperature

1 few people feel uncomfortable
2 about half of all people feel uncomfortable
3 most people feel uncomfortable
4 rapid decrease in work efficiency
5 extreme danger to human health
(*Source*: after Ruffner and Blair, 1984)

A knowledge of climate is essential for certain critical human activities, for example aviation, shipping, farming, power generation and all forms of ground transport. Improvements in climatological knowledge have been essential to guarantee the safe operation of civilian aviation, mainly due to the need to provide improved weather forecasting along flight paths. Of particular relevance to modern aviation is the presence of the so-called *jets*, strong high-level winds that occur about 6000 metres above the Earth's surface that blow from east to west at speeds in excess of 480kph. Commercial aircraft can save fuel by picking up a 'jet stream' when flying east to west, but must avoid the jet when travelling in the opposite direction.

4.4.1 Critical temperatures

The response of humans to critical air temperatures is determined by the fact that we are warm-blooded mammals and can *regulate* our internal temperature. Regulator organisms respond quite differently to temperatures than species that *conform* to the temperatures of their surroundings (see Box 4.2).

BOX 4.2

What are conformer and regulator organisms?

Plants and animals need a means of maintaining their internal temperature within a range that enables life to continue safely. There are two main ways of achieving this objective. First, the *conformer* organisms show relatively unsophisticated means of controlling their internal body temperature relative to environmental conditions in which they live (Vernberg and Vernberg, 1970). Almost all members of the plant kingdom and all animals apart from birds and mammals are conformers. Most conformers have a

fairly narrow *normal life zone* within which they operate most comfortably (see Figure 4.6a). If the organism moves to either end of its normal life zone it encounters a *tolerance zone* which can only be endured by assuming a special condition such as dormancy, hibernation or aestivation. Alternatively, the organism can migrate back into the normal life zone. Different organisms show widely different maxima and minima values to their normal life zone, and often show variations depending on the different stages of their life history. A critical temperature occurs at 0°C. Freezing of water in plant cells can cause serious damage. Well before this temperature

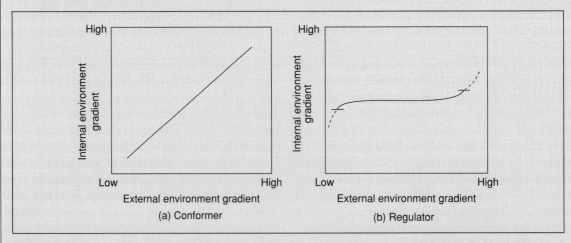

Figure 4.6 Generalised response patterns of an animal's internal environment to a fluctuating external environment: (a) conformer species: the internal environment conforms to that of the external environment; (b) regulator species: the internal environment remains relatively unchanged when the external environment changes (*Source*: after Vernberg and Vernberg, 1970)

BOX 4.2 CONTINUED

is reached plants undergo internal chemical changes controlled by internal growth hormones, substances such as auxins, gibberellins and cytokinins, and it is these that cause plants to 'conform' to the external environment. The growth hormones are responsible for triggering new growth in springtime and for the onset of dormancy in the autumn and are activated by external variables such as temperature. In addition, the growing points of stems and branches (the so-called meristems) are protected against extremes of temperature and aridity by scales, hair or deposits of lignin (Shimwell, 1971). Plants growing in the most inhospitable climatic conditions usually show greatest adaptation.

Animals such as insects, amphibians and reptiles (cold-blooded animals) often show extreme behavioural adaptation to environmental extremes. At dawn reptiles crawl into the sun to warm their bodies (see Plate 4.1). Houseflies become sluggish below an external temperature of 6°C and try to find protective cover from their enemies, while at very high temperatures (above 46°C) the metabolic rate becomes so rapid that behaviour becomes frenzied and the energy requirement exceeds the ability of the metabolic system to deliver. In all cases, if a conformer moves towards the limits of its environmental gradient, death is usually inevitable.

By contrast, species that are able to maintain or 'regulate' their bodily functions independently of a fluctuating external environment are called *regulator* species (see Figure 4.6b). This ability brings both advantages and disadvantages. On the plus side, it means that the distributional range of a regulator animal can extend across a wide thermal range of

Plate 4.1 Crocodile (conformer species) warming up in the early morning sun

conditions as the species has the ability to keep warm in cold conditions and can cool itself when conditions become warm. Regulator animals are warm blooded, i.e. mammals or birds, and regulation is achieved by means of a sophisticated blood circulation system supplemented by feathers, fur, hair or layers of subcutaneous fat to assist warming or cooling of the surface of the animal. The price to pay for this regulatory capability is that a large proportion of the food intake is used to maintain body temperature especially in cold conditions when up to 70% of food intake can be used to keep warm. Ultimately, regulators resort to hibernation or migration to withstand the most extreme conditions.

Although humans can survive a wide range of temperatures, experiments by Hardy and Dubois (1940) have shown that, at rest, naked, white-skinned Caucasians operate best at a temperature of about 27°C, while at temperatures below 15°C a naked human involved in moderate physical activity will lose heat faster than it can be generated and, ultimately, shivering and hypothermia would occur. If the living conditions coincided with windy locations (giving a wind chill effect) or wet conditions (a wet surface loses heat faster than a dry one)

then a minimum temperature greater than 15°C is required for comfort. Suitable locations providing these basic climatic requirements would have been subconsciously searched for by early hominoids and used as the preferred areas of colonisation. Apart from high-elevation areas, land between the equator and latitudes 30°N and S usually receives sufficient solar energy to maintain an average annual temperature of at least 15°C throughout the year (see Figure 4.7). Beyond these latitudes there exists a zone of varying width within which

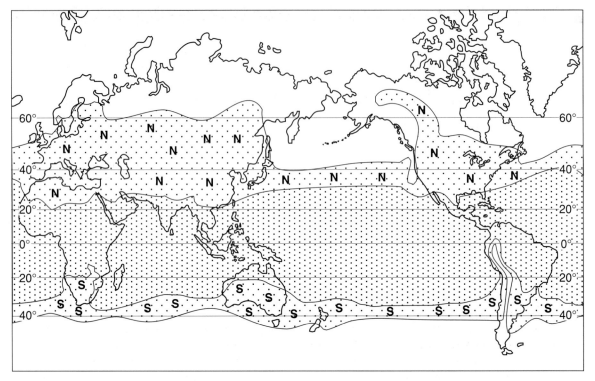

Figure 4.7 Distribution of land areas that attain an average of 15°C or above in the summer season

summer temperatures equal or exceed 15°C. Figure 4.7 shows the propensity of the greater land area in the northern hemisphere to attain this figure in summer, whereas in the southern hemisphere, the larger heat absorbing properties of the extensive oceans results in a narrower zone of temperatures at or above 15°C.

The natural ability of humans to regulate the internal body temperature has been supplemented by a wide range of technical adaptations such as protective clothing, the use of fire for heating and, more recently, by central heating, air conditioning and intelligent climate control of buildings. However, extreme temperatures still cause disruption, hardship and ultimately, loss of life, especially for the socially deprived sectors of modern society unable to afford air-conditioned accommodation or are required to work outdoors in extreme temperatures (see Box 4.3).

One other temperature threshold is indirectly of great importance for society, that of 5.6°C (42/43°F). At and above this temperature, soil bacteria

begin the process of ammonification and nitrification (see page 82). Both are essential processes in the manufacture of nitrates, vital for successful plant growth. It is unlikely that any human being is conscious of this temperature threshold, yet its importance to our survival is vital as it enables agricultural production to take place (Taylor, 1967).

4.4.2 Critical moisture levels

The effect of extreme temperatures becomes even more critical when combined with an excess or a deficiency of moisture (rain, snow, hail, sleet, dew or rime). The amount of precipitation received at a site is subject to a great number of influences, resulting in extreme variation in the amount of precipitation being recorded. Table 4.3 provides a small selection of locations that illustrate the extreme variations. Locations which experience high-pressure air masses are often characterised by very low precipitation levels (e.g. In-Salah) whereas the

BOX 4.3

Heat waves and cold spells bring human misery

The combination of excessive heat and high humidity bring particular difficulties for humans due to the inability of our bodies to achieve natural cooling. Heat exhaustion, heat stroke and death may result.

The inhabitants of the Indian sub-continent are used to coping with great heat prior to the 'breaking of the monsoon' that usually occurs each year in June. Conditions in 1998 were worse than usual and a week-long heat wave in the last week of May saw temperatures that *averaged* 44°C (111°F) across the northern and eastern plains of India. The temperature in New Delhi reached a record 46.5°C (116°F) while Agra touched 50°C (122°F). A total of 713 lives were lost with 283 dying of heat exhaustion in the eastern state of Orissa alone as a direct effect of the unrelieved heat (Bedi, 1998). Essential services failed as power breakdowns resulted in water shortages and road and rail chaos was caused by the failure of traffic lights and signals.

The USA is also subjected to periodic heat waves, especially the southern states of Texas, Arizona and New Mexico. An intense summer heat wave in 1980 killed at least 1250 people (Smith, 2001) while, during the period 1936–75, Quayle and Doehring (1981) calculate, 20,000 deaths occurred due to excessive heat. In Texas, the summer of 1998 experienced temperatures in excess of 40°C for up to 14 days, resulting in 101 heat-related deaths. The effects of the extreme heat waves on the population have been studied by the Federal Emergency Management Agency (http://www.fema.gov/rrr/talkdiz/heat.shtml).

June 1999 was marked by daytime temperatures in Moscow and St Petersburg that hardly dropped below the 31° Celsius breaking all records since 1895 (Warren, 1999). At least 140 lives in Moscow were lost, mostly men who overindulged in alcohol and then tried to cool off in lakes or rivers, only to drown. An unknown number of other Russians have died of illnesses brought on by the heat, such as heart attacks or strokes. The exceptionally hot weather brought other environmental problems, especially the risk of buying food left out in the heat for too long or of contracting skin diseases or stomach upsets by swimming in unclean water. Forest fires raged outside many big Russian cities bringing a major reduction in air quality and consequent breathing problems for city dwellers already severely stressed by the heat.

Extremely low temperatures bring different problems for humans. At the end of December 1996, much of Europe experienced a severe cold spell resulting in the death of more than 80 people (*Daily Telegraph*, 1996). Many hundreds of elderly people were taken to hospital suffering from heart and respiratory illnesses as a result of the cold weather. Temperatures fell as low as –27°C in Moldova and many of the deaths resulted from hypothermia (http://www.dnr.state.ak.us/parks/safety/hypother.htm) while other loss of life occurred at the end of the cold spell when avalanches destroyed ski resort villages. Apart from the direct loss of human life, adaptation to cold spells costs society as much as US$3 billion annually, mainly due to the cost of keeping transport links open (Smith, 2001).

opposite is true for coastal regions in the path of low-pressure systems (e.g. Cahirciveen). Distance from the sea is another major factor influencing precipitation totals, land-locked areas thousands of kilometres from the sea are invariably arid (e.g. Barnaul). Where inland areas are surrounded by high mountains and also experience high-pressure air masses then extreme aridity will occur (e.g. Denver). Finally, coastal land areas that are bounded by cold ocean currents offshore will also experience arid climates, as the onshore winds will deposit their moisture over the cold ocean current, leaving dry air to pass over the land (e.g. Arica).

It is not possible to give a precise figure for the *minimum* precipitation amount necessary for specific vegetation or agricultural land use, as it is the 'effectiveness' of rainfall that is critical and not the total amount received. For example, Walter (1973) provides examples of grass cover in southwest Africa growing satisfactorily with as little as 100mm rainfall per annum while elsewhere in the region a figure of 500mm per annum is required.

Table 4.3 Examples of different geographic locations that cause specific combinations of precipitation and temperature

In-Salah (Algeria)   Lat. 27°12'N, Long. 2°28'E
Attributes of location: approximately 1000km from nearest ocean and in the zone of descending air of the tropical high pressure cell
Long-term average annual precipitation   7.62mm
Long-term average monthly maximum temperature   42.23°C
Long-term average monthly minimum temperature   4.44°C

Cahirciveen (Ireland)   Lat. 51°49'N, Long. 9°55'W
Attributes of location: located on the coast at eastern extremity of the Atlantic Ocean and in the path of oceanic depressions
Long-term average annual precipitation   1412.24mm
Long-term average monthly maximum temperature   16.11°C
Long-term average monthly minimum temperature   4.44°C

Barnaul (Altai Republic of Russia)   Lat. 53°16'N, Long. 83°30'E
Attributes of location: approximately 2100km from nearest ocean and in the zone of the Asiatic winter high pressure cell
Long-term average annual precipitation   350.52mm
Long-term average monthly maximum temperature   26.67°C
Longterm average monthly minimum temperature   −23.34°C

Denver (USA)   Lat. 40°33'N, Long. 102°56'E
Attributes of location: approximately 1300km from Pacific Ocean and shielded by the Rocky Mountains
Long-term average annual precipitation   363.22mm
Long-term average monthly maximum temperature   29.45°C
Long-term average monthly minimum temperature   −5.0°C

Arica (Taracapá Region, Chile)   Lat. 18°18'N, Long. 70°44'E
Attributes of location: located on the coast at the eastern edge of the Pacific Ocean and washed by the cold Peruvian (or Humboldt) ocean current
Long-term average annual precipitation   7.62mm
Long-term average monthly maximum temperature   25.0°C
Long-term average monthly minimum temperature   14.45°C

Similarly, Eucalyptus forest in southwest Australia has a precipitation requirement that varies between 500 and 1000mm a year. One explanation for such variation in amount of moisture required for plant growth involves the nature of the rainfall (frequent light showers are more effective than occasional heavy downpours), percolation rate of moisture through the soil, the time of year and, consequently, the stage of growth of the vegetation, and the rate of evapo-transpiration from the leaf surfaces.

In contrast to minimum amounts of precipitation, an excess of moisture causes saturation of the soil. Permanent water-logging causes the soil to become gleyed (FitzPatrick, 1971), and anaerobic conditions lead to an impoverishment of soil fauna, an increase in acidification of the soil, and ultimately to the accumulation of undecomposed vegetation on the surface of the soil and the formation of a peat layer (see Colour Plate D). Persistent heavy rainfall or a rapid thaw of lying snow can release huge volumes of water into the ground, raising the groundwater table and, in the worst cases, the occurrence of flooding. Goudie (2000: Chapter 5) provides a thorough review of the ways in which humans have attempted to manage the problems created by too little (corrected by irrigation) or too much water (requiring drainage).

Chapter 5 examines some problems associated with the management of fresh water.

It is difficult to reach a firm conclusion whether droughts or flooding are more frequent now than in the past. Better management of rivers and their catchment areas has undoubtedly done much to reduce the occurrence of minor floods. Automatic monitoring of water levels at critical points in the system is used in flood prediction models. The construction of flood embankments and pumping schemes has encouraged development to move onto flood-prone areas and planners and developers forget that eventually the '100-year flood event' will reoccur. Flooding, when it occurs, now causes damage to underground power and telephone cables, to transport links, and to commercial and domestic properties, causing immense financial damage (see Box 4.4).

BOX 4.4

Flooding and damage to the built environment

Floods are a natural hazard and have probably caused damage and loss of life to humankind since our first ancestors existed on Earth. The first recorded evidence of a flood is the Biblical deluge recorded in the Book of Genesis, Chapter 7. Of the worst flood events to cause loss of life, six have taken place in the Far East, mainly in China, and three in countries bordering the North Sea (http://library.thinkquest.org/C003603/english/flooding/tenworst.shtml).

Coastal floods cause great loss of human life. These are often the result of intense tropical storms that sweep off the sea onto land, and are characterised by violent winds that whip up the ocean causing a storm surge that raises the level of the sea by several metres. Storm surges are usually short lasting, extending only for the duration of the high tide, yet the suddenness of the event can cause catastrophic destruction. Among the worst examples of coastal floods are those that have occurred repeatedly along the Bangladesh coastline. The storm of 9 November 1970 caused approximately 225,000 human deaths, 280,000 cattle were lost and $63 million of crops destroyed (Brammer, 1990; Burton et al., 1994).

Less damaging in terms of loss of life, but capable of causing great financial damage and social and economic disruption are river floods, especially those occurring in the lower reaches of the drainage basin where flat land allows flood water to inundate extensive areas of the flood plain. These floods are nowadays forecast well in advance and the population moved to safe ground. This type of flood builds up over several days and subsides slowly. Because of the early warning signs of impending flood, loss of human life is usually low but financial damage is considerable. The contrast that river flood damage makes on developed and developing nations is stark.

In recent years, both Europe and the USA have been subjected to catastrophic floods. Floods are the most common natural disaster to strike the USA and account for 90% of all presidential disaster declarations. In July 1993 the Mississippi and its tributary rivers turned an area extending from Missouri and Minnesota to Iowa and Illinois into a great lake with an area approximately equal to that of Great Britain. Despite an expenditure of $7 billion over 60 years on a flood management programme for the Mississippi, water levels in St Louis reached almost 15 metres above normal on 18 July 1993. Only 25 deaths by drowning occurred but the extent of financial damage reached an estimated $5–$10 *billion*. The federal government made available an aid package of $2.5 billion with $1 billion being allocated to farmers as compensation for crop losses (Münchau, 1993). Wealthy nations quickly recover from this type of flood and usually implement new safeguards to minimise further flood damage.

The autumn and early winter of 2000 saw rainfall levels in central and southern England rise to more than double the normal 230mm received between September and December. A saturated water table resulted in severe flooding in the flood plains of the main river valleys (Severn, Ouse, Thames, Trent) while hundreds of minor rivers burst their banks. Flood deaths in the main were caused by vehicle accidents and by people

---

BOX 4.4 CONTINUED

falling into swollen rivers when trying to walk home or rescuing animals. In the aftermath of the floods, the British Government re-assessed the planning policy guidelines known as PPG 25 Development and Flood Risk (http://www.planning.dtir.gov.uk/consult/ppg/pdf/ppg25.pdf) allowing building on flood plains. A new internet-based service was provided by the Environment Agency to allow people to find out the risk of flooding to their homes (http://www.environment-agency.gov.uk/subjects/flood/).

By contrast, floods that hit 100 kilometres of coastline known as *El Litoral* in Venezuela in December 1999 resulted in 50,000 deaths and made 200,000 people homeless (http://news.bbc.co.uk/hi/english/world/americas/). Torrential rain was followed by devastating mudslides from the deforested mountains onto the plain. The Worldwatch Institute had drawn attention to Venezuela's high proportion (85%) of population living in urban areas in a report published in June 1998 (http://www.worldwatch.org/alerts/000527.html) in which it highlighted the lack of consideration for safe disposal of rainwater from buildings and roads. Total destruction of the local infrastructure might take 10–15 years to rebuild and illustrates the medium-term impact severe flooding can cause in a developing country.

Flooding in Southern Africa during February 2000 caused the worst flooding for 50 years with Mozambique particularly hard hit (http://news.bbc.co.uk/hi/english/world/africa/). Because the flood built up slowly people had the opportunity to move to safety. Only 70 deaths due to drowning were recorded although the lives of 300,000 people were disrupted by moving to higher ground. Flood water rose to between four to eight metres above normal. Helicopter airlifts organised by the South African Air Force prevented further loss of life. More remarkably, the catchment of the Zambezi River again became supercharged with rainwater in March 2001 when the volume of water exceeded the storage capacity of the Kariba Dam in Zimbabwe (http://ens.lycos.com/ens.mar2001/2001L-03-02-12.html). Two of the four floodgates were opened to prevent the dam from suffering structural damage, but the consequence of releasing the water downstream to an already saturated land resulted in severe flooding in Mozambique. The National Disasters Management Institute estimated that 406,000 people were affected and 81,000 were forced to move home due to the flood water, most of whom lost everything. The UN Office for Coordination of Human Affairs launched an appeal for $30 million to bail out a country unable to rebuild the nation.

4.5 Recent changes in the atmosphere and weather

Our knowledge and understanding of the atmosphere improved more during the second half of the twentieth century than throughout the entire duration since the ancient Phoenician and Greek mariners and philosophers began studying the atmosphere. The speed with which our knowledge of the atmosphere has changed has necessitated constant revision of climatology textbooks, for example one of the main climatology textbooks of the day, *Atmosphere, Weather and Climate* by R. G. Berry and R. J. Chorley, went through six editions between 1968 and 1992. Our climate is now known to be capable of changes that occur

both at a rate and a magnitude far greater than was previously thought probable. Houghton *et al.* (1990), Kemp (1994) and Harvey (2000) provide detailed accounts of the nature of the changes, as does the Global Climate Change website (http://www.globalchange.org/).

Figure 4.8 shows the variations that have occurred in global temperature and precipitation figures over the geologic timescale. The last Ice Age (Pleistocene period) that ended about 10,000 years ago had initially been considered a unique event but geological research has revealed that every major geological time period has experienced a cooling of such dramatic proportions that a glacial epoch has occurred on at least one of the land areas of the planet (Goudie, 1992). During the Pleistocene era it is estimated that the global

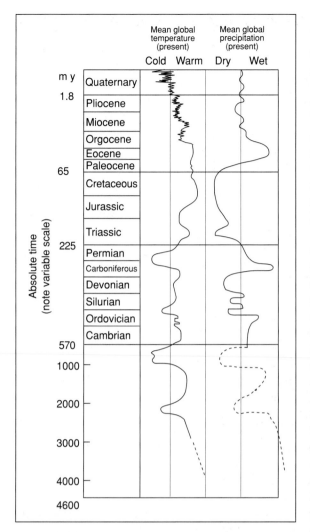

Figure 4.8 Variations in global temperature and precipitation over the geological timescale

temperature fell by about 5°C. Some climatologists believe that we are living in an interglacial period (the Holocene period) and that eventually, glacial conditions will return (Berger and Loutre, 1991; Mannion, 1992a). We can be certain that climate change is a natural event beyond the control of humankind (Goudie, 2000). However, an additional group of changes has been identified since the 1960s resulting in much scientific debate, argument and conflicting opinion on the role of humans in adding to the natural changes and on their consequent impact to the atmosphere.

Depletion of ozone molecules located between 15 and 40km altitude in the stratosphere is one of these potentially devastating changes. Atmospheric research in the Antarctic carried out during the 1970s and 1980s led to the announcement in 1987 of the depletion of stratospheric ozone levels. The cause of the depletion was attributed to the accumulation of man-made substances known as chlorofluorocarbons (CFCs) (see Box 4.5). The CFCs have resulted in small changes to the incoming transmission of short-wavelength ultraviolet radiation and to major changes in the outgoing long-wave radiation (Phillips and Mighall, 2000: 162–70). There is relatively little disagreement among scientists on the consequences of the thinning of the ozone layer. More ultraviolet radiation will penetrate the atmosphere, leading to an increased risk of skin cancer, cataracts and infectious diseases. People living at high elevation, such as the hill tribes in Nepal, Bhutan and Afghanistan, will be most vulnerable. Excessive exposure to ultraviolet light causes damage to DNA and causes cells to mutate. So great was the threat to humankind from the total loss of the protective ozone shield that the warnings given by scientists triggered a rapid and effective response from governments worldwide. The collective response on behalf of world leaders clearly shows that society can react to environmental problems and gives hope that other major problems can also be solved, provided scientists can convince politicians of the need to act.

The case against CFCs was clear. Far more usual are environmental problems for which conflicting causes can be found. Controversy still exists surrounding the accumulation of the so-called 'greenhouse gases' in the troposphere and that many scientists believe are responsible for global warming. The disagreement among the scientific community regarding the causes of global warming have been reflected in the inability of world political leaders to agree a response on how best to respond to global warming. At the centre of this problem is the question of how far the current warming of the lower atmosphere is due more to natural events and less to human intervention. This is such an important question that it is discussed in detail in section 4.6.

BOX 4.5

The CFC controversy

Chlorofluorocarbons (CFCs) were first synthesised in 1892 but had no known use and remained unused until the American chemist, Thomas Midgley Jnr., 'rediscovered' them in 1930, recognising their unique properties, being non-flammable, non-toxic, non-corrosive, stable and with thermodynamic properties. They were used first as refrigerant gases and later in fire extinguishers and finally became widely used in the 1960s as 'propellants' in aerosol spray cans, as solvent cleaners and for blowing foam plastics as used in the packaging associated with 'fast food' retailing.

CFCs comprise two main groups, fully halogenated and partially halogenated. The former are composed exclusively of carbon and halogens and have a high potential for depleting ozone molecules. Partially halogenated CFCs, for example, H-CFC-22, is now used as a substitute for fully halogenated CFCs although this substance can also destroy the atmospheric ozone layer but to a much lesser extent than the fully halogenated compounds.

In the pre-industrial atmosphere the natural level of chlorine (the main component of CFCs) was two parts per billion (ppb). By 1988 the figure had risen to 3ppb and was sufficient to cause destruction of the atmospheric ozone layer above the Antarctic (Farman, 1987). The function of the ozone layer is to protect the Earth from the Sun's ultraviolet radiation.

The seriousness of the depletion of the ozone layer was such that in September 1987, the seven main industrialised nations met in Montreal and agreed that the use of CFCs was to be limited to the 1986 level until 1990. Thereafter, a 50% reduction in consumption was planned by the year 1999. This agreement became known as the 'Montreal Protocol' (see http://www.unep.org/ozone/pdf/Montreal-Protocol 2000.pdf). So rapid was the rate of accumulation of CFCs in the atmosphere that by March 1989 the Member States of the European Union decided to work unilaterally towards a 100% reduction by 2000.

The rate of replacement of CFCs in developing countries was much slower than in the developed world. This, combined with the rapid industrialisation of countries such as India and China, indicated that unless an international CFC replacement programme could be implemented then atmospheric levels of chlorine would double to 6ppb by the year 2000. In 1989, for example, India consumed 5000 tonnes of CFCs out of a world total production of 1 million tonnes per annum. By 1999 the level of consumption in India had risen to 18,000 tonnes.

Less damaging replacements for CFCs were proposed and initially hydrochlorofluorocarbons (especially HCFC-134a) were used but these, too, caused a powerful adverse effect on atmospheric ozone. The Montreal Protocol had devised an index value of 'ozone depletion potential' (ODP) with which to compare the impact of different compounds on the stratospheric ozone layer. Under this approach HCFCs *appear* to have less effect on ozone than traditional CFCs. One kilogram of CFC-12 is 5750 times as effective as a greenhouse gas than a kilogram of CO_2 whereas a kilogram of HCFC-134a is 4130 times more potent a greenhouse gas than CO_2.

A further meeting held in Copenhagen in December 1992, attended by representatives from 93 countries, successfully implemented an accelerated programme to eliminate CFCs, halons and other ozone-depleting chemicals. It was agreed to progressively reduce the use of HCFCs by 35% (2004), by 99.5% (2020) and a total ban by 2030. In the short term, annual production of HCFCs would increase from 300,000 tonnes to 800,000 tonnes per year. This increase was required to replace the total ban on the more damaging CFCs. Some American experts claim HCFCs to be as harmful as the substances they are replacing (but cost about four times more to produce).

The hunt for alternatives involved two approaches:

- non-chemical means – improved sealants, pump-action spray cans, no-clean or abrasive cleaning agents
- alternatives – liquid sulphur dioxide, ammonia, carbon dioxide as refrigerants, alcohols or aqueous cleaning agents, hydrochlorofluorocarbons (HCFCs) for a variety of applications.

BOX 4.5 CONTINUED

Before any of the proposed alternatives can be used they, too, must be stringently tested for their impact on the environment. Three main testing schemes have been established and supported by 17 of the world's major chemical companies. Two study programmes, entitled *Alternative Fluorocarbon Environmental Acceptability Study (AFEAS)* (http://www.afeas.org) and *Programme for Alternative Fluorocarbon Toxicity Testing (PAFT)* (http://afeas.org/paft), have commenced with funding in excess of US$8m. In addition, the *Global Warming and Energy Efficiency Study* has been co-funded by AFEAS and the US Department of Energy to investigate the total equivalent warming impact (TEWI) of chemical and non-chemical alternatives to CFCs.

The proposed substitutes show significant improvements over the CFCs relative to reducing the damage to the ozone layer. Comparative ozone depletion potentials (ODPs) and global-warming potentials (GWPs) are as follows:

Substance	ODP	GWP
CO_2	—	1
CFCs	0.6–1.0	1400–8500
HCFCs	0.02–0.11	30–4200
HFCs	—	50–5200

4.6 How humans have changed the atmosphere

The first signs to suggest that the amount of CO_2 in the lower atmosphere was increasing were discovered in the mid-1970s and the assumption was made that this gas would entrap more outgoing energy and lead to a warming of the atmosphere. Because the increased level of CO_2 originated from the burning of fossil fuel (mainly coal) the problem was considered to be due to anthropogenic activity and was additional to and separate from natural global warming (Schneider, 1989).

The suggestion that the burning of fossil fuels could generate sufficient additional greenhouse gases and thus alter the natural heat balance of the atmosphere was strongly contested by some sceptical scientists. The established idea of the time was that the processes at work in the atmosphere were too complex and too massive to be altered by human activity (Bryson, 1971). In retrospect, it is difficult to understand why this attitude prevailed as Christianson (1999) gives clear evidence that by the 1870s the atmosphere above all the industrialised towns and cities of Great Britain and many in Europe had already been substantially altered by the addition of coal smoke. One hundred years later, the belief still existed that the self-cleansing processes at work in the atmosphere would prevent permanent damage from occurring.

However, by the 1950s local evidence was emerging to show that industrialisation, urbanisation and the increasing use of motorised transport was responsible for permanent, unfavourable local changes to the atmosphere and to the consequent microclimate. Tyson (1963) conducted detailed experiments into the trapping of pollution under specific climatalogical conditions in South Africa and showed that warm air lying on top of cooler air could effectively prevent dispersal of pollution. Such events, known as 'temperature inversions' have been responsible for almost every serious local pollution problem. For example, in October 1952, a high-pressure air mass became stationary for five days over London. Within this air mass a temperature inversion layer trapped a lethal air pollutant, the exact nature of which was never discovered. By chance, the Smithfield Agricultural Show was in progress in London at the time and the first indication that a serious air pollution incident was imminent was the appearance of a mysterious illness and the subsequent death of cattle exhibited at the show. Following the incident, analysis of human death certificate records for the period showed that a prolonged period of above-average mortality occurred in the second half of December and the following January and between 6000 and 10,000 extra human deaths resulted from the pollution incident (Mellanby, 1967).

Practical evidence that human activity was capable of achieving major change to the atmosphere came in the early 1960s from the rapidly growing city of Los Angeles. Each day in summer, downtown Los Angeles would become enveloped in a pall of mist that drifted over the city during the heat of the day and moved out to sea at night. The mist was the very first occurrence of the now ubiquitous *photochemical smog*, a combination of vehicle exhaust gases, rubber and tarmac particles that became chemically activated through the presence of strong sunlight. The effect of smog especially on the citrus trees, for which California was famous, was devastating, destroying vast areas of citrus groves (Heggestad and Darley, 1969). To counteract the devastating effects of the smog, Los Angeles county officials have devised the most stringent anti-pollution laws in the world and these have been copied worldwide.

4.6.1 Global warming

The threat of global warming, whether due to natural or anthropogenic causes, has emerged as one of the greatest challenges facing modern society. The disbelief, uncertainty and incredulity that surrounded scientific opinion during the 1970s and 1980s that humans were capable of causing global warming has been replaced by a more pragmatic realisation that the current consumption of fossil fuels is capable of altering the atmospheric heat balance, although some nations (notably the USA) and some organisations (for example, the American Petroleum Institute, http://www.api.org/globalclimate/) still maintain that global warming is mainly a natural phenomenon. Figure 4.9 shows the average annual global radiant energy budget. For arithmetic simplicity all values are expressed as percentages of the solar energy received at the top of the atmosphere. On average, 45 per cent of the available energy reaches the Earth's surface, 33 per cent is reflected back to space (the Earth's albedo value) and 19 per cent is absorbed, scattered and reflected as it passes through the troposphere. Human intervention in the passage of energy through the atmosphere takes place at many different stages. For example, the amount of particulate pollution (grit, dust, ash, smoke particles and aerosols) causes a scattering of incoming energy as it passes through the troposphere. Prior

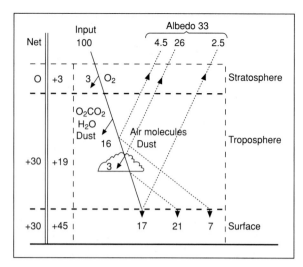

Figure 4.9 Global average annual solar radiation values, unmodified by recent changes caused by burning of fossil fuels
(*Source*: after McIlveen, 1992)

to the introduction of clean air legislation, industrial towns in England lost up to 50 per cent of their potential incoming solar energy during the wintertime because of local particulate pollution (Leicester Report, 1945). Veryard (1958) stated that on the worst foggy days in winter, central London received only 10 per cent of its potential solar radiation. Changes to the vegetation cover and increases in the amount of built-over land have changed the amount of energy reflected from the Earth's surface and, in turn, the albedo. Finally, increases in greenhouse gases in the troposphere as a result of fuel combustion have *reduced* the transmission of outgoing long-wave energy leading to a rise in temperature in the troposphere.

Figure 4.10 summarises the movement of carbon dioxide between the atmosphere, oceans and vegetation. Initial scepticism about the role of humans as an agent of global warming was in part due to a lack of long-term information about the changes in amount of CO_2 in the troposphere, and of changes in long-term average temperatures both at the surface and in the lower atmosphere. As statistical information has been collected, both from recent historical records and contemporary sources, changes in the levels of CO_2 and temperature have become apparent. Harvey (2000)

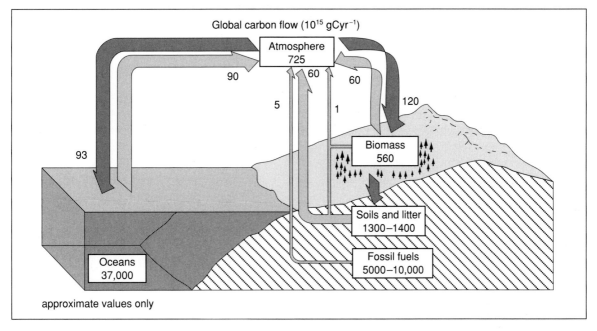

Global carbon flow (10^{15} gCyr^{-1})

Figure 4.10 A simple global carbon budget. Values in gigatonnes of carbon per year

provides information on the fluctuating level of atmospheric CO_2 from 160,000 years ago to the present. Maximum concentration reached almost 300ppmv (parts per million by volume) approximately 140,000 years ago. By 1750 the so-called 'pre-industrial' level of atmospheric CO_2 was 280ppmv but by the start of the twenty-first century, burning of fossil fuels had increased the figure for atmospheric CO_2 to 369ppmv in 2000. The 1997–98 increase of 2.87ppmv represents the single greatest yearly jump since the Mauna Loa record began in 1958. Figure 4.11 shows the rise in CO_2 since 1958 (http://cdiac.esd.ornl.gov/trends/co2/sio-mlo.htm). The saw-tooth pattern shown by the line reflects the uptake of CO_2 via photosynthesis during the northern hemisphere summer. There is no equivalent uptake in the southern hemisphere summer due to the smaller land mass and lower photosynthetic potential (Hansen *et al.*, 1981).

Average annual surface temperature values collected from around the world show there has been an *average* warming of almost 0.7°C during the twentieth century (see Figure 4.12) (IPCC, http://www.ipcc.ch/pub/spm22-01.pdf). Two complex questions arise from this graph. First, have the *rate*

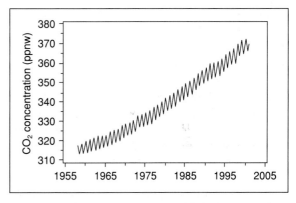

Figure 4.11 Changes in atmospheric CO_2 levels measured at Mauna Loa, Hawai'i, 3397 metres above sea level

and *manner* of climate change in the latter part of the twentieth century been due to *genuinely* different causes compared to earlier times and, second, has the warming been due primarily to natural processes or has it been due more to human actions?

Karl *et al.* (2000) reviewed recent record-setting temperatures in the context of the long-term warming trend. While confirming the overall average rate of warming of almost 0.7°C in the

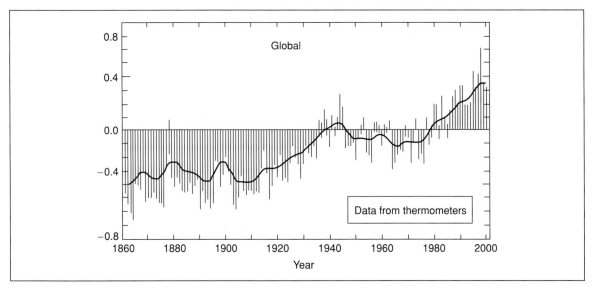

Figure 4.12 Variations in the Earth's surface temperature over the last 140 years (*Source*: data from IPCC, 2001)

twentieth century, they also identified a *change point* in the temperature trend in 1976, when the warming rate jumped to an equivalent rate of about 2°C per century. In addition, statistical analysis of 16 consecutive months of record-high global temperatures between May 1997 and September 1998 suggested that further acceleration in the order of 2.5–3.5°C per century may have occurred in 1997, although it is too soon to tell for certain. It is almost certain that in the twenty-first century global warming will take place at a much faster rate than in the last century.

Although the surface and subsurface of the Earth has warmed significantly since 1976 (Huang *et al.*, 2000), satellite temperature records from 1979 for the lower 8km of the troposphere provide a different picture and show a relatively insignificant warming trend of 0.05°C per decade. The discrepancy between temperature trends at the surface and in the troposphere generated doubts about the validity of global warming in scientific circles. However, it is now argued that both surface and troposphere temperature trends could be correct and the differences are due to variation in spatial coverage of the data. The satellite record provides global coverage while the surface record relates to specific locations. It is believed that this

could account for about one-third of the difference between the two data sets.

One particular problem concerning the authentication of global warming has been how best to integrate the many different findings from research centres around the world. Achieving interdisciplinary science that would allow an integrated environmental policy has proved very difficult to achieve. Institutes such as the Tyndall Centre for Climate Change Research based at the University of East Anglia (www.tyndall.ac.uk) in the UK have made some progress on integrating the atmospheric science that is thought to be relevant to global warming with the sub-disciplines of social sciences such as planning, assessment and policy. Society will be required to make adaptations to cope with global climate change, but knowing the timescale on which change will be necessary and the extent of change will be critical. Adapting to climate change and, in particular, global warming, will be expensive and even the most wealthy countries will be anxious to make as few changes as possible to the established economy.

Given sufficient time, the atmospheric heat balance would probably be capable of reaching a new equilibrium but not before a significant surface warming had occurred. Note that in numeric terms the changes we have caused seem to be quite small. For example, each square metre of the planet receives an average of 240 watts of energy per year while the rate at which energy is radiated back to the atmosphere is 236 watts. The surplus of four watts per metre has been sufficient to cause the current level of global warming and gives rise to what is called climate 'forcing'. If society adopted a 'business as usual' strategy, that is if we go on behaving in the future as we have done in the past, we would release ever-greater amounts of greenhouse gases that would possibly cause an out-of-control warming of the Earth's surface.

At the beginning of the 1990s industrialised countries were responsible for releasing three-quarters of all greenhouse gases and, consequently, these countries were required to make the greatest contribution to curbing output of the six main gases (Flavin and Dunn, 1998). It is probable that in the future we will be faced with other, currently unknown, climatic events that are caused by the interaction between human activities and natural processes. (Readers are recommended to visit the excellent Global Change site at http://www.globalchange.org/ for the most up-to-date information on this topic.)

4.6.2 Implications of global warming

A warming of the atmosphere will bring about many detailed changes to the way in which our societies operate. Some of the adaptations will be the result of beneficial changes to the climate while others will be due to unfavourable events. Many of the enforced changes will have both beneficial and detrimental elements. For example, a rise in the winter average temperature will result in fewer frosts, will extend the growing season for crops, will reduce the demand for energy for heating buildings and, in turn, reduce the burning of fossil fuels. On the down side, warmer winters will allow the spread of diseases that would otherwise be killed off by the cold; agricultural practices such as breakdown of clay soils by frost action will also be reduced; the lucrative skiing industry that relies on snowfall and low temperatures will suffer disproportionately as less snow lies for a shorter period of time.

The Department of the Environment (1996) has forecast that drier summers in southern Britain may result in shrinkage of clay soils leading to increasing subsidence damage to foundations of buildings. Crop yields in southeast England may also decline as lack of water for irrigation inhibits growth (Parry and Swaminathan, 1992). Tourism and recreation in southern Britain may benefit from the warmer summers. A warmer atmosphere is likely to lead to a warmer ocean, the level of which will rise due to thermal expansion of the surface layers as well as an increase in volume due to the melting of polar ice (Robin, 1986). An increase in sea level of 20 centimetres would flood 10 per cent of British coastal nature reserves (Department of the Environment, 1991) and detailed mapping using geographic information systems software by Jones and Ahmed (2000) along part of the River Clyde estuary in Scotland has identified 73 per cent of existing conserved areas at risk from coastal flooding.

4.7 Atmospheric quality, environment and human well-being

An implicit characteristic of the Earth's atmosphere is that it provides a safe environment in which plants, animals and people can thrive. However, the atmosphere is not completely benign. Atmospheric conditions that cause an excess or deficiency of any of the life-giving necessities such as too much heat, cold, drought, moisture or storminess can lead to temporary problems for species survival. In the past, humans have been affected as much as plant and animal species in this respect, suffering horrendous hardship and loss of life especially in the longer lasting episodes of drought or the short-lasting tropical storms. At present, the effect of atmospheric extremes on the human population operates at two quite distinct levels, inhabitants of low-latitude countries facing a quite different situation from mid- and high latitudes.

As our scientific knowledge of the atmosphere has improved, we have acquired the capability of predicting when and where unfavourable events occur. Also, as our technological skills have improved, it has become possible, for example, to irrigate drought-prone areas and build defence systems to prevent river and coastal flooding. Consequently, we might reasonably anticipate that extreme atmospheric events would play a smaller role in affecting our lives. For inhabitants of mid-latitude, developed world countries this is largely true. Smith (1975) has shown how modern building design and transportation systems have insulated us against many of the climatic extremes that formerly disrupted our lives. Reliance on a year-round, 'just-in-time' transport system allows supermarket companies to deliver food from around the world to the shopping trolleys of the developed world with little thought being given by the customer to problems of inclement weather affecting the farming system.

A very different situation exists in other parts of the world, especially in low-latitude, developing countries. Partly as a result of their colonial history, and partly due to the contemporary political and economic situation, low-latitude countries are struggling to cope with massive population growth rates that affect the provision of basic human services. Superimposed on the struggle to cope with the socio-economic problems faced by low-latitude countries is the problem of climatic extremes. Box 4.4 has drawn attention to the floods that devastated large parts of Mozambique in 2000 and 2001 destroying the ability of the country to feed itself for two successive years. Bangladesh is regularly battered by tropical storms with the potential to wipe out almost a quarter of million inhabitants in the space of several days. It is impossible for inhabitants of developed countries to understand and appreciate such devastation, unfamiliar as we are with the impact of the raw extremities of the atmosphere. Nkemdirim (1997) has discussed the impact of climate on life in the Caribbean Basin. The destructive force of hurricanes, of which up to six batter the Caribbean islands each year, is focused mainly on 'the underclass and marginalized persons', and causes repeated damage to the built environment and curbs the economic stability for individuals and countries alike. If global warming produces a more dynamic atmosphere then the incidence of

hurricanes could increase in the future, bringing even more hardship to an already impoverished region. Developing countries also struggle to produce sufficient basic food for their burgeoning population. Climatic fluctuations cause major uncertainties for agriculture. In Chile, annual losses to agricultural production from frost damage vary between US$12 and 20 million, while droughts add a further US$40–120 million per year (Santibáñez, 1997). A malnourished population becomes more susceptible to illness and disease; infant mortality rises and age expectancy of the population declines. Climate change could alter any of these existing conditions.

The relatively stable climate of the mid-latitudes may undergo considerable change during the twenty-first century as the effects of global warming begin to exert their influence on the atmosphere. The extent to which this happens is partly under our own control and partly down to the vagaries of the atmosphere. In 1999 the Intergovernmental Panel on Climate Change (IPCC) published its Special Report on Emissions Scenarios (SRES) (http://www.grida.no/climate/ipcc/spmpdf/sres-e.pdf). This is divided into four groups each with its own 'storyline' describing the way in which the world population and economy may develop over the next century. Data used to prepare these scenarios has been subsequently used in the calculation of future climate change. The scenarios are shown in Table 4.4.

The potential for global warming to disrupt the established lifestyle of developed nations is massive. For example, a shift in the climatic boundaries in the temperate zones would result in altered temperature and moisture regimes for agriculture. Soil organisms responsible for decomposing organic material would respond differently to the new climatic conditions and this would result in changes to the soil-forming processes. Farmers would need to adapt farming practices in an appropriate manner. New varieties of crops may be necessary to cope with changes to the heat and moisture regimes and altered growing season conditions and these may require new farming machinery, and new marketing strategies resulting in different shopping patterns by the consumer. The distribution and life histories of insect pests

Table 4.4 Narrative description of sea-level rise scenario families

Scenario name	Storyline
SRES A1	A future world of very rapid economic growth, low population growth and rapid introduction of new and more efficient technology. Major underlying themes are economic and cultural convergence and capacity building, with a substantial reduction in regional differences in per capita income. In this world, people pursue personal wealth rather than environmental quality
SRES A2	A very heterogeneous world. The underlying theme is that of strengthening regional cultural identities, with an emphasis on family values and local traditions, high population growth and less concern for rapid economic development
SRES B1	A convergent world with rapid change in economic structures, 'dematerialisation' and introduction of clean technologies. The emphasis is on global solutions to environmental and social sustainability, including concerted efforts for rapid technology development, dematerialisation of the economy, and improving equity
SRES B2	A world in which the emphasis is on local solutions to economic, social and environmental sustainability. It is again a heterogeneous world with less rapid and more diverse technological change but a strong emphasis on community initiative and social innovation to find local, rather than global solutions

(*Source*: Intergovernmental Panel on Climate Change, 1999)

and plant-pollinating insects may change while disease-carrying insects may bring new health hazards for humans. The report of the Intergovernmental Panel on Climate Change (IPCC) Working Group II examined the impacts, adaptation and vulnerability of society to climate change. The full report can be accessed at http://www.meto.gov.uk/sec5/cr_div/ipcc/wg1/WGII-SPM.pdf. The report recognises both positive and negative consequences for society resulting from climate change. Beneficial impacts include:

- potential for increased crop production in mid-latitudes that experience small increases in temperature
- potential for increased timber yield from managed forests that respond to a slightly warmer and wetter climate
- increased precipitation levels may ease the fresh water crisis in arid areas
- reduced winter mortality in mid- and high latitudes due to less severe cold spells
- reduction in winter energy consumption from space heating resulting in lower consumption of fossil fuels.

Negative impacts of climate change are generally thought to bring greater costs and disruption than the benefits. These include:

- reduction in crop yields in tropical and subtropical regions
- a general reduction in potential crop yield in mid- and high latitudes if global warming causes moderate or severe increase in temperature
- changes in precipitation patterns will, on balance, reduce the amount of available fresh water worldwide
- increase in proportion of global population exposed to vector-borne disease (e.g. malaria), water-borne disease (e.g. cholera) and an increase in heat-stress mortality
- sea level rise will inundate fertile agricultural land and flood coastal towns and cities
- increased demand for energy to power air-conditioning units in summer will offset winter energy savings.

How we respond to the possible changes brought about by climate change will dictate the quality of life our descendants enjoy 100 years

Table 4.5 Possible strategies to be used in combating the global impact of climate change

Objective	Action required	Response by society
Attempt to maintain current climate regime	Reduce production of greenhouse gases and release to the atmosphere	Cut back in economic activity until 'clean' alternatives are found for the existing processes Developed countries accept a lowering in quality of life
Accept that climate change is inevitable and beyond human control	Maintain a 'business-as-usual' strategy	Accept that there will be 'winners' and 'losers' and consequent changes in balance of world power Competitive forces predominate
Combat climate change by relying on greater use of new technology	Develop 'clean' systems. Use agricultural system based on advanced technology, e.g. GMOs, synthetic photosynthesis	Wealthiest countries gain at the expense of impoverished countries unable to afford high-cost technological solutions

Mixed strategy involving a combination of the above	Use appropriate new technology where necessary, but retain tried and tested ways until sustainable methods are developed	International cooperation, based on transfer of technology between nations
		   Transfer of surplus food to areas of shortage
		   Cooperative forces predominate

or more from now. Many different strategies of response by society can be used (see Table 4.5). Two problems, in particular, exist. First, how to assure the developed nations that the chosen action plan will not jeopardise their future development and, second, how to pay for technology transfer to the developing nations so that they participate in sustainable development.

The threat to society created by global warming is a totally new type of crisis. The consequences of global warming for society appear to have more in common with a health epidemic such as typhoid or diphtheria. The cause of the problem remains hidden and involves time, effort and money to investigate the cause of the problem before a solution can be found. Because the warming of the atmosphere is a *global* problem, it is likely that the solution will need to be applied on an international scale. If global warming is allowed to continue uncontrolled, the consequences for society are dire and the scenario painted of conditions by the end of the present century by the 'doomsters' suggest a fractious, disease-ridden society facing chronic shortages of fresh water and food. More

realistically, it is unlikely that a species such as *Homo sapiens*, so adept at making opportunities out of adversity, will allow itself to be caught off guard by the challenge of global warming.

Although global warming differs from other challenges we have faced in that it affects us all, we have already faced similar challenges, albeit not of the same magnitude. In the 1970s and 1980s acid precipitation threatened our crops, soil and fresh water. That threat has largely been overcome by imposing strict controls on the volume and concentration of gaseous pollutants we release into the atmosphere. Success in reducing the threat of acid precipitation demanded new technology, new management and new legislation that required international agreement. Similarly, the threat to the stratospheric ozone layer by CFCs has been recognised and although the problem has not yet been solved, the means to do so now exists.

Management of air pollution from urban and industrial sources has also been achieved. A succession of increasingly stringent clean air acts has been passed by almost every country in the world, the stimulus for which has been the acceptance by society that air pollution damages the health of individual citizens, causes financial damage and lowers the quality of life. Low-level ozone is estimated to cause between $2–6 billion reduction in crop yield in the USA each year. Pollution causes an unquantified loss in lowered forest growth and prevents fast-growing conifers such as Douglas fir from being planted in the UK. Pollution also kills many lichens growing on rocks and trees. The cost of controlling air pollution is generally less than the direct and indirect cost of pollution. While the main motive for pollution control is financial there are many consequential benefits, such as a more healthy and diverse natural environment.

4.8 Legislating for a healthy atmosphere

No nation has yet attempted to establish formal 'ownership' of the air space above its land area in the same way as legal jurisdiction of offshore space has been established. However, as a result of air pollution incidents and the risk of imparing the health of its citizens, many nations have enacted laws to control permitted levels of air pollution. Section 4.7 has already introduced the concept of 'clean air legislation'. The introduction of legislation restricting the type and amount of pollution into the atmosphere dates from the mid-1950s when first the USA and then Britain passed legislation designed to curb air pollution levels. In 1955 the USA passed the Air Pollution Control Act designed to encourage research into air pollution. The first legislation to control air pollution output came in 1956 when the British government passed the Clean Air Act directed at particulate matter. Control was achieved by limiting the output of dark smoke from a chimney to a maximum of five minutes within an hour. The success of the early legislation was dramatic. In Britain it led to the virtual disappearance of the notorious London smog – a foul-smelling, dirty, acid mist that caused irreparable damage to the lungs of humans and caused building materials to decay.

The clean air legislation stimulated a move away from coal fires as a means of heating buildings and the introduction of new heating technology based on oil, electricity and natural gas. Coal had supplied about 80 per cent of the world's commercial energy in the 1920s (Jones and Hollier, 1997) and coal smoke had been largely responsible for the gloomy conditions that prevailed in industrial towns and cities (see pages 97–8). As the consumption of fuel oil increased, accompanied by more efficient and cleaner methods of burning coal, the main type of air pollutant changed from particulate to *gaseous* pollution – at first dominated by sulphur dioxide, and later by a range of chemicals that included carbon dioxide, many oxides of nitrogen (the so-called NO_x pollutants), hydrogen fluoride and many trace gases.

The setting of legally safe levels for atmospheric gaseous pollution proved much harder than setting limits for particulate matter. Gaseous pollutants usually became quickly diluted once liberated to the atmosphere but the rate of dilution varied on the actual atmospheric conditions of temperature, humidity and instability. Monitoring gaseous pollution was made more difficult due to the ability of gases to undergo chain reactions (a synergistic effect) to form new gases of unknown concentrations. Most notable were the peroxyacyl acids (PANs) found in the Los Angeles photochemical smog (discussed on page 62). It was not until 1970 when US legislators passed the Clean Air Act Amendments that legally enforced gaseous air pollution levels were formally established. This legislation included two distinct standards: the National Ambient Air Quality Standards (NAAQS) and New Source Performance Standards (NSPS). NAAQS established standards of allowable concentrations in the air while NSPS set the rates at which pollution could be emitted to the atmosphere. It was the task of the newly formed Environmental Protection Agency (EPA) to enforce the standards (McKinney and Schoch, 1998). NAAQS required the monitoring of six major air pollutants: CO, NO_2, O_3, SO_2, Pb (lead) and PM_{10} (particulates smaller than ten microns in size). Taken together, their concentration constitutes the American Pollutant Standards Index (PSI), a measure of air quality that ranges from zero (best quality) to over 400 (worst air quality). If the PSI reaches 200–299, this constitutes a First-Stage Alert and the elderly and persons with respiratory diseases are advised to remain indoors. A Second-Stage Alert is reached when the PSI attains 300–399. At this point all people are recommended to stay indoors. Finally, a Stage Three Alert occurs when the PSI exceeds 400 at which point all physical exertion should stop. Congress has subsequently passed additional legislation, for example, in 1990 the Clean Air Act Amendments that focused on three specific problems: urban smog, acid rain and air-borne toxins (specifically, the carcinogenic heavy metals such as mercury and arsenic).

New technology has enabled the concentration of gases released from combustion processes to be substantially reduced. Improved burning of fuels combined with exhaust gas-cleaning devices such as catalytic converters and electronic precipitators, has led to a considerable cleaning of exhaust gases to the extent that some car manufacturers claim that the exhaust gases leaving the tail pipe of their cars is cleaner than the air entering the engine via the air filter! Unfortunately, the increasing consumption of fossil fuels worldwide has negated the undoubted clean up of gases released to the atmosphere, for example the commercial consumption of energy has increased from 1645 million tonnes of oil equivalent (MTOE) in 1950 to a peak 8500MTOE in 1997 (Jackson and Jackson, 2000: 285).

In the 1990s attention switched from the setting of safe air pollution levels for plants, animals and especially for humans to the much wider issue of preventing runaway changes in the stability of world climate. The setting of internationally ratified targets for the release of specific air pollutants has become standard practice and the Framework Convention on Climate Change, otherwise known as the Kyoto Protocol after the conference held in Kyoto, Japan in 1997, may herald a new phase in the willingness of governments to accept limits on pollution output (Flavin and Dunn, 1998).

4.9 Conclusion

In the twentieth century it became possible for scientists to study the atmosphere at first hand through the use of aircraft equipped as flying laboratories. Later, another major step was achieved with the availability, from the 1970s onward, of satellite-derived information. Unfortunately, the twentieth century witnessed the culmination of practices that had been universal throughout the Industrial Revolution – that of uncontrolled release of pollutants into the atmosphere. Clean air legislation progressively coped with particulate and gaseous pollutants but by the 1980s, evidence existed to show that far more dangerous forms of pollution were being released, CFCs and greenhouse gases. Remarkably, some countries still refuse to accept that human activity is responsible for a large part of the current global warming. Society is taking a big gamble with its own future. If we allow the processes that maintain the atmosphere within a quite narrow band of suitability for life to get out of control, we could face a situation where a severe climatic fluctuation takes place. If this occurs it will certainly bring great hardship to human society and could eliminate a large number of both plant and animal species.

Useful websites for this chapter

Alaska Department of Natural Resources
http://www.dnr.state.ak.us/parks/safety/hypother.htm

Alternative Fluorocarbon Environmental Acceptability Study (AFEAS)
http://www.afeas.org

American Petroleum Institute
http://www.api.org/globalclimate/

BBC News
http://news.bbc.co.uk/1/hi/world/africa/655510.stm

Carbon Dioxide Research Group
http://cdiac.esd.ornl.gov/trends/co2/sio-mlo.htm

Environment Agency Flood Maps
http://www.environment-agency.gov.uk/subjects/flood/

Federal Emergency Management Agency
http://www.fema.gov/nwz98/tx0805.htm

Global climate change
http://www.globalchange.org/

Intergovernmental Panel on Climate Change
http://www.ipcc.ch/pub/spm22-01.pdf

Intergovernmental Panel on Climate Change
(Working Group II)
http://www.meto.gov.uk/sec5/cr_div/ipcc/wg1/
WGII-SPM.pdf

Intergovernmental Panel on Climate Change
http://www.grida.no/climate/ipcc/spmpdf/sres-e.pdf

Lycos Environment
http://ens.lycos.com/ens.mar2001/
2001L-03-02-12.html

Montreal Protocol
http://www.unep.org/ozone/pdf/Montreal-Protocol
2000.pdf

Planning Policy Guideline 25
http://www.defra.gov.uk/environ/fcd/hltarget/
HLT12_Apps02C.pdf

ThinkQuest
http://library.thinkquest.org/C003603/english/flooding/
tenworst.shtml

United Nations Environment Programme
http://www.unep.ch/ozone/montrea.shtml

Worldwatch Institute Social and Economic Inequities
Impeding Global Environmental Action
http://www.worldwatch.org/alerts/000527.html

World Meteorological Organization
http://www.wmo.ch

5 Environmental (mis)management of fresh water

Water is the sovereign wealth of a state and its people. It is nourishment; it is fertilizer; it is power; it is transport.

Brunhes, 1920

Mismanagement of water has occurred almost since humans became settled agriculturalists and water management policies became focused on short-term solutions and not long-term sustainable solutions. This chapter examines how, in spite of the abundance of water on Earth, we have mistreated a free resource, resulting in water shortages and a lowering of the quality of supply. A number of case studies are used to show how management of water is central to the economy of a region. The chapter ends by examining a strategy for sustainable water management.

5.1 Introduction

How unfortunate that the quotation from Brunhes (1920) with which this chapter opens has been all but forgotten by society over the years since it was written. The ways in which we have mismanaged our fresh water supplies has led to the current predicament of increasing shortages, a lowering of quality and an increase in cost of fresh water that is occurring throughout the world. In the past, society neglected the value of their water supply at their peril and the once great civilization of Mesopotamia that flourished circa 4000 BC (Sears, 1972: 132) and the Inca civilization between AD 1440–1534 (Chepstow-Lusty *et al.*, 1998) are thought to have failed because the agricultural

systems fell into decline due to a breakdown in the irrigation system. The early failures to manage the water supply were, undoubtedly, disastrous for the relatively small numbers of people dependent on them. By comparison, the implications of the current challenge to provide a sustainable, safe water supply is several magnitudes greater and is of concern to approximately two-thirds of the world population – some four billion people living in almost every part of the world. The meeting of the World Water Forum held at The Hague in March 2000 and the eleventh Water Symposium held in Stockholm in August 2001 (http://www.siwi.org/menu/valframe.asp) recognised the numerous fundamental goal conflicts that exist in global water management issues and concluded that despite the fundamental necessity of water, provision of water had a relatively low priority for many government bodies. Politicians focus on the visible and acute water management crises rather than the invisible, slow changes that in the long run may undermine the structure of society. These are strong claims and this chapter investigates some of the main problems concerning water management and, where possible, suggests long-term, sustainable solutions.

5.2 The uniqueness of water (hydrological cycle)

It is impossible to study fresh water resources without first making a brief reference to the salt water oceans. Fresh water is derived from the oceans and, once used, returns to the oceans as part of the

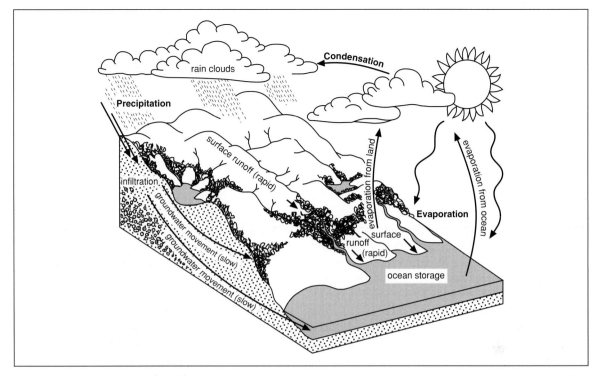

Figure 5.1 The hydrological cycle
(*Source*: after Jones and Hollier, 1997)

great hydrological cycle (see Figure 5.1). Water is the one commodity that makes our planet different, perhaps unique, from all the others in our galaxy. Water covers approximately 70 per cent of the surface of the planet and at its deepest point, in the Mariana Trench east of the Philippines, reaches a depth of 10,924 metres. The potential usefulness of water as a human living space is severely limited by three factors: the intense pressure below ten metres of the surface; the lack of sunlight below about 30 metres and the saltiness of the oceans. As a consequence of these physical and chemical restrictions, the oceans have been seen as a negative human resource. We therefore choose to ignore 97.108 per cent of all the water (the salt water oceans) on the planet and concentrate instead on the remaining fresh water amounting to 2.892 per cent. Of this amount, 2.24 per cent is locked up in the permanent icecaps and snowfields of the polar regions and at high elevations.

It must be stressed that quite substantial discrepancies occur between authors as to the precise amounts of fresh water that are held in various parts of the hydrological cycle. Where possible, data used in this chapter is based on the most recent information published by the World Resource Institute. There is broad agreement on the total amount of salt and fresh water on the planet, a total of 1,400,000,000km³ (see Gleik, 1993: 120 for detailed figures). It is difficult to imagine this quantity as the volume is so large and the unit of measurement so different from those we normally use in daily measurement of liquid (e.g. litres). Fortunately, it is unnecessary to become too concerned with the absolute numbers, as the amount of water occupying each part of the hydrological cycle constantly fluctuates. Fresh water lakes and reservoirs account for 0.009 per cent, atmospheric moisture for 0.001 per cent, rivers and streams for 0.0001 per cent and underground water for 0.61 per cent, in total a mere 0.6201 per cent of all the water on the planet. In volume terms, the total amount of renewable fresh water available is 41,022,000km³ (World Resources Institute,

1999) and of this, between 90,000 and 125,000km³ is found in rivers and lakes from which we take most of our supplies. In total, there is approximately 833,319,000km³ fresh water on land but most is locked up in ice sheets or lies deep within the planet (Arnell, 2002). These figures may require substantial revision in the light of the latest findings by Murakami *et al.* (2002) who postulated that 1000km beneath the Earth's surface, five times as much water as in all the oceans may be contained under conditions of great temperature and pressure. However, this water is also likely to be 'unavailable'.

The atmosphere and the oceans are closely coupled in terms of energy transfer. Of the 50 per cent or so of the solar energy that reaches the Earth's surface (see Figure 4.9), about nine-tenths is absorbed by the oceans. This energy accumulates in each hemisphere during the summer months and is released slowly during the wintertime, thus helping to maintain the overall temperature balance of the planet. Some of the incoming energy is used to heat the surface of the oceans and to bring about a change of state of the surface water molecules. Water in its liquid form absorbs energy and by so doing is converted into its gaseous state, water vapour. For each gram of water evaporated, about 600 calories of energy is held by the water vapour (Strahler and Strahler, 1974). Once formed, the majority of the water vapour recondenses into liquid but a small proportion continues to be heated in the atmosphere. This warmed portion is lighter than its surrounding environment and therefore expands and rises into the atmosphere. As it rises it expands further and its temperature begins to stabilise to that of its surroundings and at this point the water vapour condenses to form minute water droplets or clouds. Once again, the majority of this process takes place over the oceans, Shiklomanov (1993) estimating that 90 per cent of all rainfall occurs over the oceans.

The water vapour that does not immediately fall back as rain becomes caught up in major air movements and eventually becomes part of weather systems that move many thousands of kilometres from their origin (the source area) and by so doing, the water vapour is transported from above the oceans to above land. As the air mass moves through the atmosphere, two critical situations can occur, both of which have a broadly similar effect. First, the air mass may move from lower to higher latitudes resulting in the air mass encountering cooler conditions, or, second, the air mass encounters mountainous terrain at which point the air mass is forced to rise and as it does so, cools. Both conditions result in a release of latent energy. The energy that was extracted from the water to convert it into water vapour (600 calories per one gram of water) is now released and the water vapour condenses into one of the alternative forms of precipitation and falls out of the atmosphere onto land or over the oceans. By this means a supply of fresh water, known as *rain-fed* water, is delivered to the land. It is vitally important as it provides an immediately accessible source of water for plants and animals. Any excess of rain-fed water is removed from the land surface either as *percolation* into the soil or as *overland flow* that eventually forms streams and rivers.

There are many possible diversions and temporary storage points during the time a water droplet lands on the land surface to the time it arrives back at the sea. On average, a water droplet will take 14 days to complete the cycle from sea through the hydrological cycle and back to the sea. In many parts of the world, for example most of northwest Europe and the eastern and western seaboards of the USA, humans depend for a large part (90 per cent or more) of their fresh water on the constant availability of rain-fed supplies. In these areas, a dependency on rain-fed moisture has been sufficient to guarantee a water supply in all but extreme drought situations. Away from the coast, or between latitudes 25°–35°N and S of the equator rain-fed supplies are too irregular to provide a guaranteed supply of fresh water. In these locations resort must be made to groundwater supplies accessed by means of artesian systems or wells (see Allison, 1994: 107–10). In Figure 5.2, a water-holding rock, the aquifer, is confined both above and below by impervious materials known as aquicludes. The aquifer is recharged by percolating rainwater or snowmelt and as the water moves through the rock particles it is filtered of any impurities. The water moves by gravity to the lowest point where it accumulates and builds up

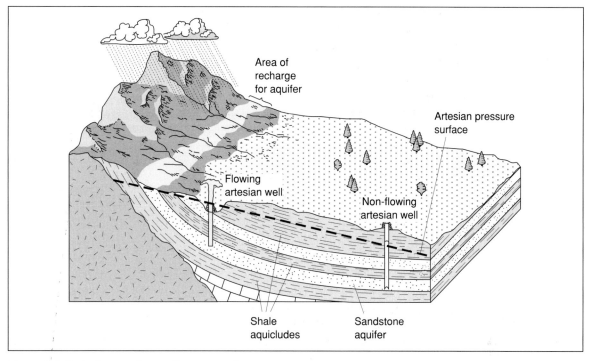

Figure 5.2 Block diagram of an aquifer
(*Source*: Monroe and Wicander, 1994)

a hydrostatic pressure that creates a water table, shown on the diagram by a dashed line. A well that taps into the aquifer will flow naturally if the well head is below the level of the water table but more usually, the well will require to be artificially pumped. Examples of groundwater use are given in section 5.3.3.

5.3 Meeting human demand for fresh water

The minimum quantity of water required to sustain every aspect of human life has been calculated as about 25 litres per day of which a minimum of five litres of liquid must be consumed (World Bank, 1994). We may not drink five litres directly as fresh water but instead absorb much of our water as fluids in the food, especially fruit and vegetables, that we eat. In modern urban-based societies most individuals consume well over five litres of fluids per day as our social meeting places usually involve the consumption of liquids such as coffee, tea, soft drinks or alcohol-based drinks. Apart from our needs to directly consume water we also use water in a wide variety of other uses (see Table 5.1), not all of which require that the quality, or purity, of water matches that of water intended for human consumption. The global per capita consumption for all uses in the early 1990s was approximately 650m^3/year (650,000 litres), equivalent to 1780 litres per person per day (World Resources Institute, 1996). Our use of fresh water has more than tripled since 1950 and between 1990 and 1995 global consumption rose sixfold – more than double the rate of population growth – and by the late 1990s the total global annual consumption had reached 4340km^3 (Goudie, 2000) while Shiklomanov (1993) predicted the figure for 2000 would reach 5200km^3. This compares with the figure of 41,022,000km^3 total amount of renewable fresh water given in section 5.1. Clearly, with such a massive difference between the total amount of available fresh water and the actual amount we

Table 5.1 Uses of fresh water by modern society

Category of use	Approximate consumption in litres	Proportion of total consumption
Domestic uses Cooking, food preparation, laundry, personal hygiene and sanitary uses, cleaning, garden uses	Up to 750 litres per person per day (this includes consumption outside the home, such as food eaten in restaurants, showering in sports clubs etc.)	Worldwide average 8%. Usually no greater than 20%, although in non-industrialised nations the proportion can reach 65%, as in Australia
Industrial uses Cooling, cleaning, diluting, use in food processing, steam production, hydroelectricity generating	Making one kilogram of paper uses 64 litres of water	Worldwide average 23%. Maximum – Norway 72%, Europe 55%, North America 42%. Minimum of 2% in Australia
Agricultural uses Consumption by animals, irrigation, washing, cleaning, diluting wastes	$2700km^3$ of water per year is used for irrigating crops worldwide. Typically, 8000 litres of water per ha per crop	Worldwide average 69%. Up to 90% in agricultural areas, as low as 2% in heavily urbanised areas. In the USA, about 33% of all water use is for irrigation
Transport Rivers, canals, lakes used by shipping for transport of bulk products	Multiuse function provided users do not pollute the water source	No direct consumption although pollution can make subsequent use impossible
Recreational uses Swimming, boating, surfing, artificial water courses, amenity use of water in built environment	Requires water of high portability due to risk of water-borne disease transmission	No irretrievable water loss. Some loss through evaporation

consume there should be no problem in meeting demand. However, most of the readily accessible supplies have been used and accessing further fresh water supplies will be two or three times more expensive (Serageldin, 1995). In practice, more and more of the world's population will experience an increasing shortage of water and a decline in the quality of available water. The most up-to-date information on water consumption can be obtained from the Sustainable Development Information Service located at the World Resources Institute website at http://www.wri.org/trends/water2.html.

Water intended for human consumption, either directly in food preparation or drinking, or indirectly such as washing clothes, showering and cleaning must be of a *potable* standard, that is it must not contain disease-bearing bacteria, viruses or parasites. For most other uses water can be of a much lower quality. For example, water released from washing machines and described as *grey water*, can be used for cooling machinery, flushing toilets or irrigating fields. However, segregated disposal and reuse of water is currently not widely practised, apart from situations of extreme water shortage (see section 5.4.3).

Water obtained directly from rainfall should, in theory, be perfectly safe to drink. Its chemical reactivity is slightly acidic, with a pH of about 5.6 (on a scale ranging from pH 0 – extreme acidity – to pH 14.0 – extreme alkalinity; neutrality occurs at pH 7.5). In heavily industrialised areas rainfall becomes tainted with pollutants from chimneys and the pH of rainfall can drop to a highly acidic 4.0 (Cunningham and Saigo, 2001). Long-term consumption of acid fresh water would lead to severe stomach and bowel illnesses in humans (Howe, 1976). It is now normal in all developed countries for rain-fed supplies to be treated by means of ultraviolet light or chemicals to kill agents of disease, to have lime added to neutralise acidity, to be filtered to remove particulate matter and further chemical additives to remove discoloration due to peat staining. Fluoride is sometimes added to strengthen bones and teeth.

5.3.1 Rain-fed water supply

The majority of fresh water used by humans is obtained directly from rivers and lakes or from renewable ground water supplies (Shiklomanov, 1993). In humid areas, these two sources are recharged on a more or less continuous basis whenever precipitation falls on the land surface. In high-latitude regions and at high-altitude sites, snowmelt in springtime is a major contributor to recharging supplies. In times of excessive rainfall and during peak snowmelt periods, the super-abundance of water usually escapes as surface *runoff*, which in extreme situations causes disastrous flood situations (see Box 4.4 for a severe example of flooding in Mozambique). As a consequence, dams and reservoirs have been employed to slow down the runoff and to store excess water for use in conditions of low recharge. The period of major reservoir construction occurred between 1950 and 1980 when global capacity increased tenfold (Gleik, 1998). Three countries dominate in terms of reservoir storage, namely the USA, the former Soviet Union and Canada, although the reservoir with the largest capacity is in Uganda, the Owen Falls reservoir, off Lake Victoria (Middleton, 1999). It is notable that the primary purpose of reservoirs is the storage of water intended for the generation of hydroelectricity (see Plate 5.1) and not as a source of human water supply.

The reliance on rain-fed supplies of water is made possible due to the highly dynamic nature of the atmospheric portion of the hydrological cycle from which precipitation originates. In the arid and semi-arid areas of the world annual precipi-

Plate 5.1 Techi Reservoir and hydroelectric installation, Taiwan. The power station has a total output capacity of 720 million kilowatt hours

tation is usually very variable and provides less than 500mm per annum. Elsewhere, precipitation occurs on a more regular basis and, for example, excluding occasional summer droughts, precipitation in mid-latitudes normally occurs every two or three days while in the tropics, rainfall is usually a daily occurrence. The annual volume of runoff is almost $47,000km^3$ (see Table 5.2), although for practical purposes the runoff from Antarctica can be discounted, as there is no permanent population to utilise the supply. The availability of fresh water from rain-fed sources is problematic due to the following reasons:

- From Table 5.2 it is clear that there is considerable disparity between the

Table 5.2 River runoff resources of the world

Location	Annual river runoff (km³)	Proportion of total runoff (%)	Proportion of total population (%)	Per capita water consumption (m³)
Europe	3210	7	12.4	625
Asia	14,410	31	60.6	542
Africa	4570	10	13.2	202
North America	8200	17	5.1	1798
South America	11,760	25	7.9	335
Oceania	2388	5	0.5†	591
Antarctica	2230	5		
Total runoff from land area	46,770	100	100	Av. per capita consumption 645m³

* combined average figure for Australia and Oceania
† includes Central America
(*Source*: after Shiklomanov, 1993 and World Resources Institute, 1999)

proportions of population and that of total runoff.

- Eighty per cent of runoff occurs in extreme northern or in equatorial regions where population figures are relatively low.
- The densest population figures are found where rain-fed supplies are most limited. Apart from Oceania, all the regions listed in Table 5.2 have a substantial land area that receives limited amounts of precipitation, for example, about one-third of Europe, 60 per cent of Asia and almost all of Australia have arid or semi-arid climates.
- The 25 largest rivers account for 40 per cent of global runoff. Only four of these rivers are in North America and two are in Europe – both of which have the highest per capita levels of water consumption.
- In most regions, the pattern of annual runoff is highly seasonal in occurrence, up to 70 per cent occurring in the period of 'high water'. Somewhat surprisingly, 'high water' does not coincide with the traditional winter period typical of the northern hemisphere. Sixty-three per cent of annual runoff occurs between May to October, caused by melting snow during the summer and to which must be added the sudden flows resulting from summer convectional rainfall (thunderstorms).

- The less humid the climate of the region the higher the variability of precipitation and, hence, the runoff.
- Population growth, especially in developing countries, causes a constant change in the proportional demand for fresh water.
- Climate change has probably already resulted in changes to the spatial and temporal patterns of precipitation, although the absence of at least 30 years of average precipitation figures for many recording stations throughout the world has made the precise changes difficult to quantify.

The average worldwide withdrawal of fresh water from rivers, lakes and reservoirs in 2000 was predicted by Shiklomanov (1993) as comprising 11.6 per cent of total surface runoff. This average figure hides the fact that in parts of southern and central Europe, in central Asia, and around the Aral Sea region, the rate of fresh water withdrawals already amounts to 65 per cent of the total available water while in parts of the Middle East and in the Arabian peninsular this figure has already reached 100 per cent.

The dilemma of managing rain-fed supplies of water will increasingly challenge utility companies and governments. Some alternatives to rain-fed supplies are considered in sections 5.3.2 to 5.3.4.

5.3.2 River basin management

In many countries of the world, guaranteeing a sufficient supply of good-quality fresh water for human consumption remains an impossibility. Towns and villages are left to fend for themselves and are forced to build small dams or dig local wells to gain access to water that is often of dubious quality. In an absence of overall planning for a supply of fresh water and the safe disposal of waste water, conflict often arises between users. Modern society has been able to survive without an integrated water management plan entirely due to the inherent natural bounty of the hydrological cycle. Often, indiscriminate use of fresh water has led to water disputes, especially where rival users or adjacent countries share the same water supply. Jones and Hollier (1997: 239–47) have outlined the main geographic areas where water disputes already occur. In particular, these occur in the Middle East. While the UN Conference on Environment and Development held in Rio de Janeiro in 1992 gave little attention to the problem of managing fresh water, one of the few recommendations to emerge from the 2002 World Summit on Sustainable Development was that action would be taken to reduce by half the number of people without access to clean water and sanitation by 2015. However, no agreement was reached on the management of rivers and without this essential step it is difficult to see how the target for 2015 can be met.

The old industrial countries (Britain, Germany, the Netherlands) were among the first to establish segregated systems for the supply of fresh water and the removal of waste water. However, these systems were designed to achieve specific aims, for example, a safe water supply for the population, or a guaranteed water supply for major industrial users. It was the United States of America that first realised the need to manage water supply on a much grander scale than that practised in the Old World. The incentive for a new approach to water management came from an environmental disaster that had its beginnings in the Great Plains in 1910 when a start was made to convert the prairies from natural grassland to grain-producing areas. Between 1910 and 1930 the mechanisation

of agriculture using tractors and McCormick reapers (the first combine harvester) enabled the acreage of croplands to rise exponentially. By the 1930s the infamous 'Dust Bowl' had been created (Coffey, 1978; Goudie, 2000: 193–6), within which topsoil was stripped off a land surface that had been overcultivated and that had undergone a process that today would be termed 'desertification'. Overfarming also occurred in other regions of the USA, notably in the Appalachian foothills of the east. The economic slump of the 1920s and 1930s saw unemployed coal miners turn to the land for a living and resulted in forest clearance and overuse of the sloping land that led to soil erosion and an almost total economic collapse of the area. President Roosevelt's *New Deal* of 1930 set out an ambitious plan to renew the physical infrastructure of areas that had been mismanaged over the preceding decades. One of the areas to benefit from the *New Deal* was the Tennessee River valley (see Box 5.1).

Much of the success of the Tennessee River Authority (TVA) has been due to the application of a coordinated management policy to a natural physical unit, the drainage basin of the Tennessee River system. Using a river drainage basin (or catchment area) as the area of management was highly unusual in the 1930s and to tackle an area as large as 200,000km^2 was unique. French scientists had established the validity of the drainage basin as a natural physical unit in the 1750s when they first discovered that sufficient precipitation fell within the area known as the Paris Basin to sustain the flow of the River Seine (see Box 5.2). Previously it had been thought that river flow was sustained by springs emanating in the headwater regions of rivers. In spite of the discoveries made in the Paris Basin it was to be more than 150 years before the drainage basin became widely recognised as the relevant hydrological unit with which to manage the flow of fresh water.

Looked at from today's perspective, it is difficult now to appreciate how radical the integrated management approach to river basin management adopted by the TVA was. At that time, water engineers saw the river as a linear feature, comprising tributaries, a main river and an outlet. Arnell (2002) provides an explanation of how the river

BOX 5.1

The Tennessee Valley Authority (TVA)

The great world economic depression that began in 1929 and continued through the early 1930s, hit the Appalachian region (comprising the states of Tennessee and parts of Alabama, Virginia, Georgia, Kentucky, North Carolina and Mississippi) of the USA hard. Deforestation, uncompetitive farming, soil erosion and flooding combined to make the area one in which people could not make a living. The US Congress created the Tennessee Valley Authority in 1933 as part of President Roosevelt's *New Deal* to develop the Tennessee River and its tributaries in the interest of navigation, flood control and the production and distribution of electricity. Prevention of soil erosion and regeneration of agriculture was also an objective, while more recently the TVA has added the rehabilitation of damaged environments and conservation of habitats to its list of achievements. The current area of the TVA is approximately 207,000km^2 (see Figure 5.3).

Central to the plan was the management of the main rivers – the Tennessee and its tributaries – and the land adjacent to the rivers. The TVA was one of the first extensive, integrated resource management plans to be successfully implemented at the river basin level of management (see Box 5.2) (Huxley, 1943; King, 1959).

At the start of the twenty-first century the TVA continues to manage the natural, economic and social well-being of the area and currently has three goals:

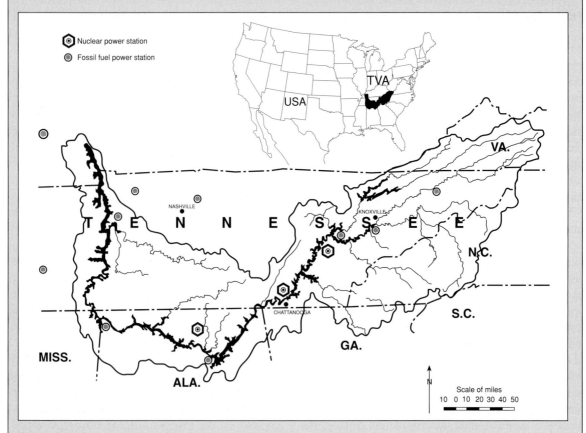

Figure 5.3 Area of jurisdiction of the Tennessee Valley Authority and location of main power plants

BOX 5.1 CONTINUED

- to supply low-cost electrical power produced by means of thermal, hydro and nuclear power plants
- to support a thriving river system, minimising flooding, maintaining navigation, supporting hydro-power generation, improving water quality, protecting public health and the environment and supporting recreational activities
- stimulating economic growth without creating environmental problems such as pollution, erosion or habitat loss.

Although the management problems facing the TVA since its inception have changed over the years, it has retained its strategy of solving problems by means of integrated solutions. Many of the solutions to problems have involved the introduction of state-of-the-art technology. From its outset, the TVA recognised the potential of becoming a major electricity-producing utility. In the 1930s and 1940s a huge dam construction programme was introduced, designed primarily to generate hydroelectric power, and also to provide local employment, to control river flow and improve navigation. The early developments did not usually include consideration of environmental issues beyond those at a superficial level. For example, new wetland habitats or breeding sites were created only as a secondary consideration of economic development of the area. The generation of electricity by the TVA power system was made self-financing in 1959 resulting in the lowest electricity rates in the USA during the 1960s. In anticipation of further demand for electricity the TVA began preparing plans for nuclear power plants but with the rise in environmental issues during the 1980s combined with new technology that resulted in major savings in energy consumption all but three of the nuclear power plants were cancelled. With 29 hydroelectric plants, 49 dams and one pumped storage scheme, the Tennessee River and its tributaries is one of the most highly controlled river systems in the USA.

From the outset, the TVA was involved with the re-vegetation of heavily eroded soils. The soil particles were washed into streams and rivers, causing siltation and also reducing the water quality, making living conditions for aquatic life more precarious. An environmental research centre has been established to develop appropriate technologies and practices that assist with the clean up of all forms of pollution in the river basin. The research centre has created many high-quality jobs and the technology developed at the centre is sold on to other states and countries that are involved with environmental management.

(The TVA website provides an up-to-date account of the organisation: http://www.tva.gov.)

BOX 5.2

The river drainage basin – a useful unit for environmental management

The drainage basin comprises the total geographical area from within which a river and its tributaries collect the water supply. The total area is called the catchment zone (Newson, 1997). The extremities of the drainage basin are delimited by the watershed, the latter forming the perimeter of the drainage basin. Precipitation falling in the vicinity of the watershed will flow into one of the two adjacent drainage basins. About 20% of the precipitation that falls on the land flows over the surface and is ultimately 'captured' by streams and rivers (McKinney and Schoch, 1998).

The shape of the river catchment is dependent on the porosity of the underlying rocks, the total amount and distribution of the annual precipitation and on the natural vegetation. A common drainage basin shape resembles that of a tree root system and is called a dendritic pattern (see Figure 5.4).

A French scientist, Philippe Buache, first presented the concept of the drainage basin to the French Academy of Sciences in 1752 and the idea of dividing the countryside into distinct topographical drainage basins was seized on by the cartographers of the day. Unfortunately, they exaggerated the importance of the watershed, often marking it with symbols or hatching that suggested the watershed was a major

BOX 5.2 CONTINUED

Figure 5.4 Example of a typical dendritic drainage basin

mountain chain, when, in fact, it could sometimes exist as a gentle rise on the landscape. By the middle of the eighteenth century the drainage basin concept of dividing up the landscape had been replaced by regional divisions that were usually based on political or economic criteria. One significant attempt was made in England early in the twentieth century to regionalise the country on a number of socio-economic criteria and were also based on drainage basins, although 'the watershed would only mark out the general trend of a boundary and not govern its details' (Fawcett, 1917). Only in France did the drainage basin remain in use as a geographical unit, being recognised as a 'natural' region (Smith, 1971). However, one is forced to conclude that attempts by geographers, particularly in the period 1900–1940, to retain the drainage basin as a significant planning unit were largely based on historical factors.

During the second half of the twentieth century the significance of the drainage basin as a planning concept gradually changed. The passing of the Tennessee Valley Agency Act by US Congress on 18 May 1933 had marked the greatest single step towards realising the value of the drainage basin as a relevant resource management area but it took a considerable length of time for the vision contained in this Act to permeate to other countries. Only when politicians were persuaded of the benefits of working at the level of *integrated* resource management was there a possibility of setting up planning zones or management corporations that were regional in extent. When it was possible to embark on problem solving in the broadest context it became possible for the drainage basin to regain its pre-eminence as a planning unit.

The use of the river basin as a natural area for resource management has found widespread worldwide use. However, a worrying trend is that interstate, or inter-regional, management only seems to take place *after* an environmental crisis has occurred. Only when planners in one region realise that colleagues in neighbouring districts or states are grappling with similar management problems such as flooding, soil erosion or river navigation problems does the value of a cross-boundary study become apparent. Gerrard (2000) has suggested the poor transfer of knowledge about resource management reflect a lack of research links between those involved in resource management, risk perception and communication studies. Despite attempts by the UNCED Conference in 1992 (United Nations, 1992) and by the Royal Commission on Environmental Pollution (1998: 51–62) to improve the pathways for information and experience transfer between countries that experience similar environmental problems, there remains little real progress in this area of environmental management.

catchment is today seen as an amalgam of three broad elements, changes to each of which can create different consequences on the behaviour of the catchment (see Figure 5.5). At the centre of this diagram are three broad types of potential changes that can occur to the catchment, each of which can be altered by changes taking place in the management pressures within the drainage basin area.

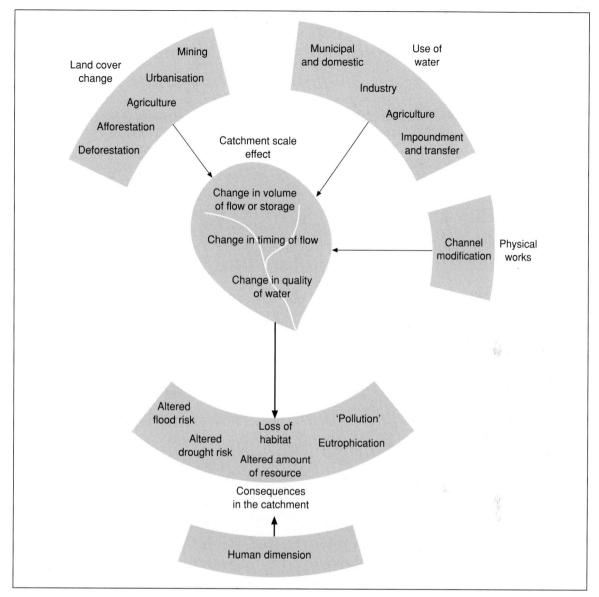

Figure 5.5 Effects and consequences of changes in the river catchment (*Source*: after Arnell, 2002)

The example set by the TVA has been copied worldwide, one of the most notable examples being that of the Murray–Darwin Basin Initiative in southeastern Australia. This scheme covers 1.06 million km² and comprises the catchments of the Murray, Darling and Murrumbidgee Rivers. Collectively, these form the Murray–Darling basin (see Figure 5.6).

Water rights issues had been a feature, particularly for the Murray River, since settlers moved into the area and the first formal discussions on installing locks to improve navigability in the dry season occurred in 1863. By 1915 a water-sharing agreement had been reached for the Murray and this agreement remained in force for some 70 years. The Murray–Darling basin area is Australia's most

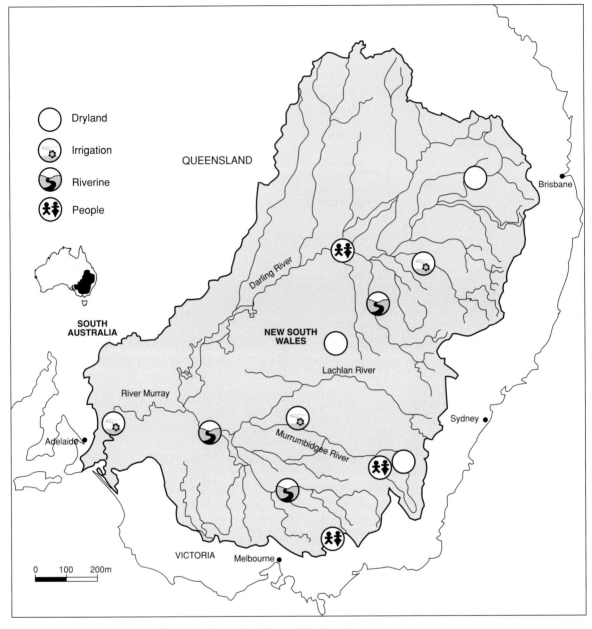

Figure 5.6 Location map of the Murray–Darling Basin Initiative, S.E. Australia

important agricultural area, accounting for about 40 per cent of the nation's gross agricultural production. It extends over approximately 14 per cent of the Australian land area yet supports 25 per cent of cattle, 50 per cent of sheep, half the cropland and 75 per cent of the irrigated land in Australia. The majority of land is in private owner-ship and successful long-term change must involve a high degree of community involvement if sustain-able resource management is to be achieved. How-ever, the maximum limit of sustainable water yield for the Murray–Darling Basin had been exceeded and substantial land degradation had occurred with soil erosion, land and water salinisation, soil

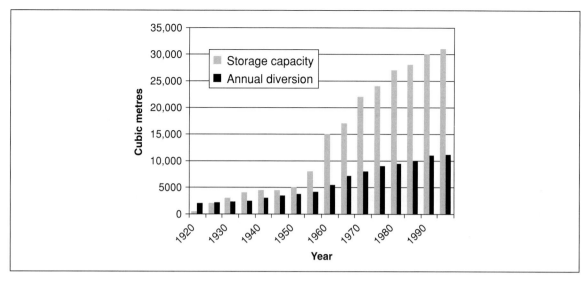

Figure 5.7 Relationship between total water storage capacity and the annual water diversion, Murray–Darling basin, S.E. Australia (*Source*: data from Blackmore, 1995)

acidification and eutrophication of rivers and streams becoming increasingly common.

In 1985 the Commonwealth government in conjunction with the states of New South Wales, Victoria and South Australia initiated the Murray–Darling Integrated Catchment Management scheme in recognition of the need to tackle the natural resource problems that had reached major proportions in the region (Blackmore, 1995). In 1996 Queensland became a member, followed in 1998 by the Australian Capital Territory.

Between the mid-1950s and 1980 the Basin area was 'managed' in the manner typical of the period – by increasing the storage capacity through dam construction, and by abstracting an ever-greater volume of water from the drainage basin (see Figure 5.7). Tension over water rights inevitably become most acute during the frequent droughts that affected the region. However, it was clear by the early 1980s that the maximum capacity of the river basin to yield further water without causing severe damage to the environment had been reached. It was also recognised that no single state government or group of private individuals could manage the catchment in a sustainable manner. Thus, between 1987 and 1998 the six partner governments established a strategic political and management framework that was intended to re-establish a balance

between the production-oriented activities (mainly agricultural) and the environmental needs of the Basin. The latter include:

- sustaining the remaining indigenous plant and animal communities
- combating salinisation of soils and reducing soil erosion
- maintaining wetland and forest areas
- sustaining the fish population and minimising blue-green algal blooms.

These environmental objectives must be set against the need to provide sufficient water for irrigation purposes and for domestic consumption. Management of all these issues must take place in a way that is sympathetic to the significant Aboriginal cultural heritage rights that exist in the area, especially around Lake Victoria.

Management has been achieved through the setting up of many individual strategic plans, for example the Fish Management Plan (1990) and the Dryland Regions Sub-programme Strategic Plan (2000). The individual plans are coordinated by means of an overarching Basin Sustainability Plan prepared in 1996 and updated in 1999 to include a Catchment Management Framework to run from 2001 until 2010. The management plan sets achievement targets that are tested at regular intervals. Information is shared between all participants of the Basin Initiative by means of an education programme. Finally, the general public and private landowners are kept informed of all the work being undertaken by the Basin Initiative.

The management of the Murray–Darling River Basin has undergone a major transformation since its foundation in 1987/88. An integrated, research-based programme in which the needs of the many different users of water are taken into consideration has replaced the piecemeal development that characterised the pre-1980 years. The river basin management programme is an additional financial burden on state finances but society now realises that an assured supply of fresh water can no longer be taken as a God-given right. Fresh water has been recognised as the most basic and essential of natural resources.

5.3.3 Groundwater extraction

A substantial amount of water exists in both a semi-permanent and permanent state at various depths below ground. The semi-permanent element is that proportion of rain-fed water that percolates into the ground and travels beneath ground either towards springs, streams and rivers and maintains stream flow or, alternatively, joins the permanent water table whereupon it can remain untouched for many thousands of years. Estimates of the amount of groundwater vary from about 0.6 to 1.7

per cent of the total fresh water supply. UNESCO calculations made in 1974 and quoted in Gleik (1993: 120) attribute the amount of fresh water held under ground as 10,530km^3.

Not unexpectedly, water engineers have turned their attention to this vast source of fresh water as a provider of a vital additional supply. Exploitation of groundwater brings two short-term problems: a technical problem of how to pump the water from depth to the surface and the associated economic problem of the cost of extracting the groundwater. In addition, there are a number of quite major longer term problems associated with large-scale use of groundwater resources, not least being the depletion of the groundwater and the subsequent subsidence of the ground (see Table 5.3). In Texas, this is contributing to coastal flooding especially in the outskirts of Houston, while in Arizona, fissuring of the ground is occurring (Cutter and Renwick, 1999).

The more immediate problems associated with groundwater use are basically economic problems. As soon as the principle had been established for the consumer to pay a realistic price for water, revenue became available to cover the technological cost associated with accessing groundwater supplies. Additional money has also been made available in the form of government-funded development grants to private water companies or through government subsidies or so-called 'tax breaks' to strategic water consumers to allow

Table 5.3 Land subsidence due mainly to groundwater removal in urban areas

Location	Maximum subsidence (m)	Extent (km^3)
Mexico City, Mexico	8.0	25
Long Beach and Los Angeles, California	9.0	50
Taipei Basin, Taiwan	1.0	100
Shanghai, China	2.6	121
Venice, Italy	0.2	150
New Orleans, Louisiana	2.0	175
London, England	0.3	295
Las Vegas, Nevada	8.5	500
Santa Clara Valley, California	4.0	600
Bangkok, Thailand	1.0	800
San Joaquin Valley, California	9.0	9000
Houston, Texas	2.7	12,100

(*Source*: data from Whittow, 1979 and Dolan and Goodell, 1986)

Table 5.4 Countries exceeding the annual groundwater recharge

Country	Annual groundwater withdrawal		% groundwater used by agriculture
	Total (cubic km)	% annual recharge	
Libya	2.0	420.0	82.5
Saudi Arabia	7.0	337.7	86.5
Mauritania	1.0	293.3	No data
Egypt	3.0	261.5	No data
Algeria	2.0	117.6	No data
Israel	1.0	109.1	No data
Tunisia	1.0	101.7	86.4

Much of the actual data used to compile this table is based on mid-1980s information
(*Source*: data from World Resources Institute, 1999).

expensive groundwater resources to compete with rain-fed supplies. In the past, the level of subsidies were often fixed too high, so that the consumer was encouraged to exploit expensive groundwater. As an example of a consumer 'free ride', the Federal Bureau of Reclamation in the USA set the repayment level for farmers utilising groundwater in the Central Valley Project in California at such a low level that only $50 million out of $931 million development costs over 40 years have been reclaimed (Postel, 1993).

The longer term problems have only recently been tackled. In a perfect situation, water would be pumped from the aquifer at a rate that is equal to the recharge capacity. This situation represents a *sustainable* use of a natural resource. However, to justify the economic cost of drilling a well, of installing a pump and distribution network of pipes, it is necessary to maximise the extraction of groundwater. This produces an *unsustainable* situation known as *water mining*, where the rate of groundwater withdrawal exceeds the recharge capacity and results in a lowering of the groundwater table. Many Middle Eastern countries are exceeding the groundwater recharge capacity in order to meet current demands for fresh water (see Table 5.4). Opie (1993) has calculated that up to 50 per cent of agricultural land currently being irrigated by groundwater supplies will have to be withdrawn from use by 2025 due to depleted reserves. (A case study for Saudi Arabia is given in Box 5.3).

BOX 5.3

Groundwater exploitation in Saudi Arabia

Saudi Arabia has no useable source of surface fresh water and, in the past, relied almost exclusively on underground supplies. Traditionally, these took the form of small natural oases (see Plate 5.2), although since the late 1960s supplies have been augmented by pumping. Only in the Asir region located in the southwest of Saudi Arabia was there sufficient rain-fed water for agriculture. Ninety-seven per cent of the country receives less than 200mm of rainfall per year. However, hydrological surveys had discovered huge reserves of groundwater that accumulated at the end of the Pleistocene period

(the Ice Age) 10,000 years ago when the climate was much wetter.

The availability of abundant supplies of very cheap oil throughout the final quarter of the twentieth century provided the necessary energy source with which to pump water to the surface. The motivation for accessing the groundwater was primarily a political one. The Saudi Arabian government decided that the country should become a major agricultural producer of food, especially grain crops, despite the fact that grain supplies such as wheat could be bought on the world market for four or five times less than it cost to produce wheat in Saudi Arabia. Between 1980 and 1992 agricultural output rose by

BOX 5.3 CONTINUED

Plate 5.2 Small desert oasis in Saudi Arabia, providing water for irrigating a few date palms

Plate 5.3 Depletion of the groundwater table by over-pumping. The well known as Al Dhalaa, near Al Kharj in Saudi Arabia, has been lowered by approximately 30 metres in the last 25 years. Note the pumping station at the top of the borehole and the exposed pipes stretching down to the water table
(Photo courtesy of Dr G. P. Hollier)

14% per annum and in 1992 alone the Saudi government paid $2 million in agricultural subsidies for wheat production (Postel, 1992). Production of wheat rose from 142,000 tonnes in 1980 to a peak of 4,124,000 tonnes in 1992 since when the figure has fallen back to 1,800,000 tonnes in 1997 (SOFA, 1997) as a result of increased pumping costs. Projected water demand suggests that by 2010, 42% of groundwater reserves would be depleted (Ali-Ibrahim, 1991). Agricultural production in Saudi Arabia is based completely on one major unsustainable resource – that of groundwater. According to the World Resources Institute (1999), 86.5% of the pumped water is used for agriculture, 5.4% for domestic use and 8.1% for industry. The total withdrawal of groundwater in 1988 (the latest year for which data is available) stood at 7 cubic kilometres, equivalent to an incredible 337.7% of the annual rate of groundwater recharge! The groundwater table has fallen from a mere 10 metres below the surface in the mid-1970s to more than 90 metres by 1991 (Jones and Hollier, 1997) (see Plate 5.3). In the not too distant future it will become technically and/or economically impossible to pump further groundwater supplies. A future generation of politicians will be forced to find an alternative source of food, not only for the Saudi population but also for countless others living in wheat-dependent Islamic states.

5.3.4 Supplying water by desalination

As 97 per cent of the water on our planet exists as salt water it is inevitable that water engineers should consider converting some of the unusable water into valuable fresh water and thus offsetting the increasingly wide discrepancy between supply and demand for fresh water.

Desalination of salt or brackish water presents few technological problems (see Box 5.4). Despite the introduction of greatly improved methods of desalination the main drawback remains that of cost. Providing a supply of fresh water by means of desalination is currently only viable for countries that experience extreme water shortage or as a temporary back-up measure for short lasting droughts in countries that can otherwise rely on rain-fed supplies. At present, it is the oil-rich states of the Middle East that rely most on desalination, with Saudi Arabia producing almost twice as much fresh water by desalination as the next country, the USA (see Table 5.5). Small islands with a large

---

BOX 5.4

Methods of desalination

In theory, it requires 2.8kJ (kilo Jules) of energy to desalinate one litre of sea water. However, the most efficient plants currently use 30 times (84kJ) this amount of energy. Howe (1978) forecast that future technology might reduce this figure to 28kJ. There are up to 11 different ways of removing salt from sea water and the choice of method is directly linked to the local cost of energy. About three-quarters of all desalination capacity uses some form of distillation process in which salt water is boiled enabling fresh water to be collected by condensation (Gischler, 1979). The remainder uses reverse osmosis (RO) based on membrane technology that separates water and salt ions using selective membranes. RO methods are being increasingly used at the expense of distillation methods.

Most desalination plants utilise finite energy resources, especially oil. The attraction of using renewable energy sources are great and attempts to develop direct thermal solar powered evaporative stills date from 1872 in Las Salinas, Chile. An area of 4500m^2 was required to produce only 20m^3 of fresh water a day. In the United Arab Emirates, mirrors have been successfully used to concentrate the Sun's energy to produce 500m^3 fresh water per day. Electricity generated by wind turbines provides an alternative energy source but the most promising sustainable energy resource appears to be the use of photovoltaic energy. Libya currently operates a 2000m^3 per day system powered by solar energy.

Table 5.5 Production of fresh water by means of desalination: top ten countries, 1996

Country	Desalination output (cubic metres per day)	% installed world capacity (not all capacity is used)
Saudi Arabia	5,006,194	25
United States of America	2,799,000	12
United Arab Emirates	2,134,233	10
Kuwait	1,284,327	10.5
Libya	638,377	3
Japan	637,900	3
Qatar	560,764	2.8
Spain	492,824	2.5
Italy	483,668	2.4
Iran	423,427	2.1

(Source: Gleik, 1998)

influx of tourists, for example Malta and the Balearics, also make heavy use of desalination plants but in volume terms their output is quite small. Globally, desalination provides about one-thousandth of total fresh water requirement.

The cost of producing fresh water by means of desalination varies greatly depending on the type of technology used, the volume of fresh water produced, the salinity of the supply and, most significant of all, the cost of the energy source used to power the desalination plant. Leaving aside the oil-producing nations for which the cost of energy is negligible, a modern high-capacity desalination plant such as that built for Santa Barbara, California in 1996 had a production cost of about $1.55 per cubic metre of fresh water, compared to a conventional rain-fed fresh water supply of $0.47/$m^3$ (Loaiciga and Renehan, 1997). Faced with an almost fourfold increase in the cost of their water, consumers in Santa Barbara changed their pattern of water use such that they achieved a 39 per cent reduction in demand making the continuous use of the desalination plant unnecessary and the operation of the plant totally uneconomic.

5.4 Water supply case studies

5.4.1 Water management in a major industrialised nation – Japan

The average annual rainfall for the four main islands of Japan is about 1700mm. This figure is twice the world average, but expressed as a per capita amount for the Japanese population of 126 million it falls to one-fifth of the world average and is insufficient to meet the freshwater needs of the country. Substantial social disruption has already resulted from a series of severe droughts during the early 1990s (OECD, 1976).

The Japanese public already face strict government control regarding planning and social policies that take account of the high risk to society from natural hazards such as earthquakes, typhoons and tsunamis. The public reaction to severe drought is therefore conditioned towards accepting the need for further environmental

management and an almost inevitable acceptance of the increased costs associated with ensuring a supply of fresh water.

First, a programme to replace old pipes was undertaken to reduce the already modest leakage rate of 19 per cent of total consumption in 1975 to 10 per cent in 1993 (in most British cities leakage remains between 25–50 per cent). Commercial and business users have been encouraged by a pricing policy to install water reuse systems. Sports stadiums store rainwater in roof tanks which is used for toilet flushing and showers. Newly designed air-conditioning units and all washing machines now use substantially less water than old units.

In a complete departure from the pricing policies of utility companies throughout the developed world, where discounts are weighted in favour of largest consumers, urban municipalities in Japan have established water rates that favour small consumers. For example, a household that limits water use to less than 10m^3 per month will pay the equivalent of 40 US cents per cubic metre, whereas a consumer using 500m^3 per month will pay almost seven times this amount ($2.64 per cubic metre). The reasons for adopting this pricing policy have been widely advertised and have been accompanied by technical and practical advice on how individual consumers can economise on the use of water.

5.4.2 Supplying water to dispersed rural communities in Africa

Provision of a reliable and safe supply of water for inhabitants of the developing world has proved to be a particularly difficult problem to solve. Signs of the impending crises that were to emerge during the last quarter of the twentieth century had been foreseen at the UN Habitat Conference held in Vancouver in 1976. Delegates urged governments to work towards the target of ensuring a supply of water to all rural and urban dwellers by 1990 and a later UN conference reinforced this decision by setting the decade 1981–1990 as the Drinking Water Supply and Sanitation Decade. The impact of these decisions proved minimal and it is noteworthy that the UN Conference held in

Johannesburg in 2002 set as one of its targets, the aim to provide all rural and urban dwellers with a reliable and safe supply of water by 2015!

The need to ensure a safe, reliable water supply for communities throughout much of Africa is founded on the following statistics: some 80 per cent of disease in Africa is related to contaminated water; 96 per cent of all infant mortality (children under five years of age) is due to their drinking contaminated water.

Access to fresh water for human use in many parts of northern and southern Africa has traditionally relied on seasonal river flow supplemented in the dry season by wells that tap the groundwater table. It is the task of women to fetch water from wells, rivers or standpipes and studies by Akintola et al. (1980) in Nigeria found that many women spent up to four hours a day fetching water, carrying up to 16 litres of water per trip. Not unexpectedly, wherever water has to be carried by humans, per capita consumption levels in Africa fall to some of the lowest levels recorded anywhere, typically about 21 litres per person per day compared with four times that amount in African cities with a piped water supply (Macdonald and Kay, 1988). The World Resources Institute (2000) has differentiated between two conditions of water stress and scarcity, although to the people who experience acute water shortages, the differences are largely irrelevant. An area is said to experience water stress when annual water supplies drop below 1700m^3 per person and water scarcity occurs when annual water supplies drop below 1000m^3 per person.

By 2025, the World Resources Institute (2000) predict that 48 countries worldwide will face water stress or scarcity conditions. Of these countries, 40 will be in West Asia, North Africa or sub-Saharan Africa. The African continent has been ravaged by some of the most extreme droughts experienced anywhere on the planet. The 1980s will be remembered as the decade when television pictures conveyed vivid images of hundreds of thousands of the mainly rural population of the Sahel region, of Eritrea and the black homelands of South Africa dying from lack of water. Many African countries, with a population of nearly 200 million people, are facing serious water short-

ages. By the year 2025, it is estimated that nearly 230 million Africans will face water scarcity, and a further 460 million will live in water-stressed countries. Among the African countries likely to run short of water in the next 25 years are Ethiopia, Kenya and Nigeria (http://www.unep.org/vitalwater/21.htm).

5.4.3 Multiple use and water reuse

An indication of how seriously water managers now take the shortage of fresh water in arid and semi-arid regions of the world can be seen through the increasing attention given to the multiple use and reuse of fresh water. These two methods of using water are based on very different principles.

Multiple use occurs when more than one user simultaneously uses a given water supply, the most common examples of which can be found from the leisure and recreation uses of water. Colour Plate E shows an urban river used for boating, as a living area for water birds, as an aesthetic feature used by town planners and as an amenity area where office workers can walk and eat lunch. For multiple use to be successful it is essential that each user does not jeopardise the quality and abundance of supply for other users or create an excessive disturbance such as noise. For this reason, the example shown in Colour Plate E precludes activity sports such as water polo or use of powered boats and abstraction for industrial or domestic use is not permitted. Multiple use of water usually involves little or no active downstream water treatment as a consequence of the 'clean' use made of the supply.

By contrast, reuse of water usually involves the chemical and/or mechanical treatment of water subsequent to its first use following which its quality is restored sufficiently to permit reuse by a downstream user. Repeated recycling of the water is possible until it eventually escapes the catchment area or contamination makes further cleansing uneconomic. Reuse usually requires the intervention of technological processes such as filtration, aeration, cooling or sterilisation prior to safe reuse. In cities with extreme water shortages, such as Jeddah and Riyadh in Saudi Arabia, cooling water collected from large air-conditioning units or waste water from commercial laundries is collected and used to water urban parkland. A traditional reuse of water is shown from the island of Mallorca, where a complex arrangement of agricultural terraces ensures that water held in mountain-top tanks is progressively reused in its journey from top-most terrace to the bottom (see Colour Plate F). This system carries the potential risk of transferring agricultural fertiliser and pesticide residues washed from the top-most terraces to those below and causes unacceptably high build-up of substances such as nitrates. Consequently, EU legislation now prevents the abstraction of water for human consumption from this traditional form of water reuse.

When the severity of water pollution was less than today, it was possible to release used water directly back into a river, downstream from where it had been abstracted. The contaminated water would be diluted to a safe level by the natural flow of clean water. Subsequent reuse of water downstream of a discharge point was possible without resort to mechanical or chemical treatment. As population density increased so the demand for water grew. In addition, the nature of water pollution changed. Addition of detergents, cleansing solvents, organic compounds, oils, as well as pathogenic organisms to waste water made it impossible to rely on the natural cleansing properties of river flow. Increasing concern with public health standards and the constant need to prevent the spread of water-borne disease has resulted in more stringent control of the effluent that can be released back to rivers. However, modern waste water treatment plants can convert the most polluted water back into a potable state and many large urban municipalities reuse polluted water as a matter of course. Two contrasting approaches are explained in Box 5.5 in which private householders are encouraged to participate.

5.4.4 Integrated water resource management – the French approach

Per capita water consumption in France has risen over the last 15 years to 665m³, becoming the second highest in the EU Member States. Only Italy is higher at 986m³ per person (World Resources Institute, 1999). Much of the increase has been consumed by irrigation of cereal crops and demand has been met mostly from groundwater supplies. Over consumption has led to serious losses in river flow, a situation made worse by a succession of severe droughts.

As explained in Box 5.2, water management in France had for many years used a different approach from elsewhere, being based on the river basin as the management unit. However, such were the new pressures placed on water managers during the post-war growth of the 1950s that it was clear that a new approach was needed to guarantee water supplies. In 1964 a highly innovative approach to water management was implemented with the passing of the Water Law that established six river basin agencies (OECD, 1998a). In addition the Water Law established the principle of *user participation* in which each river basin was controlled by a river basin committee, the so-called 'water parliament' comprising one-third each of elected officials, government officials and technical experts and consumers (mainly made up of industrial, agricultural and municipal users). The greatly increased demands for fresh water resulted in conflict of interests among the members of the water parliaments and made long-term sustainable management of water resources more difficult. In 1992 a new water law introduced four new management initiatives. These were:

- the recognition that fresh water had a value beyond an immediate economic value
- an attempt to give equal weight to all water resources (surface and groundwater)
- the requirement to manage water resources in a wider strategic manner across a longer planning period (15–20 years)
- an increase in the terms and powers of local managing bodies.

Each river basin committee has been required to prepare a long-term water use guidance document in which a dynamic approach to water use was to be developed as a result of wide consultation with all user groups. The use of financial incentives figures prominently in all stages of water management. The concepts of *user pays* and *polluter pays* govern both the consumption

BOX 5.5

Reusing domestic water: Sydney, Australia and Surrey, England

Faced with a need to reduce per capita consumption by 35% by 2010 in order to meet the total increase in demand, Sydney Water (the water company responsible for supplying the urban area of Sydney) introduced a dual piping system to a major new greenfield residential development. Each house has two sets of water supply pipes: one set supplies water for conventional indoor use (drinking, cooking, washing). Another set supplies reuse water suitable for outdoor use such as irrigating gardens. Outdoor water use in Sydney constitutes up to 25% of total household use and the substitution of grey water and storm water can save up to 50m^3 of potable water per household per year (Sydney Water, 1995).

This saving is a useful first step towards the target of 35% per capita reduction. Sydney Water has investigated the possibility of potable water reuse, that is the direct reuse of water by humans that has been previously used for all purposes apart from sewage disposal. By means of processes known as bioremediation and biotreatment that include carbon adsorption, microfiltration, ultrafiltration, reverse osmosis and ozonation, used water can be made safe for direct reuse. However, in a survey of consumer attitudes, Sydney Water found that only 9% of customers supported the reuse of potable water.

Availability of fresh water in southeast England has been in short supply especially during a series of low rainfall years during the 1990s. Planning permission for new houses is often accompanied by a condition that new development must incorporate as many water-saving devices as possible *at the design stage*. Water consumption per household is by means of a metered supply and the annual water bill in 2000 for a four-bedroom house was approximately £150. In an attempt to reduce the long-term cost of water, 400 new houses at Coulsdon in Surrey have been fitted with rainwater storage systems as standard. Rainwater collected from each 150 sq m roof amounts to an annual 105,000 litres and this is stored in a 3000 litre underground tank beneath each house. The tank holds sufficient water for an 18-day supply with the water being used for flushing toilets, supplying washing machines and for watering gardens. The rainwater is filtered before entering the tank. In times of drought the tank can be automatically topped up from the mains. Each house has a gauge which records how much water remains in the storage tank and the only job facing the house owner is to clean the filter four or five times a year. Each installation costs between £500–1000 depending on the house size and when full use is made of the stored rainwater, the household water bill can be cut by a half. However, the benefit to the community of such a system is far greater than the cost savings. Using rainwater can save scarce fresh water supplies while the storage of 3000 litres of water per household can help hold back rainwater from entering streams and rivers in times of flood, and can offset the greater runoff that occurs from built over ground compared to vegetated areas.

(An interesting website that deals with the collection and use of rainwater can be found at http://www.eng.warwick.ac.uk/DTU/rainwaterharvesting/links.htm.)

of fresh water and the return of polluted water to the drainage basin with rates set according to the volume of water extracted from and to the amount and nature of the waste water returned to the drainage basin. When a river basin committee generates a financial surplus as a result of its management policy, the proceeds are made available in the form of grants to consumers who can provide evidence of better water management, for example through introduction of new technology that generates a lower water demand and/or small pollution output.

A survey of public attitudes towards fresh water resources at the start of the 1990s showed the majority of people in France viewed water as a freely available, unlimited resource. Resources were made available to introduce a re-education programme aimed at 8–11-year-old children in the hope that in the medium term, focusing attention on this cohort would bring the greatest shift in attitudes to water use. The results of this programme will only be seen after 2001.

5.4.5 Water policy in the 'Fertile Crescent' – the Middle East

The term 'Fertile Crescent' applies to an area extending from the eastern Mediterranean Sea south to the Gulf of Arabia. It includes the countries of Palestine, Israel, Jordan, the Lebanon, Syria and Iraq (see Figure 5.8). Water management in these countries is made more difficult due to the political and religious differences that exist in the region. The situation is made worse by the unilateral action taken by neighbouring nations that often control the headwaters of rivers that pass most of their distance in one or more different country. Similarly, overpumping of the groundwater in country 'A' can lead to a drying up of springs and streams in country 'B'. Integrated water management throughout the region remains negligible and provides a classic contrast to the management conditions that prevail in the Murray–Darling River Basin (see section 5.3.2).

It is a paradox that the area covered by the 'Fertile Crescent' is now characterised by semi-desert conditions, an absence of a permanent vegetation cover, by severe soil erosion and a lowering of the groundwater table due to excessive pumping. Between 400 BC and the birth of Christ the area reached the climax of its development based on the Nabataean civilisation. The Nabataeans were highly skilled engineers and used stone to construct not only their city buildings, but aquifers and water storage cysts to manage the water supply. The region was annexed by the Romans in AD 106 and thereafter passed into decline. Whether this was due to climate change towards more arid conditions, to disenchantment of the local population due to subjugation by Rome and an unwillingness to manage the complex system of aqueducts or by exploitation of the natural resources by the Romans, it is certain that soil erosion increased, agricultural productivity declined and the area could no longer be termed the 'Fertile Crescent' (Taylor, 2001).

Present-day precipitation figures for the region show a decline from 500mm per annum in the west to 50mm in the eastern and southern extremities. Humidity during the summer is as low as 15–30 per cent, in winter it rises to about 60 per cent. Potential evaporation from free water surfaces

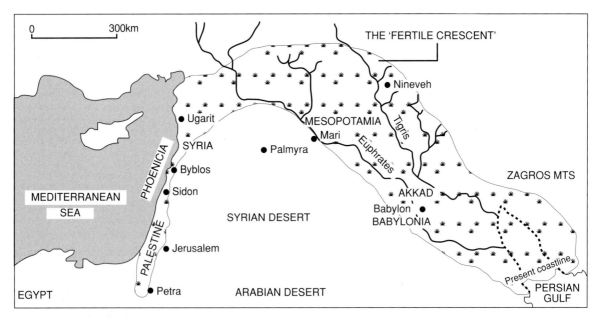

Figure 5.8 Location map of the 'Fertile Crescent', Middle East

is therefore high and in the south and east of the region can reach 4000mm per annum. Even in the more humid areas annual evaporation can reach 1600mm. Precipitation can occur in torrential storms, resulting in rapid runoff along the seasonally dry wadis which either distribute the water onto the stony low-lying areas where it infiltrates into the groundwater table or is held in small reservoirs for subsequent use.

Heavy reliance for fresh water throughout the region is placed on rainfall (Jaber *et al.*, 1997). In Jordan, about 70 per cent of surface water resources are already consumed; the remainder either evaporates or recharges the groundwater resources. The latter are currently undergoing substantial overexploitation (water mining) equivalent to 160 per cent of the sustainable yield level. Most of the groundwater supply is used for agriculture. The head waters of the main rivers flowing through Jordan are controlled by neighbouring countries forcing an ever-greater reliance on groundwater supplies. At the same time, political unrest in the region has seen a massive influx of immigrants and refugees leading to an increased demand for water. Between 1987 and 1995 the sustainable per capita water supply fell from 260m^3 to 210m^3. Assuming a similar population increase in the future, by 2015 the sustainable per capita water figure will be only 100m^3.

Per capita daily water consumption of 80 litres in Jordan is already among the lowest in the region and, to ensure safe public health, a figure of between 100–200 litres per capita per day is recommended (World Bank, 1994). The figure of 80 litres is a consequence of the frequent cessation of the public water supply and a consequent reliance on water delivered by road tanker or even more expensive bottled water. Apart from the human suffering caused by a shortage of water, large areas of the Middle East now experience a curtailment of agricultural production due to insufficient water, creating a greater demand for imported foodstuffs. The economy is also placed under stress because of the high investment and the consumption of expensive energy necessary for water pumping.

If the experience of the Tennessee Valley Authority and the Murray–Darling Basin Initiative were applied to the countries of the 'Fertile Crescent' it could bring about major long-term improvements to the water supply. Many simple steps could be taken: more efficient use of water, replacement of old pipes, reduction of pollution, reuse of grey water. While accepting the real political difficulties in establishing a transnational integrated water management policy in the region, it is difficult to propose any other method of management that would lead to an improved water supply situation. Successful resolution of the problem may await a World Bank or Food and Agriculture Organization initiative to implement a transnational integrated water management policy to the region.

5.4.6 Mismanaging fresh water – the Aral Sea

Natural fresh water lakes have always proved attractive areas of settlement for humans and some of the earliest population groups were located on the shores of lakes to take advantage of the supply of food (fish) and of the fresh water (World Resources Institute, 2000: 113–16). Overfishing and pollution have often resulted in lakes losing their advantages as settlement sites but during the twentieth century a new form of mismanagement has arisen – abstraction of so much water from the lake that it disappeared! Middleton (1999) cites the example of Owens Lake in California becoming permanently dry by 1930 due to excessive withdrawals by the Los Angeles Department of Water.

A similar fate appears to be in store for the world's fourth largest fresh water lake, the Aral Sea lying roughly in equal parts of Uzbekistan and Kazakhstan in Central Asia (Micklin, 1988). Over the last 40 years the Aral Sea has lost three-quarters of its water volume and at the current rate of contraction could disappear by 2015. Since 1968 the water level has dropped some 19 metres and the southern coastline receded by 150km. Some 33,000km^2 of former lake bed has been exposed, forming a sand-covered plain from which the wind whips up an estimated 100 million tonnes of sand and dust each year and transports it onto agricultural land. The dust is highly saline and in some areas the deposition rate reaches four tonnes per hectare, sufficient to sterilise the agricultural land on which it falls (Glantz *et al.*, 1993). The remaining area of the Aral Sea has been transformed from a fresh water lake to that of an increasingly saline, highly polluted lake. Agricultural runoff from the fields has carried pesticides, defoliant agents, fertiliser residues and sewage into the lake, killing the fish stocks that once employed 60,000 fishermen and which, in the 1970s, was still yielding 40,000 tonnes of fish per year – a vital source of food. (A website that covers the environmental problems of the Aral Sea is: http://www.dfd.dlr.de/app/land/aralsee/.)

The cause of these changes has been due entirely to what now appears to have been a deliberate mismanagement of the fresh water resource. The centrally planned Soviet economy of the 1950s and 1960s decreed that the region surrounding the Aral Sea would be used to produce cotton by means of monoculture agriculture from collective farms. Many of the environmental requirements necessary for the growth of cotton were suitable, for example the soils and the temperature regime. However, rainfall was inadequate and instead, water was abstracted from the two main feeder rivers, the Amu Darya and the Syr Darya, and used for irrigation (Saiko and Zonn, 1994). This decision was based either on a lack of understanding of the ecology of the region or on the fact the planners simply didn't care about the long-term consequences. Excessive irrigation has led to the loss of two million hectares of fertile land as a result of salinisation and the productivity of pasture grounds has been halved.

The human cost of this disaster is high. Shrinkage of the Aral Sea has led to a more extreme climate, with shorter hotter summers and longer colder winters. Diseases such as cholera, typhus and diarrhoea have increased as has anaemia and blood cancers. Infant mortality and birth defects in the areas surrounding the Aral Sea are now the highest in all the former Soviet Union.

It is unlikely that the 40 years of devastation imposed on the Aral Sea ecosystem will ever be rectified. The World Bank has worked on a three-year water management project with representatives from Kazakhstan, Kyrgystan, Turkmenistan, Tajikistan and Uzbekistan in an attempt to establish a sustainable policy that will help to avoid further deterioration. In 1997 the five central Asian states formed the International Aral Sea Rehabilitation Fund (IFAS) with the overall aim of rehabilitating as much as possible of the disaster zone surrounding the Aral Sea. The political uncertainty that exists in the region is inimical to the development of long-term, stable management plans. The rehabilitation of the Aral Sea depends greatly on international aid provided to develop and stabilise the essential components of the infrastructure – for example, re-afforestation, sustainable agriculture based on minimal use of inorganic chemicals, drip-feed irrigation and a modern waste water and sewage system. Apart from World Bank- and International Red Cross-funded projects little in the way of international interests has been shown in repairing the Aral Sea ecosystem. The successful rehabilitation of the Aral Sea through international cooperation would provide a clear signal that exploitation of the biosphere that typified the twentieth century had been replaced by a twenty-first century determination to provide an ecologically stable form of management.

5.5 Sustainable water resources management

The management of fresh water resources has occupied a prominent position in the development of urban–industrialised societies ever since it was shown that the provision of a safe water supply was an essential requirement for human health (Howe, 1976). Water management in the twentieth century concentrated mainly on the public health, engineering and technological requirements necessary to meet a fourfold increase in demand between 1950 and 2000 (OECD, 1998a). Meeting the challenge of providing sufficient fresh water has been accompanied by many instances of *environmental mismanagement*, examples of which have been given throughout this chapter. They include overpumping of groundwater supplies, excessive abstraction of river water flow, salinisation through excessive irrigation, and contamination of supply through thermal (heat), biological (algae) and chemical (inorganic fertilisers) pollution. Some of the consequences of the mismanagement of fresh water include overdrainage of wetlands, loss of watershed protection through deforestation and soil erosion, plant and animal species extinction through habitat modification and the loss of fresh water fisheries. For many countries, the overriding determinant of economic growth has involved the supply of fresh water irrespective of the cost to environmental stability and well-being. For example, the Rio Grande river (or Rio Bravo) as it flows through the state of New Mexico in the southwest USA, is reduced to a stream due to the abstraction of water to support irrigated agriculture in the state. An even more extreme example is that of the River Jordan, deprived of its headwaters that have been captured for use within Syria. According to flow rate figures for 1953 (Lowi, 1993), the River Jordan released 1250 million cubic metres (mcm) of water into the Dead Sea. By 1990 the annual amount of water reaching the Dead Sea had fallen to only four per cent of the earlier figure and amounted to approximately 50mcm. Exceptional circumstances have been responsible for this situation, not least the annexation of the Golan Heights and West Bank tributaries after the 1967 war, the damming of other tributaries within Syria, the extraction of river water upstream of the Dead Sea for use as irrigation water and a massive increase and redistribution of population in the region. The availability of fresh water in the Middle East has reached a crisis situation and fears exist that shortages could lead to acts of aggression in an area that already experiences fragile political international relations.

Faced with the real threat of a world economy slowed by a lack of fresh water, institutional and public attitudes have been forced towards adopting an *integrated* environmental and water resources management policy. This approach recognises the futility of continually striving to increase the supply of water without also minimising waste, increasing the multiple use and reuse of water, optimising water allocation to competing users and limiting withdrawals to sustainable levels. Implementing such a scheme will not be quick. It will require the application of new sustainable technology that will be expensive.

During the 1990s increasing concern with environmental issues led to the adoption of a wider interpretation of the term *competing water users*.

It is now possible to recognise the *environment* as a water user in its own right. Ensuring that rivers have minimal flow levels, ensuring that wetlands remain inundated at critical times of the year and preventing the release of polluted water back into the environment are all examples of the greater understanding of the water requirements of the environment. Examples of river catchment organisations such as the TVA and the Murray–Darling River authority discussed earlier in this chapter now include the environment, conservation and recreational use of water as legitimate users alongside the traditional water users.

Perhaps the greatest change in public attitude and in management response has been the acceptance that the availability of fresh water can no longer be considered a *free economic good*. Governments must accept a major share of responsibility in establishing a framework within which a sustainable water resources management policy must operate for it is mainly through government policy that the attitudes of the general public can be modified to accept that a greater efficiency in the appropriate use and conservation of water is necessary. Government also has a major role to play in the setting of appropriate water pricing tariffs even though the private companies will be responsible for the delivery of the water supply.

5.6 Fresh water management in the twenty-first century

Two clear conclusions emerge from information presented in this chapter. First, data relating to the amount of fresh water available for use shows very substantial variations depending on the assumptions made by the different authors. These differences become critical when trying to forecast the future supply levels, especially for countries with low natural fresh water supply levels. Problems concerning data notwithstanding, the second conclusion is indisputable: despite an overall global abundance of fresh water there is an increasing imbalance between the level of demand and the local or regional ability to provide an adequate supply of fresh water. The average annual world per capita consumption of fresh water in 1998 was an estimated 645m^3 (World Resources Institute, 1999). Assuming the annual per capita figure rises in line with recent trends, consumption will reach 1000m^3 by 2050. Because of difficulties in forecasting the exact population in 2050, and because of disagreement in the precise amount of available fresh water, it is estimated that between 20 and 66 per cent of the world's population will experience a moderate to high water deficit by the middle of the current century (Raskin, 1997; World Meteorological Organization, 1997). Developing countries face the greatest water shortages, especially those undergoing the most rapid urbanisation where a shortage of fresh water is often exacerbated by a total lack of service provision concerning segregation of fresh water and sewage have resulted in chronic levels of human ill health.

As the availability of fresh water becomes more critical, so management of the supply becomes increasingly urgent. By tradition, water has been seen as a freely available natural resource but as the cost of collection, distribution and purification of fresh water has risen, so governments have been forced to introduce a charging policy, usually in the form of a *water rate*. This charge can either be a flat-rate charge per household or a proportional charge based on the value of the property. More recently, the supply of fresh water has been made the responsibility of private companies which, in turn, are required to meet legal standards of purity and ensure that sufficient reserves are available to meet peak demand. In this situation, water as an *economic good* has become more widely accepted as increased emphasis is placed on the need to allocate water resources efficiently. However, it must also be recognised that every person should have access to clean water for personal and domestic purposes and pricing policy must be fixed so that no one is denied access to an adequate water supply (OECD, 1998b).

Expecting people to pay for a naturally occurring resource has been a difficult marketing exercise and one that is especially difficult to justify in a low-income society or developing country. Faced with an increasing scarcity of supply most consumers are eventually persuaded that payment of a small price for a guaranteed minimal supply of safe water is a price worth paying. Water man-

agers are convinced that by paying for water, the consumer will automatically value the resource more highly and not use it to excess. Such an argument may be valid in a developed country where all the consumers are familiar in the ways of commercial trading. In developing countries there may simply be insufficient money available to pay for a managed water supply.

Undoubtedly, great improvements in the management of water can be achieved although this will require investment over a period of years. The greatest single improvement would be the reduction of leakage from distribution pipes. Old cast iron pipes are notorious for their water losses through leakage. This figure can reach 50 per cent of the throughput of water! Where water is used for traditional flood irrigation, up to 75 per cent of the water may never reach the crop. Modern drip irrigation using micro-bore plastic tubes is almost 90 per cent efficient but its use still only accounts for about 1 per cent of the world's irrigated area (Rosegrant, 1997).

One of the major challenges facing society is to find ways of providing an assured supply of fresh water to meet a future world population that will far exceed the current figure of six billion. Currently, there is no clear plan of how we are going to meet the future demand for water. The only certainty is that the level of water consumption for domestic, agricultural and industrial uses will continue to rise. By diverting more and more fresh water for human use we divert water away from the remaining natural land-based ecosystems, thereby placing animals and plants under greater moisture stress. These changes are taking place in the context of a world climate that is rapidly changing as a consequence of global warming – itself a consequence, in part, of human actions.

While international cooperation has made major advances in the areas of air pollution, tropical forest destruction, biodiversity and food production, there is no single international agency to coordinate a global water resource strategy. World leaders continue to ignore the problem of fresh water supply. The World Commission on Environment and Development (United Nations, 1987) failed to include a section on water resources while the United Nations Conference on Environment and Development (the Rio Conference) paid scant attention to the provision of water management. Under the general heading of 'Promoting sustainable human settlement and development', the conference recommended provision of 'adequate environmental infrastructure facilities in all settlements by the year 2015'. To achieve this requirement, 'all developing countries should incorporate in their national strategies programmes to build the necessary technical, financial and human resource capacity . . . by 2000' (Dodds, 1992). It is difficult to understand why such superficial consideration could have been given to such an essential human resource. Preliminary press reports from the World Summit on Sustainable Development held in Johannesburg in 2002 suggested that agreement was reached to halve the number of people lacking basic sanitation and safe drinking water by 2015 but no guidance was forthcoming as to how these objectives were to be met. Once again it appears that the management of fresh water has not figured highly on the agenda of politicians and it may take the prospect of an international conflict over water rights finally to galvanise action.

Useful websites for this chapter

Aral Sea homepage
http://www.dfd.dlr.de/app/land/aralsee/

Murray–Darling Basin Commission
http://www.mdbc.gov.au/index.htm

Stockholm International Water Institute
http://www.siwi.org/

Stockholm Water Symposium August 2001
http://www.siwi.org/menu/valframe.asp

Sustainable Development Information Service
http://www.wri.org/trends/water2.html

Tennessee Valley Authority
http://www.tva.gov

UNEP Vital water graphics
http://www.unep.org/vitalwater/21.htm

Water harvesting portal
http://www.eng.warwick.ac.uk/DTU/rainwaterharvesting/links.htm

World Resources
http://www.wri.org/

6 Managing the oceans

The health of our global water system rooted in the ocean is vital to the future welfare of our planet . . . The future needs of society will be well served . . . only if we change our short-term mentality . . . and focus on long-term considerations and a sound attitude in the use of all our resources.

Cousteau, 1981

6.1 Introduction

At the beginning of the twenty-first century, an estimated 50 per cent of the world total population (amounting to slightly more than three billion) live within 320km of the coastline (Goudie and Viles, 1997). For the majority of these people, the focus of their attention is predominantly on the land and not on the sea and the oceans probably only have a *real* significance for the very small proportion of people who earn their livelihood from the sea – fishermen, mariners, oil well workers based on rigs at sea and the emergency rescue services. For all others, the only real contact may occur for several weeks a year when holidays are taken at 'the seaside'. Holiday memories are far removed from the true significance of the role played by the oceans in controlling the behaviour of the planet. Neal (1992) states that 'so vast, yet so far removed from the day-to-day lives of the majority of people of the world, the marine environment is viewed by most of mankind as everlasting and something so immense that it is immutable'. This chapter identifies how the oceans provide the life-giving properties of our planet and, as such, deserve far greater attention by society. It provides a brief introduc-

tion to those aspects of the marine environment that are currently of concern.

From the human perspective, land is the most important 'surface' in terms of providing a suitable place to live. However, when looked at from a holistic biological viewpoint, the oceans provide a far richer environment for life forms. A shallow, warm location in the primeval ocean was almost certainly the place where life on Earth began some time around 3500 million years ago (Folsome, 1979). Over the ensuing millennia, plant and animal life forms have diversified into a bewildering variety of types to suit the very different range of conditions from the warm, light and nutrient-rich surface layers to the cool, dark and nutrient-poor ocean depths.

Inevitably, human interest in the oceans is mainly utilitarian and self-focused on issues such as the tonnage of fish that can be caught from the seas or how to extract minerals that lie just below the seabed. Only in very recent years have we extended our concern to include the less tangible issues such as the possible impact of sea level rise and the consequent coastal flooding brought on by global warming (French, 1997). A full understanding of the ways in which oceans operate also requires that we include the interrelationships between oceans and the atmosphere. The two are inextricably linked by flows of incoming solar energy and outgoing long-wave energy and also by the cycling of nutrient material between the atmospheric and oceanic nutrient 'reservoirs'. Land masses play a very minor role in understanding the physical and chemical behaviour of our planet whereas the three-dimensional enormity of the oceans ensures that they are *the* major component in stabilising conditions on Earth. Without the existence of the oceans, life on Earth would probably not exist, or if it did, would have taken a different evolutionary form. As we gradually recognise the importance of the oceans in providing the very means of our existence it is imperative that we make a greater effort to understand the workings of the oceans and how human action may cause change to the operation of vital natural oceanic systems.

6.2 Why are oceans *so* important?

The beginnings of modern oceanographic studies date from as recently as the 1920s and in particular owe much to the work of Sir Gilbert Walker, Director-General of the Observatory in India (http://ess.geology.ufl.edu/usra_esse/ENSO_history.html), who encouraged mariners to record tidal patterns around the world. That it has taken so long for modern science to begin to understand the machinations of the oceans is an indication of the difficulties inherent in studying its huge and complex structures. In the oceans, the physical, chemical and biological worlds merge together, preventing scientists from studying smaller, simpler subunits. Only with the advent of powerful computers and automatic data-collecting instruments set out in the oceans has it been possible to gradually obtain the basic information we need before we can begin to understand the mechanisms at work in the oceans. The few findings we have already made are exciting in the extreme. Detecting and then substantiating the phenomenon called El Niño is one of these amazing discoveries. El Niño comprises not one event but a complex series of events linking the oceans and the atmosphere, triggered by changes in air pressure in two widely separated areas of the world (Allan *et al.*, 1996). It has only recently become understood that oceans can exert control over events such as drought and flood, periods of abnormal heat and unusual cold and indirectly exert an influence on outbreaks of disease or the level of crop production on continental land masses thousands of kilometres away from oceans. Still missing, however,

is the knowledge of *how* the oceans control the Earth's life-support mechanism (the biosphere) itself (Thorne-Miller, 1999).

Scientific knowledge of the oceans remains one of the great 'grey' areas in our understanding of the planet. At the macro level, mariners have known of the location of the major warm and cold ocean currents for many centuries. From the late nineteenth century onwards, marine scientists began to measure the chemical and physical properties of the oceans and collect information relating to the temperature, density and salinity of surface layers of the oceans. More recently, we have begun to understand the role played by oceans in the generation of hurricanes that bring so much devastation to vulnerable areas of the land masses. However, a detailed scientific knowledge of the oceans remains lacking. In part, our lack of understanding has been caused by the complexity of the oceans. This is especially true of the biological profusion that occurs in oceans. Ninety-nine per cent of all life forms and 90 per cent of all animal classes inhabit the sea and, of the 33 different animal phyla (the broadest classification of the animal kingdom), 21 are exclusive to the sea, 11 have members living both in the sea and on land and only one is exclusively terrestrial (Benchley and Gradwohl, 2000).

Our lack of knowledge of the oceans also extends to its geography. Less than 5 per cent of the ocean bed has been mapped at a level of detail given to the exploration of the Moon or Mars (Earle, 1999) and most of this relates to exploration in connection with petroleum deposits or ore nodules on the sea floor. Incentives for underwater exploration have mainly come from archaeologists and especially from teams searching for shipwrecks in deep water. Scientific research of deep water has almost entirely been connected to economic reasons, for example in preparation for laying the trans-Atlantic submarine telephone cables (Jeleff, 1999). Undoubtedly, a great deal of classified seabed mapping has been completed by the main military powers in connection with the recovery of sunken vessels, especially nuclear submarines, and the emergency recovery of stricken spacecraft that may crash in to remote areas of the ocean. (The following website provides information on the work of the seabed surveys carried out by the US Navy: http://www.msc.navy.mil/pm2/.)

6.3 Physical properties of the oceans

To understand the way in which the oceans operate requires that we work in a four-dimensional framework. Because many of us now routinely travel long distances by airplane, we recognise that the oceans have a considerable two-dimensional size. For example, a flight from Los Angeles to Auckland (a distance of 10,500km) involves a flight of about 13 hours, almost all of which is entirely over the Pacific Ocean. We also know from occasional shipping disasters that the oceans have depth. Stricken ships and submarines quickly sink to the ocean bottom, making rescue attempts difficult, for example the stricken Russian nuclear submarine *Kursk* (http://www.pulli.com/kursk). The fourth dimension is that of time. This is a difficult scale for humans to work on, limited as we are to a lifespan of 80 or so years, whereas changes to the physical environment take at least several thousand years to become evident. We are only just beginning to recognise the slow, long-term changes such as climate change or changes in the position of warm and cold ocean currents. Even more difficult to measure is the rate at which the formation of new rocks occurs on the ocean bed, leading eventually to the infilling of the oceans and the formation of new land masses by a mountain-building process known as orogenesis.

One of the most important features of the oceans is their ability to absorb vast amounts of incoming solar energy, to store and move that energy between the ocean layers and to release the energy back into the atmosphere at a slow, regulated rate. Their ability to perform this task so admirably is due to their immense physical extent and in particular to their great depth. Of the solar energy that arrives at the top of the ocean, depending on the latitude and the season, up to 90 per cent is absorbed by the dark ocean surface. By this means the average temperature of the Earth is contained with a fairly narrow range, between $15.0 \pm 0.5°C$. We know from the geological record that the average temperature fell to about 10°C during the last Ice Age while in previous hot, dry periods typified by the Permian geological period the average temperature was probably close to 20°C (Benn and Evans, 1998), but what we do not know is the role played by the oceans in allowing these changes in temperature to occur.

For most people, the role played by the oceans in stabilising the Earth's temperature is so remote as to be irrelevant! If the oceans figure at all in the lives of people, then it is events that take place in the coastal zone that are remembered. Typically, these include the twice-daily scouring of the tides bringing about wave action that may be either constructive (building up sand deposits) or destructive (causing coastal erosion). In reality, the actions that occur at the meeting of the land and sea are very atypical of what happens in the oceans in general. In the open sea, breaking waves are generally replaced by 'swell' that comprises large surges of water caused by a spherical movement of water molecules due to the gravitational effect of the Moon on the surface water layers. Depending on the phase of the Moon, the lunar impact gradually fades between 300–500 metres below sea level and instead, water movement in the ocean deep is due mainly to up-welling cold water currents although the locations at which this occurs appears to be limited, for example to the eastern side of the southern Pacific Ocean and off the coast of Newfoundland (Bradshaw and Weaver, 1995: 68–9).

If it were possible to drain off some of the sea water we would find adjacent to all the land masses a gently sloping *continental shelf* of varying width. In almost all cases the outer edge of this shelf ends at a point equivalent to approximately 180 metres below the current position of sea level. The continental shelf ends abruptly and passes into the *continental slope* that extends down to the ocean bed. Strahler and Strahler (1974) have suggested that because the ocean basins can be considered to begin at the outer edge of the continental shelf it would be reasonable to suggest that historically, sea level should be 180 metres lower and that the continental shelf is really part of the land mass. There are two opposing arguments to this suggestion. First, the continental shelf is covered by a shallow sea and is shaped by physical processes at work in the oceans. Apart from a very narrow inter-tidal beach zone there is little of this area that is directly accessible by most people. From this viewpoint, the continental shelf is part of the ocean. An alternative view is that although not *directly* accessible, modern society makes considerable use of the continental shelf by means of modern-day technology. For example, the continental shelf is dredged for sand and gravels, oil companies drill for gas and oil and the fishing industry concentrates much of its inshore fleet on the shelf. Perhaps more importantly, we legislate for the use of the continental shelf, giving legally enforceable 'rights' to countries in possession of coastline and, equally important, international treaties prevent third-party

nations from encroaching and gaining access to economic resources located on the continental shelf. Legal ownership of mineral rights and organic resources by a specific country provides the continental shelf with *de jure* status as part of a nation's land area (Jones and Hollier, 1997).

Compared to the continents, the oceans are proportionally much deeper than the elevation of the land masses. Figure 6.1 shows in diagrammatic form a comparison of the elevation and depth of the land and oceans. The mean height of all the land masses is only 840 metres compared to the mean depth of the oceans of 3800 metres (Bigg, 1996).

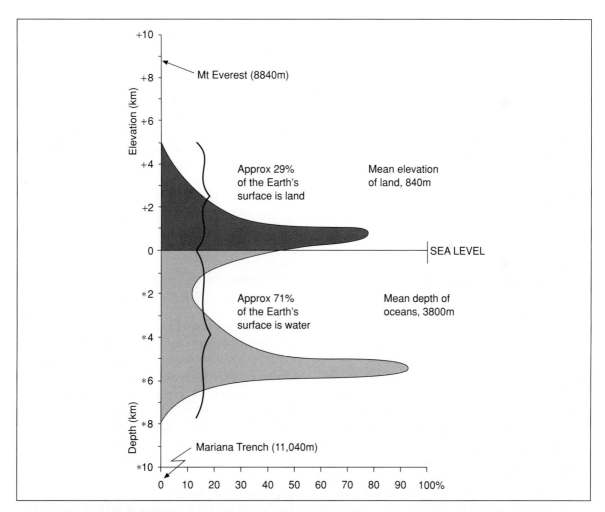

Figure 6.1 Diagrammatic comparison of the distribution of elevation of the land masses and the depth of the oceans (*Source:* after Bigg, 1996)

Away from the land masses, the oceans can be subdivided into layers based mainly on characteristics of temperature and amount of light that can penetrate to different depths in the ocean. There are four main layers:

1   The *surface microlayer* comprising at most only several millimetres of water. This layer is responsible for the transfer of energy and water vapour between ocean and atmosphere. The microlayer probably imparts its characteristics to air masses that form above oceans. It also contains the majority of the nutrients such as phosphorus, potassium and nitrogen that are transferred along with the water vapour into the air masses and which eventually fall as precipitation to the land masses and prove essential in maintaining the fertility of soils. Modern research on the surface microlayer only began in the 1960s (Open University, 1977) and our understanding of its behaviour is still rudimentary.

2   A warm *surface layer* up to 500 metres thick, which in tropical latitudes may attain a temperature of 25°C. Beyond the tropics, the warm layer thins and beyond latitudes 55°N and S may rarely exist. Within this layer, illumination levels decline rapidly and below 500 metres permanent darkness prevails.

3   The surface layer passes into the *thermocline*, a layer of water of varying depth but which is characterised by a rapid lowering of temperature. The bottom of the layer is reached when the water temperature stabilises at 5°C.

4   Finally, the ocean deep, a totally black body of water characterised by water temperatures that rarely change over time and never exceed 5°C.

6.4 Chemical properties of the oceans

Sea water comprises 96.5 per cent water by volume and the remaining 3.5 per cent is a solution of salts, the primary one being sodium chloride, the concentration of which has shown remarkably little fluctuation over three billion years of geological time (Bigg, 1996). Of the different elements combined in the salts, sodium comprises 31 per cent and chlorine 55 per cent. Traces of at least half of all the known elements are to be found in sea water as well as small amounts in solution of all the primary atmospheric gases. The proportion of dissolved salts to pure water provides a measure of the *salinity*, shown by the special symbol ‰. The average salinity of the ocean is 34.5‰ – equivalent to 34.5 grams of salt per 1000 grams of pure water (Strahler and Strahler, 1974). The presence of the salts also lowers the freezing point of the sea to below that of fresh water. Sea water freezes at or below −1.8°C.

The saltiness of the oceans has originated from two sources, one predominantly ancient and the other, mainly contemporary. The more ancient source comprised the release of vast quantities of gases from subterranean volcanic vents. Today, this source is of far less importance and only relatively small amounts of volatiles originate from volcanoes, hot springs and fumaroles that occur along sub-seaplate boundaries. The main contemporary source of salts is derived from the weathering of the rocks that comprise the continental land surface. Rivers transfer the weathered products from land to the oceans and the highest levels of sedimentation and, consequent to salinity, usually occur as a large 'footprint' offshore of the estuaries of major rivers. The salinity of the oceans has not increased over time because of two processes. First, is the constant transfer of chemical elements (via a process called *chemical precipitation*) into new sedimentary rocks that form on the ocean bed. Second, some of these precipitated elements are swept up by deep-water movement and returned to the surface by up-welling cold currents, for example the Humboldt Current off the west coast of South America. Salinity of the oceans can also be divided into layers:

1   A uniformly high saline zone occupying a shallow surface layer.

2   A transitional zone (the *halocline*) in which salinity drops.

3   A deep water layer of low salinity.

6.5 Pollution of the oceans

Marine pollution has been described by the Joint Group of Experts on the Scientific Aspects of Marine Pollution (GESAMP) as:

> The introduction by humans, directly or indirectly, of substances or energy into the marine environment (including estuaries) resulting in such deleterious effects as cause harm to living resources, hazards to human health, hindrance to marine activities including fishing, impairment of quality for use of sea water and reduction of amenities.
>
> GESAMP, 1981

From a human viewpoint, sea water is 'polluted' because of its high salinity. However, humans have added a considerable additional pollution load to oceans in the form of human sewage and industrial waste. These additional materials are mainly liquid and solids although cooling water from power stations causes a warming of water in shallow seas, resulting in thermal pollution. While the natural saltiness of the oceans has been relatively constant for many hundreds of millions of years, in contrast, pollution derived from human sources has shown a steady rise both in terms of different types and amount since the start of the industrial era. Modern pollution sources have two distinct consequences on the natural oceanic environment:

1 Pollution may reduce the quality of the marine environment. Many different types of pollution can occur, for example soluble materials (mainly chemical residues), suspended particles (materials washed off the land surface and conveyed by rivers to the seas) and physical material (wood, plastic, polystyrene, rubber). Each type of pollutant may cause a decline in quality of the marine environment resulting in conditions that are less suitable for the survival of aquatic life.
2 These may lead to a decline in the capacity of the oceans to maintain biological productivity, that is, a reduction in both the total amount of organic material and the rate of renewal of organic material living in the oceans.

Inshore, shallow and enclosed seas such as the Caspian, Black and eastern Mediterranean Seas are particularly at risk from sources of human-derived pollution where the relatively small volume of water combined with weak tidal movements allow pollution levels to accumulate close to the point of emission. Much of the physical detritus is washed up onshore where it forms an unsightly, and often dangerous, beach debris (see Plate 6.1) or as floating rafts of debris at sea. Materials such as plastics and blown foam are particular problems as they are non-biodegradable. Dixon (1990) has reported that a survey conducted on dead sea birds on Dutch beaches revealed that 92 per cent of the birds had ingested an average of 11.9 fragments of plastic debris.

In the open oceans pollution is less evident but even so, at great distances from land, sea water shows measurable amounts of heavy metals (for example, mercury and lead), persistent organics (DDT and PCBs) and artificial radionuclides – but at levels that are currently unlikely to cause adverse effects to aquatic life (Jeleff, 1999). Most of these substances are released into the atmosphere over land and are subsequently taken up in the air masses and transferred great distances until they eventually fall out into the sea either as 'dry precipitation' or are washed out by rainfall. (The following website provides a good source of information on the various forms of marine pollution: http://seawifs.gsfc.nasa.gov/OCEAN_PLANET/HTML/peril_pollution1.html.)

Pollution of the oceans takes place in a number of different ways. In most instances, modern-day pollution is exclusively the result of human activity but in some cases, notably the particulate matter described in point 3 in the following list, human actions exacerbate a natural source:

1 A wide range of contaminants (liquids and solids) are discharged from industry, commerce, transport and domestic sources into streams and rivers and subsequently flow into the ocean. The waste water conveying the pollutants may be discharged untreated into rivers or may have received various stages of 'screening' to remove physical solids and chemical treatment to remove or reduce the level of chemical pollution.

Plate 6.1 Examples of modern-day physical pollution washed up on a Mediterranean beach. The debris mainly comprises plastic granules, polythene and rubber items and includes a hypodermic needle

2 Industrial and domestic particulate matter (dust, ash, grit, soot, smoke and aerosols) derived from land-based atmospheric pollution sources are blown seaward and are subsequently removed by gravitational fallout.

3 Natural sources of particulate matter are created by dust storms, forest fires and volcanic explosion. These sources show considerable variation over time, unlike pollutants in category 2 which are more consistent in output. Pollutants in groups 2 and 3 often provide a fertiliser effect for marine plankton, resulting in a selective growth of specific plankton types, some of which may be dangerous and produce toxins that kill off other forms of plankton.

The surge in plankton growth creates algal 'blooms'. In enclosed seas, for example the Adriatic, red algae produce the so-called 'red tides', characterised by masses of floating dead algae that contaminate beaches and bathing water (Anderson, 1997).

4 Many gaseous pollutants released from industrial and domestic sources eventually end up in the oceans. Virtually every known chemical gas is now released into the atmosphere from modern industrial processes. Many gases combine in the atmosphere to form complex secondary or tertiary substances. These are removed (washed out) by precipitation and add to the chemical concentration of elements in the surface layer of the ocean.

Colour Plate A Hong Kong. A modern-day city

Colour Plate B Third World village. Botan, western Nepal in 1995. This mountain settlement had no road access, no electricity, no water supply or sewage system. It was totally lacking in any modern infrastructure

Colour Plate C Deforestation resulting in severe soil erosion on steep ground. The white area at the top of the main slope and behind the trees is bare rock, exposed by soil erosion. Natal Province, South Africa

PROFILE 2

Colour Plate D Peaty-gley soil. Permanent water-logging of the soil has resulted in the production of ferrous iron compounds characterised by predominantly grey colours in the soil profile. The darker coloured bands are layers of un-decomposed organic material, or peat. There are two layers of peat, the lower one representing the former surface of the soil, above which more mineral soil has been deposited, and finally a new layer of peat is currently forming

Colour Plate E Multiple use of an urban waterway. The River Avon as it passes through the centre of Christchurch, New Zealand, is used for recreation, wildlife and amenity purposes. The quality and quantity of the water flow are unaltered by the uses

Colour Plate F An extensive agricultural terrace system, Banyalbufar on the north of the island of Mallorca in the Mediterranean Basin. Irrigation is possible from water stored in subterranean tanks in the mountains

Colour Plate G View down the McKinnon Valley, Fiordland National Park, New Zealand. This region now forms part of the Southwest New Zealand World Heritage Area, covering 2.6 million hectares. Permanent human residents are, for the most part, not allowed in the area

(a)

(b)

Colour Plate H (a) Purple saxifrage, a late-glacial relict species growing on Ben Lawers, Perthshire, Scotland; (b) *Scilla peruviana*, growing in a fragment of soil on bare limestone, Mallorca, Spain

5 Cooling water released from thermal and nuclear electricity power stations is allowed to enter rivers and oceans. This water has usually been cooled in settling ponds but when it is released it remains several degrees warmer than the water into which it flows. As the temperature of water rises, its capacity to hold dissolved oxygen is reduced and in shallow, enclosed seas the rise in temperature can cause changes in the marine fauna and flora to occur. In open seas, no change in water temperature can be attributed to warm water from power-generation sources. The implications of global warming of the atmosphere must be recognised as a possible cause of warming of the oceans and the impact this may have on the distribution of marine life.

6 Wastes jettisoned from shipping. Until recently, it was common practice for ships to clean their holds and tanks with sea water, releasing highly polluted water (bilge water) back into the sea. This practice has now been banned by international agreement. Also banned has been the discharge of all refuse and sewage wastes from ships at sea although, undoubtedly, both practices continue to take place.

7 Release of sewage and storm water drainage. It remains common practice for domestic sewage to be released untreated by discharge pipe to a point below low watermark in coastal areas (Weale, 1998). Tidal flow can redistribute sewage onto beaches used by the public. The sewage provides a source of food for crustaceans, fish and sea birds and in some cases, accumulation of heavy metals results in the occurrence of toxic concentrations. The worst known example of human poisoning from contaminated fish and shellfish occurred in the region of Minamata in Japan during the 1960s (see Box 6.1).

8 Radioactive wastes from civilian and military installations. Because of the vast amounts of cooling water that must be discharged from a nuclear installation, most nuclear plants are located on the coast or estuaries. The majority of nuclear installations do not release any radio nuclides but occasional accidents, combined with a deliberate release of very small quantities of radioactive material to evaluate the spread of the nuclear material, has led to vociferous opposition from organisations such as Greenpeace, Friends of the Earth and residents who live locally to nuclear power plants (French, 1997).

6.5.1 The concept of the ocean 'sink'

The newly industrialised societies of the nineteenth century faced major problems of disposing of wastes generated by the new technology associated with industrial and manufacturing processes. Because of the inefficiency of early industrial technology, wastes usually accounted for very large amounts of unwanted materials, for example coal shale left from mining, 'shoddy' left over from carpet making or metal 'swarfe' (the fine metal shavings left from machining metal objects). In addition to the physical wastes there were also 'invisible' wastes, for example the products of combustion that escaped via the factory chimney or effluent flushed out of the factory by running water. During much of the industrial era, waste disposal has involved a least-cost 'dump and forget' approach, in which unwanted material has been released either into the oceans or the atmosphere. About three-quarters of all marine pollution stems from land-based sources, transferred to the seas by rivers, from drainage outfall pipes and via atmospheric fallout (Goudie and Viles, 1997). We have been able to literally wash our wastes into the ocean and then forget about it because of three reasons. One, the incessant action of tides and currents has ensured that pollutants have been dispersed, diluted and rendered safe. Second, the oceans are so vast that they appear to have a limitless capacity to absorb pollution safely. Third, the vast geochemical and sedimentary cycles that operate within the oceans capture the pollutants, locking them away for many millions of years on the ocean bed in new rock formations until eventually they are re-exposed in a future era of mountain building. These processes in which pollutants are 'lost' in the ocean depths are referred to as *sinks* (Jackson and Jackson, 2000) and they

BOX 6.1

Chemicals released into the oceans

Substances called *polychlorinated biphenols* (PCBs) are widely used in the electronics, plastics and adhesive industries. They are very resistant to change when exposed to high temperatures and are also slow to decompose if released into the environment. PCBs are fat soluble and, thus, easily absorbed in to animal bodies where they accumulate in fatty tissue. Even if an animal burns off all its fat, the PCBs would remain in a concentrated form through a process called *bioamplification* or *bioaccumulation.* High levels of PCBs can often be found in sea birds' eggs and seem responsible for preventing successful incubation. The bodies of dead seals found in the North and Baltic Seas often have high PCBs levels, implicating the substance as the cause of death.

Tributyltin (TBT) is a substance added to paints used to protect the hulls of ships against the growth of barnacles. Traditionally, marine paint contained highly toxic substances such as copper and arsenic and TBT has been used since the 1960s as a 'safe' replacement. However, by the 1980s evidence had been amassed to suggest that at even minute concentrations, TBT was not benign. Oysters growing in water in which there was high TBT levels were found to have such thickened shells that they were un-saleable. Elsewhere, female shellfish developed male characteristics, a feature again thought to be caused by TBT. Fortunately, the effects of TBT can be reversed if its presence is removed. Because of the proven effect of TBT on crustaceans and molluscs the use of this substance in marine paints has gradually been banned worldwide. Its replacement, *Irgarol*, now also appears hazardous to algae (Dahl and Blanck, 1996).

Hydrocarbons (oil-based products) enter the ocean from two main sources: tanker accidents and operational spillages during loading and unloading. Despite the often spectacular damage caused by the first of these sources, it is the second that dominates in terms of total amount of oil added to the sea. Oil contains thousands of different organic molecules and oils from different oil fields have subtly different properties thus making generalisation of their behaviour when added to water very difficult. The aromatic molecules are usually highly toxic but they evaporate quickly especially in warm sea water. The residue forms a black sticky mass known as tar balls. These slowly decompose to CO_2 and water but the process may be slow, taking up to 100 years. Oil spills away from the coast are best left to disperse naturally. Nearer land, chemicals can be sprayed to help dispersal but great care must be taken that the dispersants do not cause more environmental damage than the oil itself. Chemical dispersants used after the first major tanker disaster, the *Torrey Canyon* in 1967, caused major damage to marine life along the Brittany coastline. The amount of hydrocarbons lost to the sea shows considerable inter-annual variation depending on whether a major disaster has occurred. Figures quoted by Jeleff (1999) indicate that between 2.35 and 3.28 millions of tonnes of oil enter the marine environment each year. The Mediterranean Basin records the greatest number of individual spills per year at about 300 reported incidents (see Figure 6.2).

Heavy metals, although not soluble chemicals, are included in this section. Metallic elements such as mercury, cadmium and molybdenum are concentrated by marine organisms by bioamplification processes. The damage caused by heavy metals reached prominence during the 1960s when methyl mercury, released into the ocean from a Japanese chemical factory, resulted in fish accumulating dangerous levels of mercury in their bodies. Local fishermen, from the village of Minamata, living mainly on a diet of fish, developed neurotoxic effects, and 43 were killed, 700 disabled and a total of 2000 people were affected (Grant and Jickells, 2000).

BOX 6.1 CONTINUED

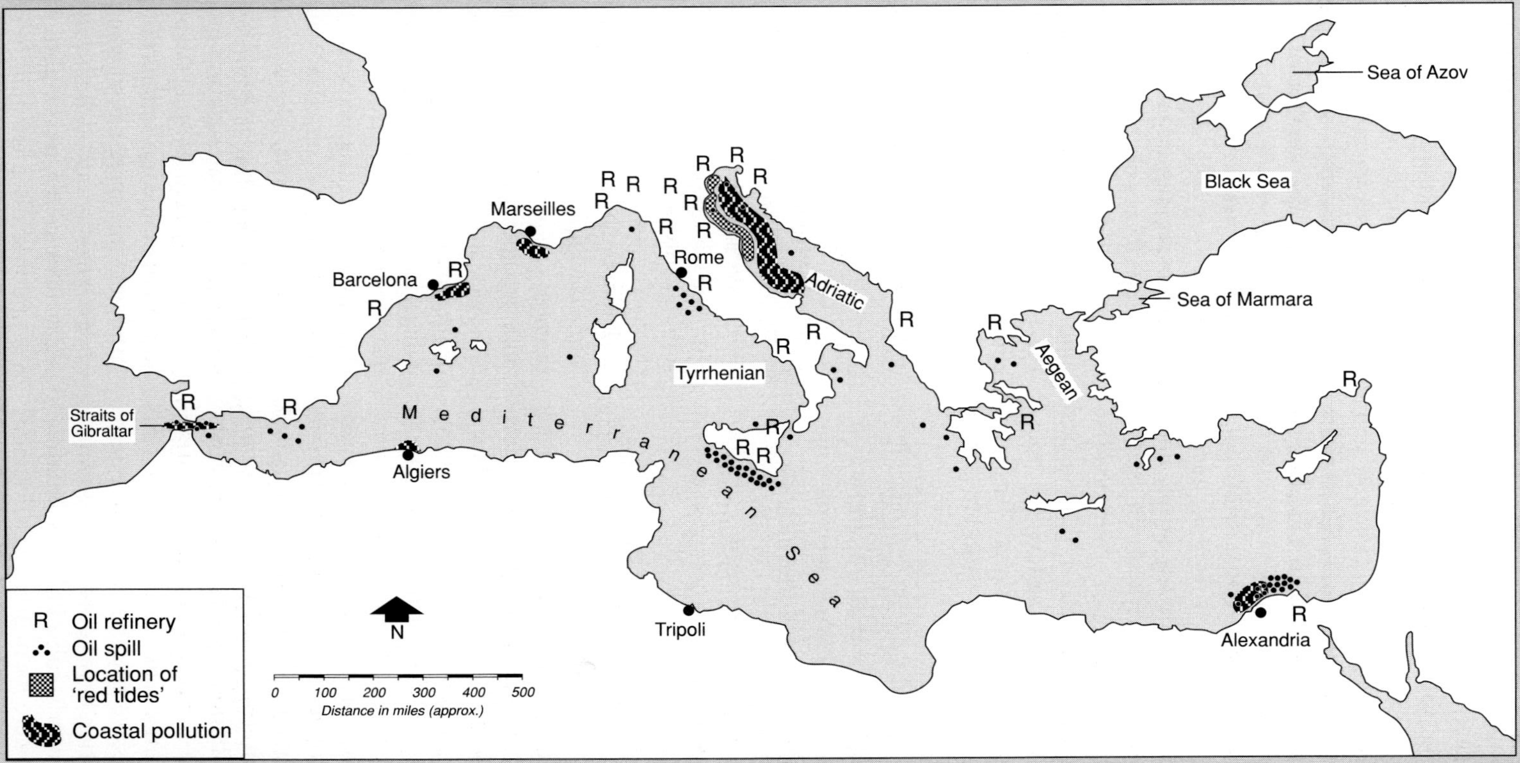

Figure 6.2 Number and location of pollution incidents in the Mediterranean, 1995–2000

Sea of Azov

Black Sea

Sea of Marmara

Aegean

Adriatic

Tyrrhenian

Mediterranean Sea

Alexandria

Tripoli

Rome

Marseilles

Barcelona

Algiers

Straits of Gibraltar

N

0 100 200 300 400 500
Distance in miles (approx.)

R Oil refinery
• Oil spill
Location of 'red tides'
Coastal pollution

occur without human intervention and with no visible, short-term economic cost.

Reliance on the unlimited capacity of the oceans to act as a 'sink' is valid only if pollutants are added to the ocean at a rate at which they can be totally absorbed by means of dilution and sedimentation into new rock formations. When this is possible, the pollutants are said to pass into an 'unavailable' reservoir where they will exist for millions of years until the seabed is raised to form new land. However, even when the most favourable of situations exist for disposing wastes in the oceans there will be a delay between the time of addition of the pollutant and its incorporation into new rock formations. This duration, the so-called *residence time*, is when damaging levels of pollution for aquatic life forms may occur. The residence time will be determined by the interaction of two main sets of factors:

1   The characteristics of the pollutant (the volume, concentration and rate of entry to the ocean). The characteristics determine the toxicity of the pollutant and its ability to cause damage to the marine environment.
2   The dispersal rate set by tidal and current movements, water temperature and width of the continental shelf across which the pollutants must pass before entering deeper water.

The problem of marine pollution has been exacerbated in recent decades by property developers concentrating their attention on coastal locations for a wide range of new development ranging from second home and time-share apartments to major harbour facilities. Despite far more stringent legislation concerning the volume, concentration and type of pollutant that can be dumped at sea, the intensification of urban and industrial land use and of agriculture has led to an inexorable rise in the total pollution load delivered by rivers and pipes onto the continental shelf. In addition, entirely new enterprises such as aquaculture (fish farms) and a massive growth in coastal-based tourism and recreation have added to the pollution load. In the USA, nearly half of all new buildings between 1970–89 were located in coastal regions but these areas account for only 11 per cent of

the country's land area (Goudie and Viles, 1997). Despite the threat of a rise in sea level of as much as 49cm by 2100 as a consequence of global warming (Meteorological Office, 2001) population numbers continue to increase in coastal regions at a rate that is far greater than the rate of global population growth. For example, in the Mediterranean Basin the number of people living in the coastal zone in 1985 was calculated to be 133 million (38 per cent) of the total population of countries bordering the Mediterranean. By 2025 corresponding figures are predicted to rise to 200 million or 55 per cent (Grenon and Batisse, 1989). The increasing effect of pollutants derived from land-based activities and entering shallow seas and the continental shelf is currently the single major cause for concern. Table 6.1 shows the main types of chemical pollution and the impact on humans.

6.5.2 Chemical pollution of the seas

Until the 1950s the volume and complexity of chemical waste disposed of in the seas posed only local problems. Since that time, the increasing use of complex industrial chemicals and of agrochemicals has led to a substantial deterioration of water quality, particularly above the continental shelf.

Dumping highly toxic chemical materials at sea began during the Second World War when a large tonnage of merchant and military shipping was sunk by enemy fire resulting in pollution from oil tanks and munitions. At the end of the war, deliberate dumping of unwanted munitions or experimental substances such as mustard gas took place especially around the European coastline (Meredith *et al.*, 2000). Since 1945 the quantity and complexity of materials dumped at sea has increased many fold. There is considerable uncertainty as to what substances have been dumped, the quantity of materials and precisely where they have been released. Despite the plethora of national and international legislation relating to marine pollution it is still uncertain whether fewer substances of less toxicity are deliberately dumped at sea today than in the recent past. Recent legislation has undoubtedly attempted to restrict toxicity of materials dumped at sea but problems remain due

Table 6.1 Chemical contaminants in coastal sea water and related health hazards for humans

Inorganic contaminants	Effects on humans
Arsenic	Cancer of the liver, kidneys, blood, and nervous system damage
Cadmium	Kidney damage, anaemia, high blood pressure
Lead	Headaches, anaemia, nervous disorders, birth abnormalities, mental retardation, especially in children
Mercury	Damage to central nervous system and kidneys
Nitrates	Respiratory problems particularly to the newborn and chronically sick
Synthetic organic substances	
Benzine	Anaemia, leukaemia, chromosome damage
Carbon tetrachloride	Cancer of the liver, kidney and lung. Damage to the central nervous system
Dionysian	Skin disorders, cancer and genetic malfunction
Ethylene	Cancer and male sterility
PCBs	Liver, kidney and lung damage

to indiscriminate dumping and from the occurrence of marine accidents. For example, on 31 October 2000 the Italian owned vessel *Ievoli Sun*, carrying 6000 tonnes of toxic chemicals, sank in the English Channel. The cargo comprised 3998 tonnes of styrene monomer, classed as 'harmful to human health', 1027 tonnes of methyl ethyl ketone and 996 tonnes of isopropyl alcohol. French politicians in particular, claimed that if legislation proposed after the sinking of the oil tanker *Erika*, on 12 December 1999 had been passed, the *Ievoli Sun* would not have been sailing. The same Italian regulators had passed both the *Erika* and *Ievoli Sun* as seaworthy.

Examples of some of the most damaging chemicals that still find their way into the sea are given in Box 6.1. It has proved remarkably difficult to verify the potential damage of many chemicals added to the oceans. Results from laboratory testing are rarely of relevance when applied to the real world and regulation of discharges have been made on a more or less arbitrary basis with emphasis being placed on those toxic substances known to be prone to bioaccumulation.

6.5.3 Disposal of human sewage at sea

Although the practice of releasing human sewage into streams and rivers is a particularly distasteful one it is still widely carried out throughout the world. Provided the quantity of sewage discharged into the water source can be adequately diluted and dispersed, this practice is not detrimental to the aquatic ecosystem. However, when sewage works began releasing large volumes of sewage that had received only limited pre-disposal treatment into rivers and estuaries it became only a matter of time before the effluent reappeared on beaches. The most common response to this problem has been the construction of long sea outfall pipes that pass well beyond the position of the lowest tide. Effluent is then 'treated' by wave action, helped by the saltiness of the oceans that act as a disinfectant while the ultraviolet rays contained in sunlight help kill some of the sewage-borne diseases (Grant and Jickells, 2000). Middleton (1999: 91) has reported how, prior to 1800, the ocean contained between five and ten times the amount of nutrients than runoff from the land to the offshore waters in the northeast of the USA. Today, that situation has been reversed with sewage outfalls providing five times more nitrogen than originates from within the ocean itself. Such major changes in the chemical cycling between the land and the waters of the continental shelf area have resulted in many undesirable changes in the population dynamics of aquatic environments.

The sludge that is released from sewage plants can be of three types: *raw* (untreated), *digested* (fermented until decomposition begins) and *activated* (in which digested sewage undergoes a secondary fermentation that removes harmful pathogens – but not heavy metals). The first of these causes the greatest potential harm in that the untreated sewage *scavenges* oxygen from the water and can result in aquatic life forms becoming asphyxiated.

Neal (1992) has claimed that the motivation for disposal of raw and partly digested sewage at sea is entirely economic with little attention paid to environmental issues.

6.5.4 Examples of marine pollution incidents

The addition of pollution to rivers from sewage and industrial outfalls takes place on a continuous basis and adds millions of litres of polluted water each year to the oceans yet attracts relatively little adverse public attention. By contrast, the occurrence of accidents to shipping regularly captures media attention. While the break-up of a giant oil tanker results in catastrophic pollution, its impact is localised to a specific geographic area and the incident lasts only for a short timespan. Fortunately, maritime accidents are relatively rare and clean-up is usually completed within six months of the incident. Marine incidents are usually quite straightforward and involve collision between vessels, break-up caused by severe storms or run-

ning aground resulting in a loss of all or part of the cargo. Crude oil now constitutes the greatest tonnage of cargo transported with approximately 22 million barrels of oil per year passing along well-defined shipping routes (http://www.eia.doe.gov/emeu/ipsr/t47.xls) and when a supertanker is ripped apart the environmental consequences are severe. These are discussed in more detail later. Other sources of marine pollution originate from the deliberate jettisoning of rubbish, disposal of bilge water and from biocides from anti-fouling paint used on ship's hulls, as described in Box 6.1.

Land-based accidents leading to marine pollution are much more common, release far more pollution to the oceans and often go unreported. Pollutants originating from land-based sources show an almost infinite variety and up to 100,000 man-made chemicals are now in widespread use and many of these enter the sea via river runoff (Jeleff, 1999).

6.5.5 Oil spills from tanker disasters

Although the shipwrecking of large, fully loaded oil tankers occurs infrequently, when incidents happen they have potential for causing untold biological and economic damage. The impact of half a million barrels or more of oil-based products spilling into the marine environment from each accident bring the condemnation of the world press but preventing these catastrophes appears impossible (see Table 6.2).

Table 6.2 Major oil spills resulting from tanker disasters, 1967–2002

Date of event	Vessel and amount of oil lost	Location
1967	Torrey Canyon, 119,000 tonnes	Scilly Isles, England
1975	Jacob Maersk, 80,000 tonnes	Oporto, Portugal
1976	Urquiola, 108,000 tonnes	La Coruña, Spain
16.3.78	Amoco Cadiz, 227,000 tonnes	Brittany coast, France
24.3.89	Exxon Valdez, 37,000 tonnes	Prince William Sound, Alaska
1990	Mega Borg, 500,000 gallons	Texas
1991	Haven, 140,000 tonnes	Genoa, Italy
1992	Agean, 72,000 tonnes	La Coruña, Spain
5.1.93	Braer, 86,300 tonnes	Shetland Isles, Scotland
15.2.96	Sea Empress, 72,000 tonnes	Pembrokeshire, Wales
12.12.99	Erika, 15,000 tonnes	Brittany coast, France
19.11.02	Prestige, 12,000 tonnes	Galicia coast, N.W. Spain

Although major oil tanker disasters cause very serious damage to the marine environment, reference to Table 6.2 shows that the frequency of major spills is, fortunately, an infrequent event. Modern tankers have double hulls, reducing the risk of a major rupture to the oil tank in the event of an accident, although single-hull tankers will be allowed to transport oil until 2020. Coping with a tanker disaster is made slightly easier in that the oil spill is focused on a single point source, the accident point, and allows the emergency clean-up campaign to be concentrated at that location. By contrast, it has been estimated that the cumulative loss from minor, often unreported incidents, that occur at loading and unloading points and from refuelling (bunkering) are at least as significant in terms of the damage caused to the marine environment and often go untreated due to their 'unofficial' status and widely scattered occurrence.

The impact an oil spill has on the marine environment will depend on the type of oil, the weather conditions immediately following the spillage, dispersal by tides and currents, the vertical dispersal of the oil in the sea water and the abundance of marine life (plankton, kelp, molluscs, fish, sea birds and sea mammals) (see Box 6.2). In addition, oil

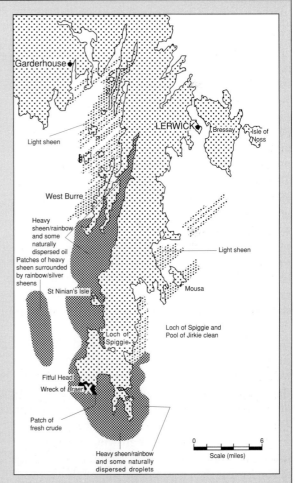

BOX 6.2

The Braer *oil spill*

Mid-morning on 5 January 1993 the Liberian-registered tanker *Braer* lost all engine power and was driven onto rocks near the Scottish islands of Shetland. The tanker was loaded with 84,700 tonnes of Norwegian Gullfaks crude oil bound for Canada along with 1600 tonnes of fuel oil. Southwesterly gales of force 10–11 on the Beaufort Scale delayed implementation of counter-pollution measures. Initial fears were that an environmental disaster of major proportions was about to unfold. With more than twice the oil lost compared with the environmentally disastrous *Exxon Valdez* incident, these fears were justifiable.

Despite the damage caused to fisheries and farmland the consequences of the spill were very different from previous incidents. A report on the *Braer* incident by the Marine Pollution Control Unit (Department of Transport, 1993) showed that two factors prevented the occurrence of severe damage to the marine environment. First, the spilt oil was light Norwegian crude that broke up quickly, and second, in the severe weather that followed the accident the oil was quickly dispersed by wave action into many tiny droplets. The area contaminated by oil reached its maximum extent four days after the wreck occurred (see Figure 6.3). Six days after the incident heavy seas had greatly reduced oil quantities all along the coast. Currents and tides helped distribute the oil at a rate of up to 10km a day. Normally, such a rapid distribution would have hindered clean-up operations but the tiny droplets of Norwegian crude

Figure 6.3 Area of the Shetland Isles coastline affected by the *Braer* oil spill, January 1993

BOX 6.2 CONTINUED

oil fell to the seabed within 15 days of discharge where they were incorporated into the sediments and became broken down by natural processes.

The incident occurred in an area of international environmental importance. Prior to the spill both the marine and onshore environments were characterised by an almost total absence of pollution. Although less than 1% of the oil cargo came ashore the amount of dispersed oil in the sea was very high and caused major damage to inshore fisheries and to salmon fish farms. A fishing exclusion zone covering more than 1000km^2 of sea and including 11 salmon farms was imposed. Income from salmon farming prior to the oil spill amounted to £35 million a year. Losses resulting from the imposition of a ban on salmon sales resulted in a major economic blow to the local economy. Some sections of the marine ecosystem also suffered damage. Bottom-feeding fish and shellfish experienced large losses in numbers due to the amount of oil that sank to the seabed. Sea birds

were especially affected by the oil. A total of 1549 dead birds of 28 species were recovered. The numerous sea otters that live in the area appeared to suffer no damage. Information relating to seals was not available.

The strong winds that followed the incident resulted in large amounts of oil-laden spray being transferred inland, covering an area of up to 50km^2. People working in the open had to resort to wearing face masks, farm animals had to be kept inside, field crops were condemned and some houses were so badly stained with oil that they required extensive external redecoration. The wind also caused dispersant chemicals that had been aerially sprayed onto the oil slick to be blown ashore and caused fears of chemical damage to people and animals. The clean-up and recovery cost of the *Braer* incident was £2.5 million. The British government has a policy of recovering as much of these costs as possible from the company that operated the *Braer*.

---

that is washed ashore coats the inter-tidal zone, covering rocks and sandy beaches and destroys any coastal tourist industry for at least one season. Damage to sea birds is especially evident as these animals are often washed ashore in a seriously weakened state. Oil destroys the waterproofing capabilities of the feathers, while the action of preening causes the birds to ingest the oil globules, so poisoning them. The Royal Society for the Protection of Birds (RSPB) has shown that the location and timing of the oil spill appears critical to the damage caused to the sea bird population. The massive oil spill from the *Amoco Cadiz* off the Brittany coast in March 1978 killed only 5000 sea birds, as many had flown to their summer feeding areas, whereas the much smaller spill from the *Erika* in the same area in December 1999 affected at least 200,000 over-wintering sea birds. Oil that forms a film on the surface of the sea may prevent fish eggs from growing, while the oil itself can poison adult fish. Commercial fishing grounds can be decimated, although recovery can occur provided that a thorough dispersal of the oil occurs. Research by the Swedish Environmental Protection Agency (SEPA) suggests that

petroleum hydrocarbons (PHCs) absorbed by marine animals can be converted into harmless substances and do not accumulate in the body (http://www.internat.environ.se/).

6.5.6 Disposal of nuclear wastes at sea

A major problem associated with generating electrical power by means of nuclear fission is how to safely dispose of the spent uranium^{235} fuel or high-level radioactive waste from nuclear power plants. During the 1960s and early 1970s, it was the practice to dispose of low-level nuclear waste by vitrifying it or encasing it in concrete before sealing in a steel jacket. These 'packets' of waste would then be buried on the ocean bed. Dumping of low-level radioactive wastes at sea was practised at a small number of licensed sites but generated such public concern about the long-term problems associated with nuclear waste that an international ban on open-sea dumping of all nuclear materials was imposed in 1983. Although some nuclear waste packages dumped at sea are known to have become damaged, no discernible effect on the ecology of the deep-sea areas concerned have been identified

(Meredith *et al.*, 2000) but clearly, careful checks on the safety of these sites will be required for the foreseeable future.

Depositing nuclear wastes at sea was initially considered to possess more benefits than disadvantages. Assuming that initially, all practical care was taken to encase the wastes in long-lasting containers, in the event that a container became damaged the vast size of the oceans, combined with the constant movement of waters, would ensure that any release of nuclear waste would become so diluted that it would become lost in natural background radiation levels. Such an argument ignored the rights of an individual nation that had adopted a nuclear-free policy from becoming contaminated with nuclear pollution from another country. If the source of the nuclear contaminants could be traced to the country or company of origin, the grounds for compensatory payments would be strong. Also, proponents of deep-sea dumping overlooked the 'worst case scenario' in which a number of containers became damaged, leading to a serious radioactive contamination of the seabed. It is now certain that the scientific and technical knowledge surrounding deep-sea dumping of dangerous materials in the 1960s and 1970s was incomplete. Gradually, as new knowledge was gained, the body of opinion moved against deep-sea dumping. However, two other situations existed whereby radioactive materials could enter the oceans: discharge pipes from nuclear reactors located at coastal locations (see Box 6.3) and military accidents (see Box 6.4).

BOX 6.3

Liquid nuclear waste discharge from outlet pipes to sea

Routine discharge of low-level liquid radioactive waste from a 3km pipeline into the Irish Sea, operated by British Nuclear Fuels Limited (BNFL) nuclear power and reprocessing plant at Sellafield in the English Lake District has resulted in the Irish Sea becoming the most radioactively contaminated sea in the world (Hughes and Goodall, 1992). BNFL has been repeatedly accused by NGOs, research laboratories, and also by concerned Irish politicians, that it was guilty of contaminating the Irish Sea with radioactive wastes. In 1997 a report from the Bedford Institute of Oceanography, Halifax, Nova Scotia claimed that discharge from Sellafield was producing a plume of radioactive sea water of an order of magnitude greater than the background level from nuclear weapons fallout and that the plume had reached the northwestern shores of Canada. According to a *New Scientist* report (*New Scientist*, 10 May 1997), Sellafield releases radioactive isotopes of caesium-137 and iodone-129 into the Irish Sea from where they are transported northeastwards to the north of Scandinavia where the plume splits in two, one branch continuing eastwards along the northern shoreline of Siberia, the other, turning westwards to flow around the tip of Greenland and into Baffin Bay in northern Canada. The level of contamination has been assessed as between two and three times that of the contamination caused by the Chernobyl reactor explosion in April 1986.

In contrast, the independent Radiological Protection Institute claim that radioactivity levels in the Irish Sea are falling and are now dwarfed by naturally occurring radioactivity. The Institute claims that the identification of nuclear wastes emanating from Sellafield is a testimony to the extreme scrutiny and efficiency of modern radiation-monitoring techniques and that the levels of radioactivity now being detected in northern Canada were to be expected from the peak discharge levels from Sellafield permitted during the mid-1970s. Current discharge levels were only 1% of the peak levels.

The cause of the problem appears to be that while the *levels* of nuclear waste released occur within the permitted limits, the *amount* of contaminated water released from the reprocessing plant has resulted in the *cumulative* radiation exposure for aquatic life to be sufficient to cause scientific alarm. Opponents to nuclear energy claim that the intentional discharge of low-level nuclear waste to the environment is totally unacceptable while proponents of the nuclear industry claim the discharge levels are safe. It is certainly the case that 'safe' levels of discharge have been drastically reduced in recent years and such is the concern over unknown and long-term effects of radiation damage that politicians now err on the side of extreme caution with all forms of nuclear waste.

Military accidents and disposal involving nuclear materials

On 2 August 2000 the Russian nuclear submarine *Kursk* suffered two major explosions that incapacitated the vessel and eventually resulted in the death of all 116 crew. In addition to the submarine's nuclear power plant, submarines of the Oscar 2 class of which the *Kursk* was a member, would have been armed with up to 24 cruise missiles (http://www.museum.navy.ru/kursk-e.htm). The incident is the last in a series affecting Soviet submarines dating back to 1970. The environmental impact that the sinking of the *Kursk* will have on the area of the Barents Sea where the disaster happened is not yet known. Military confidentiality will undoubtedly mean that we may never know how much radioactive material was on the submarine. Neither can scientists predict how quickly the radioactive material will be released or its distribution pattern in the Barents Sea as little or no detailed research has been undertaken.

Of even greater concern must be the general state of the former Soviet nuclear fleet that once numbered 243 vessels. In 1996, Greenpeace revealed plans (reported in the *Daily Telegraph* on 26 April 1996) that the Russian government intended to scuttle up to 20 of its elderly nuclear submarines at sea. The Russian navy has inherited a major submarine disposal problem with about 170 submarines decommissioned and awaiting disposal. At present, only two submarines can be decommissioned each year. The newspaper report, based on an interview with a British nuclear expert, claimed that the central problem was that of money. An estimated $240 million is needed to decommission all the nuclear submarines.

The former Soviet navy dumped six nuclear submarines in 1988 off the island of Novaya Zemlya, some in as little as 20 fathoms of water. Since 1993 the Russians have abided by international agreements on nuclear decommissioning but the worldwide ban on dumping nuclear wastes at sea does not cover military waste and submarines were specifically excluded from the START agreement. It is believed that the Russians are dumping liquid nuclear waste at sea. Britain has already assisted with some nuclear decommissioning and the Americans have offered to build a treatment plant at Murmansk to handle nuclear waste from six nuclear ice breakers. Greenpeace has claimed that if nuclear submarines were dumped it would create the worst environmental disaster to face the global marine environment.

Although the international treaty banning the dumping of nuclear materials at sea has been in force, and as far as can be verified, adhered to since 1983, the Organisation for Economic Co-operation and Development (OECD), has promoted the Seabed Working Group under the Radioactive Waste Management Committee. The objective of the Seabed Working Group, which comprises all the main industrialised nations of the world, is to provide scientific and technical information to enable international and national authorities to assess the safety and engineering feasibility of seabed disposal. The findings of the Working Group have been published in a series of technical reports, Volume 8 of which considers the likely impact a container of high-energy waste placed 30–40 metres *beneath* the seabed would have on its immediate environment (Lanza, 1988).

The report stresses that high-energy nuclear waste has never been placed beneath or on the seabed but *if* such an event occurred the intense heat generated by the radioactive waste would be sufficient to induce the water contained in the surrounding sediment pore spaces to move approximately one metre in 1000 years. Tests on how far the waste container could be expected to migrate during its lifetime suggested a maximum distance of 40cm. Extensive corrosion tests were conducted on a variety of canister materials and thickness. The life expectancy of both titanium alloy and carbon steel canisters was sufficient to allow the wastes to cool to a safe level before the canister disintegrated.

A quotation from Simon (1996) expresses a view that typifies the cornucopian outlook to many aspects of environmental management but which in the context of nuclear waste represents an *extremely* optimistic viewpoint:

> Scientists and engineers will be producing a stream of ideas about how to handle the [nuclear] waste even better, and indeed, will probably soon find ways to put the waste to such use that it becomes a commodity of high value.
>
> Simon, 1996: 211

Public concern over the safety of nuclear wastes has increased in the light of a chain of terrorist attacks starting with events in the USA on 11 September 2001. Fears that nuclear waste could be stolen by terrorist organisations now means that even greater security will surround its disposal and dumping at sea allows too great a possibility of its being hijacked by mercenaries.

6.5.7 Deterioration of water quality in the Black Sea

Relatively small, shallow inland seas such as the Black Sea have become particularly vulnerable to degradation from land-based sources of pollution. Section 5.4.6 described the almost total devastation of the Aral Sea, once a valuable fresh water resource, but now a highly saline and polluted lake. While not as severe, the mismanagement of the Black Sea provides yet another example of the total disregard we have for marine ecosystems. The Black Sea forms part of the much larger Mediterranean Sea ecosystem (see Figure 6.2), the environmental deterioration of which has been ongoing since the 1970s (Grenon and Batisse, 1989). The capacity of inland seas is often impaired by the slow natural exchange of water with adjacent seas, for example the Mediterranean Sea exchanging only half its volume every 50 years with the Atlantic through the Straits of Gibraltar.

Decline in the water quality of the Black Sea was first recognised in the 1960s (see the Black Sea environmental website for details http://www.blacksea-environment.org/) and a major part of the Black Sea is now highly eutrophic. Eutrophication is a condition in which nutrient enrichment of the water occurs due to the accumulation of effluent mainly from land-based agricultural systems. The River Danube contributes 60,000 tonnes of total phosphorus and a colossal 340,000 tonnes of nitrogen per year. The addition of raw sewage probably adds an equivalent tonnage of phosphorus and nitrogen. Eutrophication has resulted in a serious decline in Black Sea fisheries and also a decline in the aesthetic quality of waters in a region long famed for its role as a therapeutic holiday resort for the population from countries of the former Soviet Union (Mee, 1992). Interestingly, the decline in quality of the Black Sea has also been attributed to the so-called 'Iron Gates' dam between former Yugoslavia and Romania. This has caused a two-thirds reduction in the silica content of the river and is thought responsible for the dramatic decline in phytoplankton species in the Black Sea (Humborg *et al.*, 1997). An action plan to deal with the environmental catastrophe of the Black Sea was inaugurated in 1992 when adjacent countries signed a Convention for the Protection of the Black Sea (see the Protocol on Protection of the Black Sea Marine Environment at http://sedac.ciesin.org/entri/texts/acrc/BlackSeaLBP.txt.html).

The political instability that has characterised the region throughout the 1990s has not been conducive to progress and the fear remains that the rescue plan for the Black Sea will not be implemented.

6.6 Sea water quality and human health

Ever since the beginning of the nineteenth century the importance of providing a potable (that is, 'safe') water supply for humans has been a fundamental requirement for all developed nations. Environmental health regulations require that fresh water intended for human consumption does not contain the disease-carrying pathogenic micro-organisms that impart diseases such as cholera and dysentery to humans. By the late 1970s public health concern turned towards the quality of coastal bathing waters. Paradoxically, many of the countries that had established the safest fresh water supplies discovered that their coastal salt waters contained dangerous levels of bacteria. The rise in water sports, particularly surfers and bathers who spend long periods of time in the sea, led to a rise in the incidence of gastro-intestinal problems (diarrhoea), ear, eye and skin infections. These problems were thought to be a result of sea water contaminated with human sewage. In addition, Jeleff (1999) reports that outbreaks of cholera and infectious hepatitis have occurred more frequently among coastal populations made ill through eating infected, undercooked seafood. On a longer timescale shellfish can, under certain circumstances, accumulate heavy metals (zinc, cadmium, mercury) and, if these form a large proportion of the human diet, can cause metal poisoning and even cause death. The most notorious example of metal poisoning has been that which occurred in Minamata in Japan (see Box 6.1) (Grant and Jickells, 2000).

Parker and Frost (2000) have identified the earliest research from America in 1953 with the US Environmental Protection Agency setting a standard of 200 faecal coliform bacteria per 100 millilitres of water as the acceptable safe standard in 1968. In spite of repeated attempts to link incidence of disease to polluted sea water, no connection between the two has ever been universally accepted. The EC Bathing Water Directive (http://europa.eu.int/water/water-bathing/index_en.html) has set a safe level of 2000 faecal coliforms per 100ml – a level ten times higher than the US level! Somewhat confusingly, the EU directive also advised a *guideline standard* of 100 faecal coliforms per 100ml. Research has suggested that an acceptable level of pollution was only 35 faecal coliform per 100ml. In the UK over 1.4 billion litres of raw and partially treated sewage are released to the sea each year making sea water pollution inevitable. In an attempt to provide a clear guide to the general public, seaside tourist resorts throughout Europe and South Africa have collaborated in a scheme known as the Blue Flag Campaign, which awards a Blue Flag symbol to those beaches that consistently pass the official standards for permitted levels of coliform bacteria in sea water (see Box 6.5).

BOX 6.5

The Blue Flag campaign

The Blue Flag campaign is an exclusive eco-label awarded annually to beaches and marinas that reach high environmental standards as well as good sanitary and safety facilities. The scheme began in France in 1985. The 'European Year of the Environment' in 1987 provided an opportunity for the then European Commission (EC) to adopt the campaign and extend it to other European countries. An independent, non-profit organisation, the Foundation for Environmental Education in Europe (FEEE), was set up to manage the Blue Flag scheme. In 1987, 244 beaches and 208 marinas from ten countries were awarded a Blue Flag. By 2002, these figures had risen to 2093 beaches and 729 marinas in 24 countries throughout Europe and South Africa. The FEEE became a global organisation in 2001, changing its name to the Foundation for Environmental Education (FEE) and in 2002 plans to implement the Blue Flag scheme to Caribbean countries and the USA and Canada were underway.

The findings of Eurobarometer surveys (a regular opinion poll of key issues organised by the EU Commission, see http://www.gesis.org/en/data_service/eurobarometer/) reveal that the quality of bathing water ranks highly on the list of issues of greatest interest to EU citizens. To meet the EU bathing water standard, 95% of water samples must be free of sewage and chemical pollutants. Samples are taken every two weeks and an additional sample taken two weeks before the start of the locally defined bathing season. In 2001, 11,500 beaches throughout Europe were tested and 97% reached the basic standard of cleanliness, a considerable improvement since 1992 when only 85% of beaches reached a similar standard (see Table 6.3).

BOX 6.5 CONTINUED

Table 6.3 'Blue Flag' awards in 2001

Country	Beaches 2001	Marinas 2001	% of beaches meeting basic EU standards	% ± change in beaches 2000–2001
Belgium	18	5	100	157
Bulgaria	16	0		78
Croatia	33	15		106
Cyprus	35	0		3
Denmark	200	79	95.8	6
Estonia	3	3		50
Finland	6	30	98.8	0
France*	284	85		
Germany	38	204	96.8	15
Greece	354	8	98.8	1
Ireland	75	4	98.3	0
Italy	160	40	95.6	4
Latvia	4	0		33
Netherlands	17	30	98.7	−6
Norway	2	6		100
Portugal	144	7	92.2	8
Slovenia	5	3		25
Spain	423	85	98.1	8
Sweden	60	74	96.0	0
Turkey	127	11		28
United Kingdom	83	26	94.4	51
Total	2041	713	Average = 97%	Average = +30%

* Figures for France relate to 1998

During the 25 years in which the Blue Flag scheme has existed, the nature and extent of water pollution has changed considerably and, in December 2000, the European Parliament published a discussion paper to develop a new bathing water policy (http://europa.eu.int/scadplus/leg/en/lvb/128007.htm) that would maintain and improve bathing water quality in the face of changing water pollution. As shown in the main text of this chapter, there has been considerable debate between the USA and EU surrounding the setting of water quality criteria and reaching an international agreement will inevitably prove time consuming.

The organisation 'Surfers Against Sewage' (SAS) have been most active in conducting research into human illness associated with bathing and water sports. Parker and Frost (*op. cit.*) report SAS findings (http://www.sas.org.uk) that people entering the sea to paddle, but not bathe, were 25% more at risk from infection than those remaining on the beach. For swimmers this figure rose to 31% and for surfers to 80%! Only when all sewage receives full treatment, including UV light treatment, will our bathing waters be totally safe.

Prevention of water pollution as well as cleaning up past pollution has proved expensive. Britain spent £13.7 billion between 1989 and 1992 on new sewage plants. In the USA, greater use of technology is seen as the way to reduce water pollution. This method, called MACT (maximum available control technology), may prove too expensive and provides no guarantee of success.

6.7 Political management of the oceans

There is an international interest in the effective management, beneficial use, protection and development of the world's oceans. The natural, commercial, recreational, ecological, industrial and aesthetic resources of the oceans has resulted in competing demands being made on them. Indiscriminate use has resulted in the loss of living marine resources, a reduction in nutrient-rich areas, permanent and adverse changes to marine ecological systems and to shoreline erosion. In an effort to preserve, protect, develop and, where possible, to restore or enhance the resources of the oceans, governments have made great efforts to establish internationally agreed principles with which to manage the oceans.

6.7.1 International maritime treaties and territorial limits

Two specific issues (navigation and fishing rights) have traditionally dominated attempts to achieve a legal framework for the successful management of the oceans. Since the 1960s, two additional international issues have emerged, namely ownership of mineral rights on the ocean bed and the discharge of pollutants to oceans. Together, these four human uses of the world's oceans have attained immense political and economic significance especially for countries in possession of coastal boundaries.

By tradition, a nation's right to make use of the ocean was based on the principals contained in Grotius' *Mare Liberum*, published in 1609. In one of the first legal attempts to establish an international agreement to avoid conflict over the use of a natural resource, the *Mare Liberum* allowed a country that possessed a coastline unrestricted sovereignty over the resources within a three-mile territorial limit of high water. Beyond that limit, absolute freedom of the high seas existed for all users. By the 1940s a number of countries had entered into bilateral and multilateral agreements negotiated primarily in connection with fishing rights. After 1945 legislation relating to specific maritime issues became more common-

place. Among the first problem to be tackled was that of oil pollution, debated at the International Convention for the Prevention of Pollution of the Sea by Oil in 1954. This conference was to prove the first of a long series of international conferences intended to solve the newly emerging maritime problems.

Progress in reaching agreement over the use of the oceans has been fraught with difficulty. One particular problem has been the unwillingness of maritime nations to relinquish the concept of *res nullis* under which the ownership of the oceans belonged to no one and was therefore open to claim (Buzan, 1976). However, during the 1960s, conflict over access to offshore fishing rights, for example between Iceland and Britain, had reached an impasse. Depletion of inshore fish stocks around the British Isles in the 1960s led to British boats travelling to the waters around Iceland in search of cod. Icelandic fishermen objected to the invasion of their fishing grounds and imposed a unilateral 200 mile (320km) exclusion limit on Icelandic waters. Tensions between the two sides eventually led to universal adoption of the 200-mile exclusion limit in place of the antiquated three-mile limit. Of far greater concern was access to lucrative minerals located on and beneath the seabed. The economic value of fishing and seabed minerals ensured that 'ownership' rights relating to maritime areas began to rise up the attention list of governments and their legal advisers.

The 1960s was also the time when deep sea scientific exploration of the oceans began to take place. Advances in diving technology enabled a new generation of submersible vessels to probe parts of the ocean never before visited while research carried out in connection with the laying of long-distance submarine telephone cables brought further discoveries about the seabed. The work of research institutes, for example the Scripps Institution of Oceanography in California (http://www.sio.ucsd.edu/about_scripps/) and CSIRO Marine Research for Australia (http://www.sio.ucsd.edu/about/oceanview.cfm), added greatly to our knowledge of the oceans.

Throughout the third quarter of the twentieth century, ocean 'management' still comprised individual nations applying a 'sectoral' approach to specific problems as and when they occurred. For example, the overfished North Sea and the Northern Approaches to the northwest of Britain resulted in a sequence of measures intended to safeguard fish stocks. British fishermen were required to practise an early form of conservation through the imposition of a minimum mesh size of nets to ensure immature fish would not be caught. When these measures failed to prevent further decline of fish stocks, quotas on the tonnage of fish that could be caught in designated areas were imposed and, ultimately, 'closed seasons' were introduced in an effort to allow fish stocks to recover. Considerable mistrust existed between trawlermen, marine scientists and government officials. The fishing industry claimed that the scientists had insufficient information on fish stock numbers to call for fishing restrictions while politicians were criticised for not taking action to safeguard fish stocks because of the implications on employment in fishing villages and towns often located in remote areas of Britain. Hutchings (2000) has shown that, with the exception of herring and related species, 90 other species of commercially exploited fish have not shown a recovery of numbers even after 15 years of fishing restrictions.

The use of a selective and incremental management approach to regulate fishing around the coastline of Britain has been both unpopular and unsuccessful. Failure to restrict non-EU vessels to the same restrictions as home vessels meant that large gaps appeared in the management strategy. Also, the setting of fishing restrictions on small geographic areas was doomed to fail, due to fish shoals migrating to different locations depending on the season of the year. Lack of information on the life history of the different fish species was also notably lacking in the 1960–80 period.

A different set of arguments applied to marine mineral resources. Here it was possible to map the location of resources and to set territorial limits accordingly, but it was soon shown that geographical limits set in isolation from ecological knowledge could result in further disaster for the fishing industry. The habitats of many demersal fish (seabed dwellers) could be destroyed by dredging for sands and gravels. Damage to coral reefs by dredging and pollution resulted in major loss of aquatic biodiversity. Research showed that coral reefs form the nurseries for many of the species of economic fish stocks and coral reef destruction was often followed by a collapse in adult fish numbers. The unsatisfactory nature of territorial limits as a method of managing marine resources quickly became apparent and strong scientific arguments emerged for an *integrated* management strategy that could be applied to the oceans *and* to associated river basins and terrestrial environments. This approach has been accepted as inevitable by national governments. However, the relinquishing of some of the national 'ownership' rights by governments with territorial boundaries that adjoin the oceans and their replacement by 'international' management agreements has been an extremely difficult and slow process. Implementation of a management plan for the oceans has become one of the major objectives for the United Nations and negotiations under the heading Convention on the Law of the Sea (UNCLOS) have continued almost continuously since 1958. The United Nations web pages entitled 'Oceans and Law of the Sea' (http://www.un.org/Depts/los/index.htm) provides an historical account of the resolutions and conventions that have been agreed on ocean management.

The modern era of international law relating to the seas dates from 1972 when the Stockholm Conference on the Human Environment resulted in new international agreements for limiting the dumping of materials other than oil and on the protection of wetlands. However, the Stockholm Conference illustrated the wide division that existed between the industrialised and the non-industrialised nations of the world regarding the management of the oceans. Industrialised nations argued for the setting of minimal controls, afraid that more restrictive legislation would damage their economic growth. Some industrialised nations refused to ratify the Stockholm agreement. Much of the detail agreed at the Stockholm Conference made little practical difference to the way in which the management of the oceans took place. However, the Stockholm Conference marked a turning point for the management of the oceans. Since 1972, legislation relating to the use of the oceans has been progressively tightened and, of greater significance, been accepted by almost all countries.

6.7.2 The UN 'Constitution for the Oceans'

Negotiations to build on four early Geneva conventions (agreed in 1958) relating to international agreements on use of the oceans began in 1970 and culminated in the adoption of the United Nations Convention on the Law of the Sea (UNCLOS) in 1982. This legislation is sometimes called the UN 'constitution for the oceans'. It entered force in 1994, making the duration between the beginning of negotiations and its successful implementation the lengthiest of any international legislation. The legislation has been described as one important step in a process that has slowly gathered pace over the years. A decade after UNCLOS had been accepted, the adoption of Agenda 21 in 1992 by the United Nations Conference on Environment and Development (UNCED) followed in 1995 by the Global Programme of Action for the Protection of the Marine Environment from Land-Based Activities signalled a crucial turning point in the management approach of the oceans (Jeleff, 1999). Table 6.4 provides a list of the main international conventions relating to the oceans.

The UNCLOS agreement was necessary for three main reasons:

1 A realisation that the oceans should be managed on a systematic rather than traditional sectoral basis.
2 The increased capability of human society to influence the ecological integrity of the oceans made it imperative to introduce new protective legislation.
3 A recognition that marine resources lying outside the traditional territorial limits should be recognised as the 'common heritage of mankind' and as such required management under legal terms known as *res communis* and not as the established conditions of *res nullis*.

The driving force behind UNCLOS legislation was a desire to show that problems concerning use of the oceans were interrelated and should

Table 6.4 Major international legislation relating to use of the oceans, 1969–1998

Date of agreement	Convention
1969	Tanker Owners' Voluntary Agreement Concerning Liability of Oil Pollution
1969	Brussels International Convention on Civil Liability for Oil Pollution Damage
1969	Brussels International Convention Relating to Intervention on the High Seas in Cases of Oil Pollution Casualties
1971	RAMSAR Convention on Wetlands of International Importance
1972	UNESCO Convention Concerning the Protection of the World Cultural and Natural Heritage
1972	Convention on the Prevention of Marine Pollution by Dumping of Wastes and Other Matter
1972	Oslo Convention for the Prevention of Marine Pollution by Dumping from Ships and Aircraft
1973	London International Convention for the Prevention of Pollution from Ships
1977	London Convention on Civil Liability for Oil Pollution Damage from Offshore Operations Resulting from Exploration and Exploitation of Seabed Mineral Resources
1982	UN Convention on Law of the Sea (UNCLOS)
1989	Basle Convention on the Control of Transboundary Movements of Hazardous Wastes and their Disposal
1991	UN Convention on Environmental Impact Assessment
1995	Global Programme of Action for the Protection of the Marine Environment from Land-based Activities
1998	UN-designated 'Year of the Ocean'

(*Source*: after Jeleff, 1999)

be considered as a whole. The main premise on which UNCLOS was based was that the rights and obligations of coastal states must be balanced with the rights and obligations of all states to use the oceans and their resources in a rational way. The UNCLOS agreement established an exclusive economic zone (EEZ) of 200 miles (320km) for coastal states with rights over living and non-living resources, scientific research and environmental protection. In return for these rights certain obligations were imposed, the main one of which included the protection of the entire marine environment.

Despite the almost constant flow of United Nations legislation in support of integrated management of the oceans, it must be conceded that many of the initiatives have failed to achieve their objectives. The rhetoric of the UN conference halls quickly fades when the realities of implementing the legislation become apparent. As usual, money has proved the main stumbling block to success. The cost of the Recommendations contained in Agenda 21 of the UNCED Conference on how to achieve marine management have been estimated at about US$12.9 billion for the period 1993–2000. The inability of developing nations to meet their contributions has led developed countries to make their payments conditional on additional financial resources being made available to developing countries from the Global Environment Fund, administered by the World Bank. However, this has increased the indebtedness of developing countries and the consequent repayment figures and has resulted in further inappropriate development on land that causes further environmental deterioration to the oceans!

Compounding the inability (or an unwillingness) of some nations to pay for environmental management of the oceans illustrates a widespread practice of non-compliance with international legislation. Reasons for non-compliance include the use of delaying tactics, invoking of 'special circumstances' and exploitation of loopholes in the legislation.

Faced with this unsatisfactory situation, some groups of countries are returning to the practice of making local or regional arrangements. The EU nations have discussed the setting up of an EU maritime agency which would assemble multi-disciplinary scientific and technological skills to support European industry and policy makers. Such an agency would cover a wide range of multiple uses and functions of European oceans, landlocked seas and wetlands. The agency could be involved with issues as diverse as declining fish stocks, revitalising former ship building regions, pollution control and fish farming. While the 'go-it-alone' policy of the EU maritime agency reflects a frustration with the tardiness of UN negotiations on ocean management there is a danger that unilateral policies will focus more on national self-interests and less on development of an integrated policy. The conflict of interests appears once again to reflect the desire by elected politicians to achieve a result within the timespan of an elected government rather than the longer timescale needed when basic scientific research is required to identify the cause of a problem (for example overfishing, protection of whales, porpoises and dolphins) and establish a solution that is agreeable to all users of the ocean resource.

6.8 A new approach to ocean management

At about the same time as the UN began work on UNCLOS legislation, the USA began concentrating on legislation covering the continental shelf and the immediate areas landward of high water. The Coastal Zone Management Act of 1972 established new standards of management around the shores of the USA, principles that were copied by many other maritime nations of the world. The original Act has been amended to the Coastal Zone Protection Act of 1996.

6.8.1 Integrated management of the oceans and coastal regions

The enormity of the oceans, covering 360 million square kilometres of the surface of the planet, has disguised the need that even this immense natural resource requires careful and appropriate management. But as with all natural resources the challenge has been to agree on a sustainable management strategy that satisfies all potential users.

Of the 225 or so countries of the world (the exact number changes with surprising frequency as democracy spreads to some of the formerly monolithic nations) only 40 are landlocked. Obtaining agreement from the other 185 countries, many of which make widely varying uses of the oceans, has proved one of the greatest challenges for environmental lawyers working on behalf of sea-faring nations and the relevant non-government organisations (NGOs).

Not unexpectedly, it is possible to recognise all the different human responses to the use of ocean resources, ranging from cornucopianism to ecocentrism, as discussed in Chapter 2. For much of the industrial era we have adopted an 'act-and-forget' approach towards the way in which we use the ocean and its resources. When applied to ocean resources as opposed to other smaller and more immediately vulnerable natural resources such as rivers, forests and soil, the cornucopian approach becomes more understandable. The size and the bounty of the ocean resource made it appear that the seas were truly infinite in their capacity to yield resources and absorb pollutants. In reality, the vast oceans have proved just as vulnerable to ecological and environmental damage as terrestrial and atmospheric ecosystems.

The long list of disasters that occurred during the twentieth century include incidents such as depletion of the fish stocks off the Newfoundland Banks, pollution of the North Sea with industrial and human waste stemming from the rivers flowing out of Germany, the Netherlands and eastern Britain and the widespread dumping of toxic chemicals particularly by the old industrialised nations. The vastness of the oceans was used to justify (if justification were ever called for) such mistreatment of the ocean ecosystem. Because the oceans are so vast in size it has been especially difficult to convince society of the need to manage oceans in a sustainable manner, believing them to be so large that it was impossible for human actions to cause permanent change. It is surprising, therefore, to discover that of all the planetary habitats, it has been with the oceans that we have achieved some of the greatest successes in international management agreements. It is significant also that success has been uniquely based on the legal concept of *res*

communis, namely a resource that is owned by all nations and managed by means of a strategy designed neither to favour nor disadvantage any nation whether they possess a coastline or not.

The idea that management was necessary for *coastal* areas as opposed exclusively to land-based areas or to maritime areas dates from 1973 when the Economic and Social Council of the United Nations began to describe coastal areas as 'valuable national possessions' and that 'proper management and development' was necessary for these areas in order to ensure their long-term contribution to society. A report on coastal area management and development (United Nations, 1982) established the concept of *integrated resource management of oceans and coastal regions* as symbolising a means of holistic development it recommended for coastal regions. The UN publication recognised a long list of potential problems that could result from uncontrolled or unsuitable development and accepted that difficulties had arisen because of the confusion due to the different management strategies that were necessary to cope with renewable and non-renewable resources existing side by side in the oceans. Until recently, coastal management has taken place within the highly restrictive confines imposed by the need to ensure that coastlines remain 'fixed', that is the coastline does not undergo undue erosion or accretion thus causing problems to the built environment. French (1997) has claimed the narrow engineering viewpoint of coastal zone management has probably caused as many problems as it has solved! Instead, French advocates the adoption of a holistic management approach, looking at the coastal area as a whole and extending the boundaries of management both inland and seaward.

The name given to the new approach to coastal zone and marine management has varied since the mid-1970s, initially *coastal zone management* (CZM), subsequently *integrated coastal area management* (ICAM) and most recently of all *integrated coastal management* (ICM). These different names refer to the same basic concept – an integration of management policies that link coastal waters and adjacent land areas within a common planning unit. ICM has been defined by Cicin-Sain and Knecht (1998) as 'a conscious management process that acknowledges the interrelationships among most coastal and ocean uses and the environment they potentially affect'. ICM involves processes that enable rational decisions to be reached concerning the conservation and sustainable use of coastal and ocean resources and space and, in ways similar to river basin management plans described in Box 5.2, overcomes the separation that becomes inevitable through using a single-sector management approach. An ICM process should include all relevant resource uses in a locality, such as fishing, aquaculture, agriculture, forestry, manufacturing industry, waste disposal and tourism. All these uses must be examined in the context of the requirements of the communities in the study area (see the NOAA National Ocean Services web pages for a detailed explanation of the ICM process: http:// icm.noaa.gov/story/icm_def.html). Proponents of the ICM process claim that it does not replace single-sector strategies but is an additional management overlay that optimises the often varied and intertwined strands that typify complex coastal zone issues.

In order that an ICM approach makes real progress it is necessary that 'trade-offs' between competitive uses are made on an equitable basis. Compromise between rival users of the coastal zone can be made as a result of mutual agreement although this process can be time consuming and often results in none of the participants being totally satisfied with the end result. Arbitration by an independent body is also possible. A third alternative is to use a geographic information software package to compile a layered database of the study area and allocate 'value' to each data layer. Jones and Ahmed (2000) have used this approach to identify the management options necessary to cope with the impact of sea level rise on areas of different conservation value in the River Clyde estuary. GIS allows rapid and accurate identification zones of different value and as more knowledge becomes available the actual values attached to the data layers can be easily changed and new scenarios calculated. A further example of the use of ICM is described in Box 6.6.

BOX 6.6

Disaster management in Japan: a case study using ICM

Japan faces almost constant threat from natural disasters. The area surrounding the Greater Tokyo Bay area experiences about 7000 earthquake tremors each year! From the statistical record it appears that a major earthquake occurs every 75 years, the last one being the Kobe Earthquake of 17 January 1995, causing damage worth an estimated $100 billion. The origin of almost all Japanese earthquakes is located offshore along the crustal plate boundaries and thus the impact of earthquakes is almost always experienced first (and often most strongly) in the coastal zone. Earthquakes are not the only hazard the Japanese have to contend with. The coastal zone surrounding Japan suffers major damage from torrential rain, flooding and high waves (from typhoons) and tsunamis (giant, fast-moving waves associated with earthquakes), all of which have contributed to coastal erosion that is occurring at the rate of one metre per six years. Approximately one half (15,932km) of all the coastline of Japan requires protection against coastal erosion.

Because so much of the Japanese land area is mountainous, development has concentrated along the flatter coastal region. Japan comes second to Denmark in the length of coastline per square kilometre of total land area (Denmark 150 metres per square kilometre; Japan 91 metres). Almost half the population live in the narrow coastal zone, 77% of all commercial sales occur in the area and almost all the industrial development is located near the coast. Such concentration of development has resulted in major destruction of the physical and natural environments, often leading to a loss of natural protection of the land provided by shingle beaches, coastal lagoons and sandbars. Landfill of shallow coastal areas has proceeded apace since 1978 causing further disruption to the natural coastline.

BOX 6.6 CONTINUED

It is against this background that attempts have been made to use an ICM approach to future coastal developments and also to apply it retrospectively to some of the poorest examples of coastal zone development. Application of ICM in a Japanese context has four objectives:

- to preserve the remaining ecological integrity of coastal zone ecosystems
- to prevent excess material damage and loss of life from natural disasters
- to aid appropriate human utilisation of coastal areas
- to integrate the separate efforts of these three objectives, to create an attractive, safe and vibrant coast that is ecologically stable.

Unlike the USA, Japan does not yet have a legal requirement for coastal development to be based on ICM principals. In the USA, coastal development requires assurance of 'no-net-loss', that means any development in the coastal zone must be accompanied by restoration, repair or creation of new coastal environment that may have been altered by development. Managing Japan's coastal zone represents a major challenge to the technological prowess of the Japanese economy. The successful integration of the separate interests of developers and the conservationists would represent a major achievement for Japanese science and technology.

(*Source*: based on information obtained from Masahiko Isobe, University of Tokyo: http://icm.noaa.gov/country/Japan.html

6.9 Ocean 'futures'

Despite the almost universal acceptance of UNCLOS legislation, very considerable misuse of the oceans and their valuable natural resource base remain. Practical implementation of international maritime conventions depends as much on the willingness of a local fisherman as on the skipper of a supertanker to comply with the detail of the convention. It is still all too common to find examples of mangrove forest being ripped out to make way for shrimp fisheries, for illegal fishing methods to be used against endangered species or for dumping of rubbish from deep-sea shipping. Application of the ICM approach outlined in the previous section is still in its infancy. Countries such as the Netherlands and the USA have made good progress in the application of ICM strategy and benefits are already apparent. In the UK, a more cautious application of ICM has been made with considerable variation between England (moderate uptake) to Scotland and Wales (slower uptake). The vast majority of other countries have yet to make any progress with ICM, relying instead on a competitive single-sector approach. Cicin-Sain

and Knecht (1998) provide an excellent account of the application of ICM in many different situations.

In some instances a lack of uptake of ICM or non-compliance to UNCLOS legislation is due to ignorance, lack of education, lack of suitable alternative technology or to a simple lack of financial resources. Those officials charged with implementing legislation are often overtaxed in terms of their responsibilities as targets become more demanding and deadlines ever tighter. One argument suggests that a temporary moratorium on further legislation may be required to allow implementation of recently introduced treaties but a counter-argument suggests that the decline in the ocean environment is so severe that no let-up can be considered. It is possible that further advances may be achieved by so-called 'soft laws' comprising guidelines, codes of practice, and action plans and led not by governments and their scientific officers but by non-government organisations working through advisory groups. Ways must also be found to finance both the existing and any future legislation.

As with so much of contemporary management involving environmental issues, much of the attention focused on the oceans is concerned with retrospective management, such as remedial actions taken to repair or reduce past misuse or malpractice. More recent international conventions on ocean management have been far more anticipatory and precautionary in their concerns and have attempted to predict future areas of concern. This approach has become inevitable as concern with climate change and the role played by the oceans as sink areas for accumulated heat energy and carbon become more apparent.

Useful websites for this chapter

Black Sea Environment Programme
http://www.blacksea-environment.org/

Eurobarometer Survey Series
http://www.gesis.org/en/data_service/eurobarometer/

European Union Bathing Water Directive
http://europa.eu.int/water/water-bathing/index_en.html

CSIRO Marine Research for Australia
http://www.sio.ucsd.edu/about/oceanview.cfm

Electronic Telegraph (search for specific oil tanker disasters)
http://www.telegraph.co.uk

European Union New Bathing Water Policy
http://europa.eu.int/scadplus/leg/en/lvb/128007.htm

International Maritime Organization
http://www.imo.org

International Petroleum Imports and Exports (Trade) Information
http://www.eia.doe.gov/emeu/ipsr/t47.xls

Japan Integrated Coastal Zone Management
http://icm.noaa.gov/country/Japan.html

Marine Pollution One
http://seawifs.gsfc.nasa.gov/OCEAN_PLANET/HTML/peril_pollution1.html

Protocol on Protection of the Black Sea Marine Environment
http://sedac.ciesin.org/entri/texts/acrc/BlackSeaLBP.txt.html

Russian submarine *Kursk* tragedy
http://www.pulli.com/kursk

Scripps Institution of Oceanography in California
http://www.sio.ucsd.edu/about/overview.cfm

Sir Gilbert Walker
http://ess.geology.ufl.edu/usra_esse/ENSO_history.html

Special Mission (US Navy ocean surveys)
http://www.msc.navy.mil/pm2/

Swedish Environmental Protection Agency
http://www.internat.environ.se/

Surfers Against Sewage (SAS)
http://www.sas.org.uk

United Nations Oceans and Law of the Sea
http://www.un.org/Depts/los/index.htm

What is integrated coastal zone management?
http://icm.noaa.gov/story/icm_def.html

7 Land degradation

We abuse land because we regard it as a commodity belonging to us. When we see land as a community to which we belong, we may begin to use it with love and respect.

Leopold, 1949

7.1 Introduction

The need for land is a fundamental requirement for all terrestrial species and especially so for the most materialistic of all species, *Homo sapiens*. Although we share the land with all the other land-based plants and animals, our species has succeeded in removing a large proportion of the competing species so that the land can be converted into productive agricultural land, as space for our towns and cities, for industry, for extractive industries and for communication links. The conversion of natural ecosystems into ones that had a greater utility for humankind probably started soon after the appearance of hominoids. However, it is only in the last 200 years that clear signs have emerged to suggest that the ways in which we are utilising the land surface have caused major problems for other components of the biosphere, causing problems of soil erosion, loss of biodiversity and of chemical pollution of land, water and air. All these trends can be grouped under the heading of 'degradation'. This chapter is concerned with one particular form of degradation, that of land degradation which usually involves the erosion of soil from the land. The role of humans as modifiers of the land surface will also be examined.

Most governments and all major international organisations such as the FAO (Food and Agriculture Organization, 1999) and the United Nations, through the Global Environment Facility (see UN website: http://www.unep.org/documents/default.asp?documentid=74&articleid=1054), confirm that land degradation is a major world problem. In spite of unanimous international agreement and concern over land degradation, it is impossible to give a precise figure of how much land has been degraded. This is because degradation occurs across a widely fluctuating scale of intensity, much of which is invisible to human perception. An estimate of the amount of degraded land is given in section 7.5 but the reader must be aware that statistics on degraded land are notoriously inaccurate. Data on erosion rates becomes rapidly outdated as new methodology in new geographical areas reveals further examples of erosion. Some limited success at slowing down soil erosion and land degradation has been achieved and section 7.14 examines some successful land rehabilitation schemes.

7.2 What is 'land degradation'?

Land degradation is the result of a vast number of natural and human causes. It takes many different forms, some of which are easily recognised while others remain hidden from view. In addition, land degradation operates at many different scales, affecting the individual person, depleting the economies of countries and ravaging vast areas of entire continents (http://www.ciesin.org/TG/LU/

process.html). Recognising land degradation can be difficult partly through lack of knowledge, partly due to a refusal to accept that it is happening and partly due to our ability to compensate for the effects of degradation through the use of new technology and fertiliser application that artificially and temporarily boosts production from the land. The very poorest farmers recognise that degradation is occurring but at the same time it is a fact of life that can be ignored. Peasant farmers must grow enough food to feed their families irrespective of whether their farming methods cause soil erosion or not.

A satisfactory understanding of land degradation must combine subjective judgements that can differentiate between degradation caused by human actions and those of natural physical processes responsible for causing erosion of the landscape. A wealth of information exists to help measure land degradation, for example data detailing the quantity of top soil lost from an area or changes to biological or agricultural productivity figures. But statistics alone cannot adequately define the human misery that results from land degradation. For about one billion people, mainly living in the developing countries, land degradation means the inability to sustain a traditional lifestyle marked by a gradual loss of income, a constant fear of hunger, scarcity of water and fuelwood.

Degradation of land is normally said to occur when the soil becomes impoverished, eroded and unable to sustain vegetation or agricultural production. The processes leading to land degradation can be the result of natural erosion processes (for example water, wind or ice erosion) or to human causes such as deforestation or processes involving excessive or inappropriate treatment of the land such as ploughing, draining or hedgerow removal. Degradation often involves a combination of natural and anthropogenic processes. Attributing the precise responsibility to one or other is both difficult and irrelevant. Whatever the reason, once land degradation begins, it is very difficult to stop.

Calculating a definitive statistic to measure degradation is a complex process because of the different levels of severity that can be recognised. A subjective scale of measurement is often used in which degradation is described as 'slight', 'moderate' or 'severe'. Table 7.1 shows the criteria used to allocate degraded land; furthermore, the extent and the severity of land degradation varies considerably between region (see Table 7.2). Because degradation affects the ability of a nation to feed its population, it is usual to express the extent of degradation as a function not of the total land area but of the amount of degraded agricultural land. Table 7.2 shows that despite the publicity given to soil erosion in sub-Saharan Africa, it is Central America and Mexico that experiences the greatest land degradation. Table 7.3 shows both the natural and human processes responsible for degradation.

Table 7.1 Subjective scale of measurement for three categories of degraded land

Category of degradation	Characteristics of the category
Slight	Productivity declines by 10–25%
Moderate	Productivity declines by between 25–50%
Severe	Land on which most of the top soil has been washed or blown away leaving the parent material exposed. Land beyond restoration or requiring major engineering (such as terracing of hillsides) to restore productivity. 15% of the world's cropland falls into this category, equivalent to 86 million ha. Productivity declines by more than 50%

Table 7.2 Share of agricultural land degraded, by region, 1945–1990

Continent	% share degraded
Australia	16
Europe	25
North America	26
Asia	38
South America	45
Africa	65
Central America and Mexico	74

(*Source*: after Gardner, 1997)

Table 7.3 Main physical, biological and human causes of land degradation and possible methods of control

Land degradation due primarily to physical causes	Appropriate methods of control
<ul><li>Adverse climate Heavy rainfall or snow melt Hot, dry summers Strong, dry winds Freeze-thaw loosens soil surface</li><li>Loss of top soil by water and/or wind erosion</li><li>Landslips or slides</li><li>Gullying</li><li>Choking of streams with sediment</li><li>Erosion of stream banks</li><li>Coastal erosion</li></ul>	<ul><li>Minimise cultivation of the soil especially before winter time</li><li>Use of contour ploughing</li><li>Cutting of terraces on steep ground</li><li>Preparing sedimentation ponds to collect fine material</li><li>Use of netting or stone embankments along stream courses</li><li>Slowing down runoff by construction of *diguettes* – lines of stones placed across the slope to catch soil carried away by flash floods</li><li>Avoiding fallow land, planting of leguminous cover crops</li></ul>
Land degradation due primarily to biological causes	**Appropriate response**
<ul><li>Deforestation</li><li>Excessive use of fire</li><li>Overgrazing by wild and domesticated herbivores</li><li>Removal of woody plants for use as firewood</li><li>Removal of hedgerows</li><li>Monoculture</li><li>Preventing the return of humus to the soil</li></ul>	<ul><li>Protection of the watershed areas of river drainage basins by planting trees</li><li>Restricting burning to one year in five</li><li>Reintroduction of wild migratory herbivores</li><li>Restricting grazing density of domesticated animals</li><li>Strict control of goat population</li><li>Replanting hedgerows and shelter belts</li><li>Reinstating mixed farming systems</li><li>Reinstating the use of animal manure as fertiliser</li></ul>
Land degradation due primarily to socio-economic and political causes	**Appropriate response**
<ul><li>Wealth-based system based on cattle and controlled by male members of society</li><li>Large family size – extensive use of child labour</li><li>Fragmentation of land holdings</li><li>Migration of males to urban areas leading to lack of able-bodied labour</li><li>Political instability and interference</li><li>Inappropriate technology gifted by donor countries</li><li>Stranglehold by multinational seed and fertiliser companies</li><li>Direct export of crops and timber with no downstream processing to generate a local economy</li><li>Overuse of chemical fertilizers</li><li>Creation of larger field sizes</li><li>'Robber economy' that uses resources in an unsustainable way</li></ul>	<ul><li>Empowerment of females in the economic cycle of farming</li><li>Introduction of no-interest cash loans for development of local cash earning initiatives</li><li>Small-scale industries to employ local young males</li><li>Consolidation of land to most efficient local farmers</li><li>Introduction of fuel efficient wood-burning stoves</li><li>Planting of rapid-growing trees for use as wood fuel</li><li>Encouragement of indigenous knowledge systems (IKS)</li><li>Involvement of local population in all aspects of local management plan</li><li>Maintenance of biodiversity by retaining vegetation diversity</li><li>Development of agro-forestry systems</li><li>Voluntary control of population numbers</li><li>Economy based on sustainable enterprises</li><li>Real economic return to local area from bio-prospecting successes (drugs, pharmaceutical products, medicinal products)</li></ul>

7.3 Early stages of land degradation

The earliest hominoids probably had space requirements similar to those of the other large mammals they lived alongside. Territory was needed from which to obtain food, water, shelter and security by all the family clan. Anthropologists are continually reinterpreting the distribution and behaviour of our earliest ancestors as new discoveries are made. For example, remains discovered in Chad have shown that hominoids had already become established there between 6 to 7 million years ago (Brunet *et al.*, 2002). It had previously been assumed that the origin of hominoids had been in the East African Rift Valley where numerous finds, also dated to about 6 million years ago, had been believed to be our earliest ancestors. It is clear that, even at the earliest stages of hominoid evolution, our ancestors had the ability to undertake lengthy migrations in search of suitable habitats. What impact these migrating groups of hominoids would have made on the landscape is unknown. Based on current knowledge it is assumed that little in the way of land degradation would have taken place for these people neither farmed the landscape nor had implements with which to cut down trees; neither did they have the use of fire.

Neanderthal people were almost certainly territorial in their behaviour (Leakey, 1979). Family units or 'clans' of between eight and 20 people probably occupied a territory of a radius of no more than 50 kilometres. A distance greater than this would take them too far from their home base. It would become too strenuous to carry large items of prey home to their base or too difficult to return home quickly in times of bad weather or if injured. In addition, a radius of 50 kilometres probably defined an area of land that was sufficient to provide food for these hunter–gatherers – there was simply no need to extend their boundaries further afield.

It is easy to imagine that our ancestors were at one with the landscape they inhabited, existing in some sylvan state, hunting the abundant herds of large herbivores, catching fish from the teeming rivers and collecting a plentiful supply of berries, nuts, herbs and fungi from the forest. This image is probably false. Early humans were ruthless predators, consuming the most accessible food sources, gorging in times of plenty and often going hungry for much of the year. Their small and scattered numbers and their technical ineptitude resulted in their inability to cause too much environmental degradation.

The contrast between the demands made on the environmental resource base by our ancestors and that of *Homo sapiens* in the twenty-first century could hardly be greater. Our present relationship with the environment and its resources differs in almost every aspect from that of our ancestors. Not least is the pressure exerted on the environment by a population greater than 6 billion worldwide and increasing at a rate of 1.3 per cent per annum (United Nations Development Programme, 2000). For 60 per cent of the world's population, difficulty in obtaining food on a daily basis is, in a relative sense, as precarious as it was for Neanderthal people (Brown, 1997). While it has proved extremely difficult for the FAO to determine precise rates, undernutrition and dietary deficiency are thought to affect 25 per cent of the population in Africa, South America and the Far and Near East (Food and Agriculture Organization, 1992) and also the urban poor in developed world towns and cities (Jones and Hollier, 1997). Thirty-one per cent of all children worldwide are estimated to be below normal body weight (United Nations Development Programme, 2000).

While our planet and its resources remain finite in terms of the absolute content, the resourcefulness of humans has allowed a significant increase to be made in the proportion of the resource base that is accessible to modern society. The very same finite natural resource base that supported a world population of perhaps 100,000 Neanderthal people now provides a much larger range of resources for the current population and is testimony not only to the inventiveness of humankind but also to the capacity of our planet to sustain its resource output. However, ecologists fear that the ever-growing demand for land, from which we take many of our physical resources, is resulting in a decline in the quality of the environment as well as in numbers of species and individuals other than *Homo sapiens* (Daily, 1997). The loss of forest, especially tropical forests, the reduction in biodiversity, the erosion of soils by overcultivation and the pollution of water and air by the contaminants released by industry and transportation all result in a progressive reduction in the quality of the environment. When these changes specifically affect the land then land degradation has occurred.

7.4 How humans cause land degradation

Land degradation occurs specifically when 'excessive' use is made by humans (and more rarely by other animals or plants) of a part of the landscape (Blaikie and Brookfield, 1987; Johnson and Lewis, 1995). Defining 'excessive' use and identifying the threshold between it and 'sustainable' use remains a major problem for land managers. Ecosystems have a natural resilience and can accommodate change. But how much change can natural ecosystems withstand and at what rate? Geographers and ecologists have concerned themselves with ecosystem changes such as deforestation (Sioli, 1997), soil erosion (Morgan, 1995), soil nutrient depletion (Albrecht, 1971), salinisation (Thomas and Middleton, 1997) and desertification (Saiko, 2001) for many years. More recently, other problems such as acid precipitation (Howells, 1990), loss of biodiversity (Wood *et al.*, 2000a), and the spread of genetically modified species (Huckle and Martin, 2001) have become further cause for concern, bringing new implications for degradation of the land.

Over time, the impact of humans has been to modify natural ecosystems so that these modified ecosystems become more suited to supporting humankind. Unlike most other animals that occupy a specific place in the ecosystem, humans are capable of taking food from most levels of the food web. We have eliminated many competitors and are no longer subject to the threat of being preyed on by other animals in the food chain. Our success has gone far beyond merely being another super-efficient animal, for we have developed the capacity to modify natural ecosystems so that they provide the materials that we find most useful in our day-to-day lives. We have replaced natural ecosystems by the agricultural system in which many of the energy and nutrient movements are managed, a situation described by Billings (1969) as the formation of a 'synthetic ecosystem'.

At first glance, the dividing line between 'natural' and 'managed' ecosystems is clear, for example as between a woodland area located next to an agricultural field. It is possible that although the woodland appears natural, in reality it is not. It could be managed to provide suitable habitats for birds, insects and small mammals and also to provide a source of timber. Despite its 'natural' appearance, the woodland could be as intensively managed (but in different ways) as the agricultural land. Human influence on ecosystems is now ubiquitous and we justify our intervention on the basis that the modified ecosystem has greater 'utility'. Such a trade-off is acceptable as it enables us to grow more food in the intensively managed agricultural ecosystems while at the same time allowing other less intensively used ecosystems to retain an element of biological diversity (World Resources Institute, 2000: 16–17).

However, we have come to recognise that there are limits to the extent to which we can utilise ecosystems. Ultimately, there comes a point when the use of specific parts of the ecosystem exceeds the level of their sustainable output. When that level is exceeded, degradation occurs. Unfortunately, we usually fail to recognise the early stages of degradation. This may involve the reduction in numbers of inconspicuous members of the natural ecosystem, for example insects or organisms that live in the soil. Only when a major species disappears do we realise that the ecosystem has undergone permanent change. Even in these cases, we rarely change the level of management, as the trade-off still favours the production of resources that are useful for humans. Ultimately, degradation of the ecosystem base occurs: the soil shows an increased rate of erosion and the capacity of the ecosystem to produce goods and services in terms of quantity and quality becomes reduced from its previous level.

Initially, land degradation was considered to be the result of physical processes that had become out of control. For example, geomorphologists such as Sparks (1969) persuaded a generation of physical geography students that 'process' was fundamental in causing landscape change. It was assumed that ways of controlling land degradation could be achieved by controlling the physical processes that had caused that degradation. Only later did a broader approach emerge, for example the work of Chisholm and Dumsday (1987), in which socio-political factors were shown to be at least as responsible for soil erosion and salinisation in Australia as were the more traditional ideas of physical process. Even more forceful in their argument, Blaikie and Brookfield (1987) began their book *Land Degradation and Society* with a clear statement: 'Land degradation should by definition be considered a social problem and thereby cannot be considered to be due to natural physical causes.' They emphasised that land degradation comprises a complex set of processes involving a combination of physical and human factors that makes any study of land degradation an interdisciplinary issue.

Barrow (1991: 1) has summarised the process of land degradation as 'the loss of utility or potential utility or the reduction, loss or change of features or organisms which cannot be replaced'. The first part of Barrow's definition implies a reduction in the usefulness or profitability of the land, which is a hallmark of the way humans evaluate land – namely, its ability to yield resources and generate wealth. The concept of land degradation therefore becomes a consequence of the way in which we apply economic values to natural resources. In the absence of humans, a loss of utility cannot occur, as there is no yardstick by which to measure *utility*. To test this theory we would need to find a contemporary landscape that is completely free from human involvement (see Colour Plate G). Given the dominance of the landscape by humans this is a difficult task, but one such landscape that is almost completely free of human interference is examined in Box 7.1. Alternatively, it is possible to reconstruct an earlier time period when human numbers were small and impact on the landscape was minimal. To follow this approach we must travel back 10,000 years before present. At this time, much of northern Europe was recovering from the effects of the last major glaciation. The deglaciated land surface

---

BOX 7.1

Natural landscape evolution in Fiordland National Park, New Zealand

Fiordland is the largest national park in New Zealand, consisting of 12,116 sq km (4678 sq mi) of subtropical woodland, southern beech (*Nothofagus nothofagus*) forests and alpine landscape extending across fiords, headlands and mountains. It was designated a National Park in 1952, becoming a World Heritage Site in 1986. In size it is equivalent to a little more than half that of the principality of Wales and yet has a current permanent population numbering less than 20! Historically, population numbers have always been low. Maori tribes settled in the area during the 1700s but conditions must have been difficult for when the first Europeans began exploring the area at the end of the 1800s only a few impoverished villages were found at the head of the fiord inlets.

Fiordland is one of the wettest areas in the world, receiving up to 7000mm of rainfall per annum. The contemporary landscape is the result of severe glaciation that formed huge U-shaped valleys rising to 1000 metres above sea level (see Colour Plate G). Precipitous valley sides are covered in thick natural southern beech forest. Periodically, the weight of the trees (the *biomass*) exceeds the holding capacity of the thin, saturated soil and large slices of valley side comprising the forest and the soil tumble downslope exposing bare rock beneath. Vegetation succession begins anew on this exposed surface and gradually the scar is covered. Although the landscape is undergoing rapid and constant change it cannot be described as a *degraded* landscape for two reasons. One, there is no human land use to suffer from the changes, and second, the landscape can regenerate itself quickly, leaving no permanent degradation. The changes that occur can best be described as *interruptions* to the natural succession. Following the interruption, succession begins again on a similar, but not identical pathway.

passed quickly through many stages of physical development characterised by rapid and highly active weathering, erosion and transportation (Summerfield, 1991). Over the course of several thousands of years, the climate fluctuated rapidly between cooler, drier conditions and milder, wetter conditions. During the wetter climate phases, the soils gradually become more acid in reaction (that is, the pH become lower) and at the end of each growing season an accumulation of partly decomposed vegetation remained on the surface of the soil. This material eventually formed into peat. As the peat thickened, a change occurred in the vegetation that grew on the peat. Species such as *Sphagnum* mosses and sedges thrived whereas trees found it difficult to establish a firm anchorage in the soft peat. Climatic conditions that favoured the development of extensive peat deposits are known to have occurred throughout much of the European land mass during a period called the 'Atlantic' era, so named because of the dominant influence of that ocean (Pennington, 1969). Atlantic conditions reached a climax about 5000 years before present (BP).

Although humans were beginning to become more numerous in northern Europe at this time, it is probable that their actions still caused only local changes to the vegetation. For this reason it is possible to say that the sequence of events that occurred between about 10,000 and 5000 years before present can be considered as *natural*. The expansion of peaty conditions resulting from climate change undoubtedly caused a reduction in the biological productivity of the landscape, but so few people were present to extract *utility* from the landscape that it is unlikely that the combination of climate change and human pressure caused *degradation* to occur.

While northern Europe still struggled to recover from the physical conditions of the post-glacial period, southern Europe and especially the Middle East presented a much more favourable environment in which human development could occur. Diamond (1997, 2002) has charted the rise of settled agriculture, suggesting that the transition from the hunter–gathering lifestyle to that of settled farmer would have been accompanied by extremely difficult conditions. Farming brought with it an obligation to tend the fields and the animals and resulted in more work and a loss of freedom. Early agriculture would also have been associated with frequent crop failures leading to food shortages and a rise in human diseases as people now lived in close contact with each other and with animals. Cohen and Armelagos (1984) claim that archeological evidence shows that the stature of early farmers was smaller than that of their predecessors, suggesting poorer diet and reflected a harder working lifestyle.

Agriculture developed at a number of separate locations mainly within an area known as the 'Fertile Crescent' in southwest Asia (see Figure 5.8) and also in China at approximately the same time (11,000 years before present), and made use of the locally available plants to create the *cultivars*, those species that eventually formed our crop plants. Zukovskij (1962) has recorded genetic details of the 140 or so genera of cultivated plants and his work clearly shows that a diverse number of plants were used to create the cultivars. Even more remarkably, very few new cultivars have been created since the beginning of agriculture. Domestication of animals began about the same time (Price and King, 1971) and was largely the result of selective breeding programmes.

The earliest agricultural communities existed in those areas now occupied by Iran, Iraq, Israel, Jordan, Syria and Turkey; others existed in southeast Asia, in Thailand, along the River Nile in Egypt and in Europe along the River Danube and in Macedonia. Elsewhere, early centres of agriculture have also been identified in the Huang He (Yellow River) area of China, the Indus River valley of India and Pakistan and the Tehuacán Valley of Mexico. One of the first examples of settled agriculture has been identified in the region of Jarmo, a site in present-day Iraq (http://ancientneareast.tripod.com/Qalat_Jarmo.html). From this site, evidence shows that the inhabitants had started to make use of the utility of the land on which they lived. The people relied on the annual flood of the two great rivers of the region, the Tigris and the Euphrates, to deposit fertile silt on the land. However, after several centuries of largely successful development, this early agricultural civilization declined rapidly (Jacobson and

Adams, 1958). We can but surmise that its eventual collapse was due to the operation of either one or more factors such as climate change, to salinisation of the soil, to population growth that placed excessive strain on the primitive agricultural system or from raids by more aggressive neighbouring tribes. Whatever the cause, the system began to disintegrate, resulting in one of the first known examples of *land degradation* of the natural resource base.

7.5 How does land degradation occur?

Geomorphologists have traditionally described the wearing down of a land surface as involving processes of degradation. These include the effect of running water (fluvial erosion), erosion from moving ice (glaciation and deglaciation), wind action and rejuvenation of the landscape owing to a change in sea level due to isostatic readjustment, volcanic activity or disruption of the land surface by earthquake activity. Because all of these processes are *natural* they cannot strictly be defined as 'degradation' in the same sense as the changes brought about by humans. From about 1750 onwards the industrial age provided humankind with a new capacity to alter the landscape. First, steam-powered excavators were responsible for reshaping the land surface and later these were replaced by more powerful and versatile diesel-powered machines. Such was the capacity of humankind to modifying the landscape that it soon became necessary to reinterpret the meaning of the term *landscape degradation*. For the first time, humans could be credited with possessing a greater role in shaping the surface of the Earth than the traditional natural processes. Changes to the landscape have been carried out with the intent of bringing benefits for humankind such as new roads, draining swamps or reclaiming land from the sea. The almost unlimited capacity of modern civil engineering has brought about landscape change at such speed and with such magnitude that it has often led to land degradation. It is a combination of human initiative and ability to cause landscape change that has placed us at the forefront of causing land degradation.

To understand natural degradation processes it is necessary first to understand something of the great material cycles that take place within and on our planet and, at the same time, to understand the role played by energy in the process of natural change. When our planet was formed some 450 billion years ago, it inherited a finite amount of matter (the materials from which our planet is constructed) and an infinite supply of energy. The energy comprised two forms: a slowly declining proportion of radiogenic heat derived from the internal radioactive energy of the materials that made up the Earth (Strahler and Strahler, 1974) and an infinite supply of solar radiant energy that constantly arrives from the Sun. At the present time, the energy ratio is about 0.1:99.9 in favour of solar energy but when our planet was in its infancy, the ratio was probably more evenly balanced. When the Earth's surface began to solidify, the newly formed rocks contained part of the radiogenic energy. Over time the radioactivity has decayed and the minerals making up the rocks have weathered, decomposed and disintegrated. The materials have been reused many times over and, on each occasion, the amount of radioactivity has decreased until today the surface of the Earth is largely benign of radioactivity. Despite the passage of 450 billion years, sufficient energy remains to provide the rocks with a physically recognisable form. The minerals comprising the rock are held together by energy but, once exposed to the forces of nature, weathering of the minerals occurs. A freshly exposed rock is resistant to weathering. It is said to have a high energy content and its structure is well organised, that is, it has low entropy. Over time, the rock begins to disintegrate into its basic materials of which quartz (silica) is usually a major component. As the rock is weathered, it loses energy and passes from low to high entropy. Its structure passes from highly organised to poorly organised and eventually the rock becomes reduced to sand. On a lifeless planet, the surface would become a thick layer of completely degraded dust. However, on Earth, the loss of entropy is counteracted by life forms, especially green plants that are capable of rebuilding the basic components into more complex forms via photosynthesis. The natural process of disintegration is part of the much larger cycling of materials including the essential elements of life: carbon, nitrogen and phosphorus, as well as the minor elements such as sulphur, manganese and boron. Without the constant cycle involving landscape disintegration and rebuilding, our planet would eventually become inert and unable to support life forms.

The rate at which natural degradation occurs is dependent on the prevailing climatic conditions, on the vegetation, on the lithology (the nature of the rocks) and on the topography of the land surface. Degradation (weathering) is fastest in tropical latitudes where a combination of heat and moisture encourages chemical disintegration of the rocks. By contrast, arid areas experience some of the lowest rates of chemical decomposition, although mechanical shattering of the rock by the Sun's heat often produces a layer of rock rubble. Where chemical degradation is active, the products of degradation pass into solution and unless stored in the vegetation and soil are carried away in runoff. The surface of landscape is said to undergo 'chemical rotting' and over the millennia, the topographic features will be reduced to a flat or rolling landscape. In reality, most land surfaces are subjected to rejuvenation (uplift) or surface warping that results in the weathered layer being transported down slope by water, wind or ice movement. A fresh rock surface is exposed and the cycle of degradation continues.

Accurate measurement of degradation can be made at trial plots but the results are probably atypical of reality. Boardman and Evans (1997) have examined the rate at which sediment is removed from different mid-latitude land surfaces and calculated an average loss of 0.5 tonne per hectare per year from 56 river catchments with mixed land use in Britain. This figure corresponds to the lower range of soil erosion as indicated by World Resources Institute (1986: 53) which quotes 'average' soil erosion figures of between 0.5 and two tonnes per hectare. This rate of soil erosion corresponds very broadly to the loss of one vertical centimetre of topsoil per year. A figure of two tonnes per hectare per year is a representative figure for all crop land in the USA. This figure rises to an average of 8–12 tonnes per hectare per year in the corn belt states of Missouri, Illinois, Indiana and Ohio (Cutter and Renwick, 1999) while for the state of Iowa, a figure of 35 tonnes per hectare per year has been recorded. An extreme figure of 296 tonnes per hectare per year has been quoted by Herweg and Stillhardt (1999) for Ethiopia.

Superimposed onto a landscape shaped by natural degradation, our ancestors gradually developed technical skills that allowed deliberate and complex changes to be made to the landscape, mainly with the intention of increasing the amount of food produced from a finite area of land. Shifting cultivation (collector–gatherers) was replaced by settled cultivation (primitive agricultural societies). At first, settlement was temporary and each summer time the population would migrate, often to higher ground, returning to the valley bottom in winter time. Gradually, as agricultural skills improved, so reliance on a transhumance lifestyle was abandoned and the so-called *anthropogenic factor* began making an increasing impact on the landscape (see Box 7.2). The operation of entirely natural processes of land degradation became altered as vegetation was cleared, crops planted and the soil surface exposed to the direct action of water, sun and wind. Degradation rates became faster and soil was eroded more quickly than it could be renewed.

It is paradoxical that the advent of settled agriculture allowed humans to achieve a substantive leap forward in their development yet this very achievement contained within it the seeds of its own destruction. Wherever agriculture occurred on sloping ground or in areas with high rainfall, the increased rate of erosion that accompanied disturbance of the land surface would eventually strip away the topsoil, lowering the agricultural productivity and forcing the farmer to search out new ground. Once again, we find evidence to show that when faced with adversity, humankind can adapt and improvise to find ways around the problem. New management techniques such as the construction of terraces or crop rotation were capable of slowing the rate of erosion. But as fast as new land management techniques were introduced, farmers intensified their farming systems, offsetting the potential reduction in erosion by using the

BOX 7.2

The anthropogenic *factor*

In the 1950s and 1960s, geographers debated whether people were responsible for causing changes to the very conditions that enable civilisation to exist. The changes could range from small-scale changes such as clearing small areas of woodland for agriculture to the total removal of vegetation and its replacement by towns and cities. Clearing a small space within the forest usually makes little overall impact on the forest as a whole but felling many thousands of hectares of forest usually destroys the ecosystem. Both the scale and the speed at which changes are made to the landscape are therefore of critical importance in determining the potential for landscape recovery. Collectively, these actions are responsible for creating the *anthropogenic* impact, or the 'human imprint' on the landscape. The main methods for bringing about change are:

population growth
deforestation
agricultural practices including:
    ploughing
    draining
    scrub clearance
    use of fire to clear land
    irrigation
    use of pesticides and insecticides
    elimination of wild herbivores

clearance of land for urbanisation
clearance of land for industrialisation
land pollution from extractive industry
water pollution
air pollution
disturbance by transportation
global warming
loss of biodiversity
problems caused by globalisation
release of genetically modified organisms (GMOs) to the wild
mismatch of environmental and human timescales.

Russian geographers such as Isachenko (1974), Nesterov (1974) and Gerasimov (1976) argued that landscape change should not be seen as degradation but as *constructive geography* – literally 'a planned transformation and management of the natural environment for the sake of the future of mankind'. They recognised the ability of the natural environment to withstand changes brought about by mankind, but beyond a certain point those changes would cause permanent and irreversible changes to the environment. How to identify the critical point was not known, but it was thought that by means of *appropriate* technology the level of landscape transformation could be safely extended.

Some years earlier, a major symposium entitled *Man's role in changing the face of the earth.* (Thomas, 1956) had marked both a high point and a turning point in geographer's interest in the human impact on the landscape. Thereafter, concern over the human impact on the landscape passed to the environmental scientists rather than geographers in what Goudie (2000) observed as a retreat by Anglo-Saxon geographers from examining relationships between human agency and the physical environment.

new management techniques to extend agriculture onto more vulnerable physical locations. Better management of the land has delayed the time when land degradation reached such proportions that commercial agriculture was no longer possible. On several occasions in the past, writers such as Malthus, Goldsmith and Ehrlich forecast imminent decline in agricultural production and ecological catastrophe, but were proved wrong as new technology was introduced. Even though the earlier predictions of disaster have been proved wrong, it is difficult to avoid the conclusion that the main achievement of all the previous 'new' technology has been merely to delay the point when conventional agriculture will fail to deliver sufficient food to feed the world population. One reason for ultimate failure is because substantial areas of land have become degraded to such an extent that they will no longer be able to support conventional commercial agriculture. The problem has been intensified due to the application of constantly evolving technology that has enabled a larger human population to be supported. If our agricultural systems ultimately fail to deliver sufficient food because of some unsolvable critical shortage, the potential for a human disaster will be substantial and will have reached a position similar to that shown in Figure 1.3 when a population 'crash' decimates the population.

7.6 Agriculture as a prime cause of land degradation

The majority of all ecosystems pass through plant succession that leads towards a 'closed' format, that is, all available living space both above and below ground becomes occupied by vegetation. Occasionally, along river banks or on steep mountain slopes, instability causes landslips to occur, from which soil erosion can be severe until vegetation succession recolonises the scar. Fires that start naturally, wild fires, can also destroy the vegetation leaving a landscape that remains bare for at least one growing season. Soil erosion can be locally severe from such natural gashes on the landscape, but these sites apart, erosion is rarely a serious problem in undisturbed ecosystems. On the contrary, beneath mature, closed ecosystems, soil accumulates often to a depth of more than one metre. It was the anticipation of cultivating the deep forest soils that lured the early agriculturalist to expend so much energy on removing the forest. But the very act of opening up the forest, by a combination of burning, cutting and grazing of domesticated animals, resulted in an inevitable erosion of the very resource that was the prize.

Because soils form the essential base of agricultural production, any degradation of the soil from its original state represents a serious threat to the profitability of agriculture. Hardin (1968), in his now classic essay *The Tragedy of the Commons*, has identified how land degradation came about:

> Picture a pasture open to all. It is to be expected that each herdsman will try to keep as many cattle as possible on the commons. Such an arrangement may work reasonably satisfactorily for centuries because tribal wars, poaching, and disease keep numbers of both man and beast well below the carrying capacity of the land. Finally, . . . the long-desired goal of social stability becomes a reality. At this point, the inherent logic of the commons generates tragedy.
>
> As a rational being, each herdsman seeks to maximise his gain. . . . He asks 'What is the utility to *me* of adding one more animal to my herd?' This utility has one negative and one positive component.
>
> 1 The positive component is a function of the increment of one animal. Since the herdsman receives all the proceeds from the sale of the additional animal, the positive utility is nearly +1.
> 2 The negative component is a function of the additional overgrazing created by one more animal. Since, however, the effects of overgrazing are shared by all the herdsmen, the negative utility of any particular decision-making herdsman is only a fraction of −1.
>
> The rational herdsman concludes that the only sensible course for him to pursue is to add another animal to his herd. And another; and another. . . . But this is the conclusion reached by every rational herdsman sharing a commons. Therein is the tragedy. Each man is locked into a system that compels him to increase his herd without limit – on a world that is limited.

Real-life examples of *The Tragedy of the Commons* have been evident since permanent human settlement first existed. Jacobson and Adams (1958) provide evidence to show that crop yields from Mesopotamian agriculture *circa* 4400 years before present were already beginning to decline probably due to the salinisation of soils. This was probably the result of applying too much irrigation water to the fields during the hot summers. As the water evaporated from the soil surface it brought up salts from within the soil profile. Worthington (1977) has estimated that 50 per cent of the land area that was previously farmed by the Mesopotamians has now seen its agricultural potential lost through salinisation. A similar extent of the former fertile Euphrates valley has also been salinised. In contrast, Chepstow-Lusty *et al.* (1998) have shown in remarkable detail how the Inca civilization between AD 1440–1534 successfully adapted their agro-forestry system to the natural resources of the central Peruvian Andes. The arrival of the Spaniards placed sudden and increased demands on the agricultural system and resulted in a severe famine for the local community. Many of the old management techniques used by the Incas have been 'rediscovered' in the twentieth century and found to be relevant to the modern farming conditions.

In section 7.5 a figure of 0.5 tonnes of soil loss per hectare per year was given as a 'normal' rate of degradation in temperate latitudes. This figure can now be compared with the rate of soil loss from sites that have been disturbed by agriculture, forestry or from building sites. Irrespective of the cause of disturbance, the appearance of bare soil at the ground surface profoundly increases the potential for soil erosion. Removal of the protective vegetation cover allows a wide range of natural agents to disturb the soil surface. Rainfall disturbs the soil particles, carrying away fine particles in the overland flow. The sun and wind dry out the land surface allowing the wind to transport soil particles in a complex bouncing motion known as 'saltation'. In winter, ice crystals form at the soil surface, loosening the soil to a depth of several centimetres in a process called 'freeze-thaw', making the soil more vulnerable to the influence of gravity in moving small particles downslope. In addition, the mechanised activities involved in preparing a land surface for crop planting, such as ploughing, harrowing, draining, removal of large stones and hedgerows all add to the disturbance of the soil and make it more vulnerable to natural erosion events. Loss of the binding capabilities of the roots of natural vegetation is another major reason for the increased vulnerability of land converted for agricultural use.

Because of the great variation in the management practices applied to agricultural land, soil erosion figures exhibit so much variation that to quote 'average' values is probably misleading. For example, Boardman and Evans (1997) calculated that losses from agricultural land in temperate European latitudes rise to as much as 12.2 tonnes per hectare per year, a figure confirmed by the earlier work of Cooke and Doornkamp (1990), while Morgan (1995) recorded a figure of 17.7 tonnes per hectare per year on steeply sloping sandy soils in eastern England. Figures for soil erosion from agricultural land in the USA are usually much higher than those for European sites. The problem

is greatest in the Mid-West and between 1940 and 1956, research scientists began to develop a quantitative procedure for estimating soil loss. It was recognised that a soil loss equation could have a great value for farm planning. Several factors were used to construct a soil loss equation, in which slope and land management practice were considered important. Wischmeier and Smith (1965) developed the universal soil loss equation (USLE), later modified to the revised universal soil loss equation (RUSLE). RUSLE has the same formula as USLE, but has several improvements in determining factors. (The RUSLE On Line Soil Erosion Assessment Tool website explains the details of the equation: http://www.iwr.msu.edu/rusle/.) Using the equation, farmers can predict the soil erosion loss in advance of introducing new land management techniques. RUSLE has encouraged farmers to use contour ploughing, the planting of wind breaks and to avoid managing the land on the occasions when climatic factors (such as heavy rain or strong wind) would cause major soil erosion problems.

Both RUSLE and USLE can be expressed as follows:

$$A = R * K * LS * C * P$$

where:

A = estimated average soil loss in tonnes per acre per year
R = rainfall runoff erosivity factor
K = soil erodibility factor
L = slope length factor
S = slope steepness factor
C = cover management factor
P = support practice factor

Soil erosion adds enormously to the costs of agricultural production. Pimentel (1976) estimated that for the whole of the United States, direct and indirect losses from land degradation during the 1970s exceeded $44 billion per annum! Soil washed from fields enters the waterways and adds about three billion tonnes of sediment each year causing offsite damage of $6.1 billion (World Resources Institute, 1986).

Data relating to soil erosion must be treated with great care and it is especially the case whenever comparing data that has been compiled by different organisations and under different socio-economic conditions. Further errors can be introduced when local, small area data is rounded up and applied to the whole country. Local erosion figures can show a variation of at least a factor of ten especially if some of the local studies have been made in arid areas from which soil losses are known to be greater than from humid areas with full vegetation cover. In addition, land subjected to recent burning will record high erosion levels as the land surface is left in a desiccated state, leaving it vulnerable to wind blow and water erosion. Despite the uncertainty surrounding some of the data, it is certain that throughout much of the US Mid-West, cultivation has caused soils to erode at a rate considerably greater than that at which they can form. Myers (1988) states that the *average* rate of soil loss is 16 times greater than the rate at which it can be formed and in the arid areas of California this figure rises to 80 times! Myers has calculated that: 'Since the development of agriculture . . . soil erosion may have ruined 4.3 million km^2 of agricultural lands and that the amount of agricultural land now being lost by soil erosion, in conjunction with other forms of degradation, can already be put at a minimum of 200,000km^2 per year.' It is unlikely that the intervening years since Myers made this calculation have seen any reduction in the rate of land degradation.

Despite a greater understanding of the problem of soil erosion, the intensification of agriculture and the increasing use of machinery has led to a continuing increase in the rate of soil loss in recent years and has prompted Barrow (1991) to state that 'on a world scale, erosion rates seem out of control'. The following list, based on the findings of Walling and Quine (1991), represents a summary of the main farming practices responsible for soil erosion:

- ploughing old grassland on steep slopes to increase the area of arable crops
- more frequent use of larger, heavier agricultural machinery leading to soil compaction
- removal of old hedgerows, making field sizes suitable for larger machinery. The larger

fields have a longer slope length allowing a greater risk of wind and water erosion
- declining use of organic fertilisers (animal manure) and a concomitant increase in chemical fertilisers leading to a reduced aggregate stability
- use of more powerful machinery allows cultivation to occur up and down slopes instead of along contours. It also permits cultivation to take place under adverse weather and on land previously considered unusable
- use of powered harrowing machines permits preparation of finer seedbeds, increasing chance of wind erosion
- greater use of autumn-sown cereals give higher yields, but exposes bare soil to extremes of winter weather. Spring-sown crops have the advantage of winter stubble to protect the soil.

7.7 Urbanisation and construction as causes of land degradation

Since 1800, the trend towards growth of urban and industrial areas at the expense of agricultural or undeveloped land has been relentless. In 1800 only about one person in 40 lived in towns of 20,000 people or more. By 2000 that figure was one in two (50 per cent) (Haggett, 2001). Urban expansion requires land not only for houses, but also for schools, hospitals, shops, industry and transportation. All new construction involves some breaking of the ground necessary for preparing foundations, laying service pipes and roadways. Major construction work requires regrading the entire land surface, flattening small hills and filling in depressions. Streams and rivers may be diverted and former industrial land 'capped' to prevent harmful wastes from reappearing at the surface. All these changes result in high rates of erosion. Earth moving vehicles churn the land surface leaving extensive areas of bare ground that can be stripped of the topmost layers of soil by wind or by rain. Goudie (2000) quotes American data in which soil loss from a construction site reached an equivalent of 550 tonnes of sediment per square hectare per year. Vice *et al.* (1969) reported erosion figures from a construction site in Virginia ten times greater than from an agricultural site, 200 times greater than for a grassland site and 2000 greater than from a forest in the same area. Once construction has ceased and an urban landscape established, erosion rates fall back to those of an undisturbed site.

In an attempt to restrict the nuisance caused by air-borne dust and drains blocked by high-sediment load in the vicinity of construction sites, it is often a condition of planning permission that site managers take every possible step to minimise erosion during the construction work. Open-ground areas must be regularly sprayed with water during dry weather, soil storage areas must be grassed and settling pools built to collect sediment from the runoff from the site. Unlike in agriculture, the increasing use of powerful machines can reduce the loss of soil by speeding up the time during which excavations remain open. Cash claw-back penalties for companies that exceed the agreed contract times ensure that building and road construction now take place under strict time management arrangements and work is completed as quickly as possible, minimising soil loss.

7.8 Taking control of soil erosion and land degradation

Many methods have been devised to limit the loss of soil from the land. Some of these methods are almost as ancient as settled agriculture itself while others involve the latest in technological developments. The earliest methods included the laying of small branches over bare ground to prevent the wind blowing away the soil particles or the cutting of small, flat terraces into sloping land to minimise the downslope movement of soil. However, apart from recognising the problem of erosion and how to minimise it, early farmers were also thwarted by the limitations of their own inadequate human muscle power, lack of durable hand tools and the absence of wheeled vehicles with which to transport earth and stones. Gradually, these problems were overcome as iron implements and horse- and oxen-drawn carts came into

widespread use. It was only after the First World War that petrol-driven tractors became available in the USA, Canada and Europe. The flat midwestern plains of the USA provided ideal terrain for the new tractors, the steel ploughs and the early threshing machines and the extent of ploughed land increased exponentially from 1920 to the early 1930s. Interest-free loans provided by the US government stimulated the uptake of new technology and agriculture spread into the more arid midwestern states of Kansas, Nebraska, Colorado and Oklahoma. A sequence of years with below-average rainfall combined fatally with the increased area of ploughed land and led to a drying out of the top soil and created the infamous Dust Bowl covering more than 13 million hectares at its maximum extent (Goudie, 2000: 191–6).

At approximately the same time as degradation occurred in the Mid-West, in the Appalachian Mountain region of the southeast, deforestation and the spread of cultivation onto steeply sloping and unsuited soils had created another area of severe land degradation and absolute poverty. The events coincided with the great economic depression of the late 1920s and 1930s when conditions prevented the payout of social benefits to entire communities deprived of their livelihood. Under the leadership of President Franklin D. Roosevelt (1933–45), a series of new government agencies were established with the express task of halting soil erosion and land degradation. Agencies such as the Tennessee Valley Authority (see Box 5.1), the Soil Erosion Service (later renamed the Soil Conservation Service) and the Agricultural Adjustment Administration (renamed the Agricultural Stabilization and Conservation Service) began primary research into ways of preventing land degradation. President Roosevelt's administration recognised that the causes of land degradation involved not only management of the physical environment but also of the social and economic environment. As the initial results of the research programmes became known it was evident that the causes of land degradation were primarily those of poor farming techniques, spurred on by poverty and inappropriate government incentives. Unfortunately, much of the excellent work undertaken at the beginning of the Roosevelt era took a back seat from January 1941 as the requirements of a nation at war took precedence over the recovery of the land.

7.9 Preventing land degradation

Cutter and Renwick (1999) have suggested that modern society is particularly inept at solving complex, multi-factorial environmental problems. Managing soil erosion has proved to be a particular challenge, involving a complex interaction between natural physical causes and socio-economic reasons. As the previous section explained, some methods used to prevent soil erosion are ancient, while others are the result of modern techniques. Research institutes throughout the world have searched for ways of preventing soil erosion yet despite the millions of dollars spent, only two simple conclusions can be made. First, it is far easier to manage the land in ways that prevent soil erosion from starting in the first place than to devise ways of stopping degradation once it has become established. Second, simple methods that can be used in the field by local landowners and farmers, with little reliance on technology or money, have greater chance of success especially in developing countries.

In Table 7.3, the three main causes of land degradation have been identified. Of these, it is the socio-economic and political factors that are usually the main cause of land degradation. Mismanagement of the land by landowners caught in the trap of needing to grow more food to feed an ever-increasing population is undoubtedly a major cause. Poverty, lack of education and exploitation of the rural population by their political leaders are other important factors. Rural farmers are stymied, believing that if they can produce 'one more crop' they will finally make enough money to leave the land and make their fortune in the city. In this way, farmers exploit their land to the utmost before finally abandoning a worthless resource, moving to the urban slums and adding to the misery of the countless millions of dispossessed people living in total despair. While land degradation is usually thought of as a problem associated with rural economies, engaged mainly in agricultural land use, the migration of people no longer able to gain a living from the land causes major political problems for governments of developing nations as landless people move into towns (Fellmann *et al.*, 1995: 108–110).

7.10 Determining when land becomes degraded

In order to determine when land has become degraded it is necessary to assess the change in land quality over time. If it is possible to show that a loss in the *utility* of a unit of land has occurred then it may be possible to state that degradation of the site has also taken place. However, before we can make such a confident statement, we must possess the means of evaluating the quality of the land at a point in time and from which we can measure any change in quality.

Wasson (1987) and Stocking (2000) discuss problems associated with the detection and measurement of land degradation. As the role of economic and social factors in soil erosion have become prominent, so methodology has changed to include these factors as well as the more traditional measurements of physical and biological change. Whereas mid-twentieth-century methodology recognised degradation, for example in terms of fertility loss, soil structure change or increasing acidification of the soil resulting from acid precipitation, the approach of the late twentieth century took account of emerging economic, social and technical circumstances. For example, government subsidies can be used to encourage the opening up of land that was previously considered marginal for agriculture; grants can be paid to individuals or to companies prepared to plant woodland; and landowners can be paid to withdraw environmentally sensitive land from agriculture use. But it is technological advances that have been responsible for most of the changes in land potential. New agricultural machinery, improved knowledge of fertiliser use, better use of irrigation and drainage schemes and, most recently of all, genetically modified crops have allowed land previously considered unusable to be cultivated effectively and efficiently. Previously, land was considered to offer a finite number of possible uses (Meadows *et al.*, 1972; Blaikie, 1985) whereas today, technology can shape the land to the use that is most desirable. Farmers no longer rely on the inherent fertility qualities of the soil to support crop growth but use the soil as a sponge into which they place the chemical fertilisers necessary for specific crop demands. In addition, if improvements in technology continue to bring new land into productive use at an economic cost that is *less* than the cost of rehabilitating degraded land, Wasson (1987) asks whether land degradation really matters. Greater soil erosion might be more acceptable in the future than it is today if soil fertility losses through erosion can be offset by new technology.

Such a radical perception of the role of soil as a medium for supporting agricultural production brings new problems in deciding what constitutes land degradation. Instead of the amount of soil lost by erosion, the amount of fertiliser residues in the groundwater table might become critical. Nitrate accumulation in drinking water as discussed by Goudie (2000: 239–45) has reached critical levels in areas of high fertiliser use, especially in the Netherlands where application of nitrogenous fertilisers have increased almost 20-fold since 1920 (Ministerie van Volkshuivesting, 1992). Nitrate levels in English rivers were up to 400 per cent higher in 2000 than 20 years earlier, an inevitable consequence of increases in the application rate of nitrogenous fertilisers from about 30 kilotonnes (kt) in 1930 to 1200kt in 1980. Of equal importance is the impact of inorganic fertilisers on the soil fauna and the bioaccumulation of residues throughout the food chain. Changes in technology and in social values we now place on the environment mean that degradation of the landscape can take on a totally new significance as the meaning of landscape degradation is reinterpreted.

Any process allowing the evaluation of degradation should possess the merit of being simple to implement in the field. It should be useable by farmers and foresters, land economists and land managers as well as pedologists, ecologists and conservationists. Ideally, it should be based on a set of standardised and easily measurable land characteristics. It should also be relatively cheap to implement, as the process of evaluation may need to be applied on two or more occasions and to the non-urbanised land of entire countries. If possible, the evaluation process should be capable of incorporating advances in new technology (most notably, the incorporation of data derived from satellite surveillance, from chemical analysis of samples and the storage of data in digital form allowing subsequent computer analysis). New data must be capable of use alongside historic records. This last requirement is necessary so that long-term changes in land quality can be identified. The methodology should be clear and avoid the use of subjective, value-based judgements of land use quality. Use of subjective terminology makes comparison of results between different surveyors more difficult and results between regions or countries open to criticism. The evaluation methodology should ensure that, for example, a unit of land classified as 'severely degraded' in the context of the Highlands of Scotland should be directly comparable to a unit of land in southern England with the same designation. Whenever possible an objective set of criteria should be used to assess land quality. Suitable criteria are shown in Table 7.4.

One of the most basic ways of evaluating the land is in terms of its ability to generate agricultural products that are either directly or indirectly used to support animals and people. For example, agricultural land can be assessed in terms of the number of head of animals it will support per 100 hectares or the tonnage of agricultural crops it will produce per hectare per year. An alternative would be to convert the agricultural product as it leaves the farm to a standard financial value, the so-called 'farm gate' value before transport, processing and marketing costs have been added. It is often difficult for an economic evaluation method to cope with the very different values of a commercial and subsistence farming economy.

A very different approach to evaluation would involve the measurement of agricultural output not in terms of economic value but expressed as the biological productivity of the farming system (Jones, 1979). In this approach, it would be necessary to establish a carefully controlled field experiment, in which the total amount of animal protoplasm or the total quantity of crop growth would be measured over a specified time span. This

Table 7.4 Objective criteria used to assess the quality of land

Land use capability criteria

- Wetness
- Soil limitations
- Gradient and soil pattern limitations
- Liability to erosion
- Climatic limitations

Depth of the topmost mineral soil horizon (A horizon)

Type and amount of humus material

Soil water-holding capacity

Number and type of soil fauna

Quantity of soil washed into collecting trays after heavy rain

Number of layers of vegetation

Presence/absence of key indicator plant species

Soil pH

Soil cation exchange capacity

Net primary productivity (npp value)

Agricultural production value (biomass value)

Economic value of the crop

Number of workers and their families supported by a specific area of land

Table 7.5 Typical biomass levels for the major natural biomes and a comparison with agricultural systems

Vegetation type	Biomass value (tonnes/hectare)
Equatorial rainforest	600–800
Tropical forest with dry season	420–460
Savanna (tropical grassland)	20–600
Temperate latitude forest	101–384
High-latitude coniferous forest	80–350
Agriculture	4–120
Wheat (UK)	4.38 (12.72)*
Barley (UK)	5.21 (10.86)*
Rice (Philippines)	4.18 (17.50)*
Maize (Uganda)	3.69 (18.66)*
Sorghum (N. Nigeria)	4.81 (13.70)*

* figures for agricultural crops: first value = grain yield, figure in brackets = recoverable biological yield (straw, fodder, fuel)
(*Source*: data from Jones, 1979 and Holliday, 1976)

might correspond to an agricultural 'growing season' as delimited by the time an animal is first allowed out to graze and the time it is sent for slaughter. For crops, the time period is usually the time between sowing and harvesting. An advantage of this approach is that comparisons of production levels are possible between different agricultural types and with natural vegetation systems (see Table 7.5). Table 7.5 shows that agriculture production compares unfavourably with natural or semi-natural vegetation production. Two reasons account for this poor agricultural performance. First, agricultural production measures the proportion of the total 'crop' that is useable by humans whereas the production values for natural systems measure the total biomass, that is all vegetation both above and below ground *and* the animal life which feeds on it. Second, agricultural systems usually operate for only part of the year,

i.e. the *growing season*, and will take several weeks or possibly months to begin active growth whereas the natural systems normally have the advantage of starting growth from a pre-existing growth stage attained in the previous growing season.

7.11 Evolution of land use patterns

Apart from the most arid deserts and the highest summits, almost all land areas have an inherent ability to provide resources of use for humans. The simplest of land use patterns must ensure that we have access for land on which to grow food (agricultural use) and for building our houses (which ultimately become towns and cities). As society becomes more sophisticated, we also require land for specialist purposes, for example for religious celebrations (churches) or for extraction of raw materials (mines and quarries), land on which industrial activities take place and land for communication links to enable easy movement between the different land use areas. There are other specialised requirements, for example for water storage (reservoirs), for depositing wastes (landfill), for burying our dead (in cemeteries), tertiary uses such as park and recreation areas and, nowadays, specialist retail areas, business and science parks.

Apart from one or two notable exceptions, there has been little attempt in the past to segregate the different types of land use from one another. Exceptions include religious sites that were often located in places considered sacred, while the chieftain or head family usually lived at a central, often elevated location. Ease of transport, in an era when all movement was time consuming and difficult, was a major consideration in determining land use patterns. Indeed, in the nineteenth century it was seen as an advantage to locate workers' houses close to factories and mines, so reducing their 'journey to work'. These early attempts at designing an optimal layout for industrial towns and villages were oblivious to the many forms of pollution that occurred near to heavy industry (see Plate 7.1).

Neither was any consideration given to ensure that the land most suited for a specific use was reserved exclusively for that use, for example from an early point in the growth of urban settlements, buildings were allowed to extend onto good agricultural land (Davidson and Jones, 1986).

An understanding that specific areas of ground might be best suited for particular land uses was slow to develop. Making the link between the yield of a crop and the site on which it was grown was a difficult concept and one that could not be made until an appreciation of the science that underpinned plant growth was understood. As long as new land remained available onto which agriculture could expand, there always existed the belief that more food could be grown simply by bringing more land into production whenever it was needed. In addition, by the mid-1700s farmers became

Plate 7.1 An industrial landscape typical of industrial Britain in the early twentieth century. Preston, Lancashire, circa 1950. Note the high-density terraced housing in close proximity to the cotton mills
Source: Aerofilms.

caught up in the new technical discoveries that allowed significant improvements in agricultural to take place, resulting in an increase in agricultural output. The new agricultural practices more than compensated for the spread of urban areas onto what had previously been good farm land. Only when special circumstances caused by the First World War, resulting in maritime blockades on shipping and food supplies, did governments began to realise that certain categories of land, especially those that were scarce, justified protection from transfer to non-agricultural use.

The concept of planning the use of land to ensure that land of a specific type was retained for a particular use had occurred prior to the early 1900s on the more enlightened estates where, through trial and error, specific crops were found to grow best in particular locations. Making the link between quality of land with specific land use was not a concept that was arrived at easily by our ancestors. Understanding the 'capability' of the land in terms of what it could best deliver in terms of useful resources represented the first stage in optimising land use productivity. The second, and far more difficult stage, was to ensure that society was able to respond positively to the first step. Land was often fragmented into tiny patches, 'common' land was owned by everyone in the village and obtaining unanimous agreement to make a change would have been difficult. Technical skills were very limited, making large-scale land use change virtually impossible. Not surprisingly, it usually required an edict from the king, landlord or governing body to enforce land use change. In France, where land was traditionally inherited from their fathers by all male children, including illegitimate sons, land holdings become increasingly fragmented and unable to produce enough food for even one family. Eventually, in a process called *remembrement*, land reform was forced through and small land units were exchanged and amalgamated into economic units. Even in this example, there appears to have been no attempt to group land into units of similar capability, although it is possible that adjacent land units would have been broadly similar in their physical characteristics. In southern England, one preferred method of land distribution was to divide farm land into strips or fields that ran from low ground to higher ground. In this way, the landowner would be ensured of some good land on the valley bottom and poorer land on the higher ground as the influence of climate and angle of slope became greater. While this arrangement worked well when farming was predominantly of a subsistence nature, when produce began to be traded during the Middle Ages, most farmers were severely restricted in what they could grow because of the limited amount of good land. Only when wealthy land owners began buying up land was it possible to reorganise field patterns into a system based on land capability. Once the best land in an area fell into single ownership, that farmer was then able to prosper at the expense of most others.

7.12 Land classification systems

Despite the undoubted advances made during the Agricultural Revolution in England between 1760 and 1830, what we would now recognise as modern agriculture did not begin until the second decade of the twentieth century. Section 7.8 has explained how the arrival of petrol-driven farm machinery allowed a new phase of agricultural expansion, opening up land previously considered unworkable. The stimulus provided by two world wars and the consequent blockade on the free movement of food by shipping made the European countries re-examine their land use strategies. Acute food shortages in Britain during the Second World War resulted in the appointment of a geographer, Dudley Stamp (later knighted for his contribution to his work on land use mapping), conducting a land use survey at the scale of one inch to the mile (1:63,360) across the entire country. This became known as the First Land Use Survey and was the first real attempt to discover the land use of Britain since the Domesday Book, a written record of a statistical survey of England ordered by William the Conqueror in 1086. Stamp's land use maps were used to identify areas of Britain that were suitable for conversion into productive agriculture. In southern England, areas of ancient grasslands located on chalk downland were ploughed and planted with grain crops, coincidentally, destroying valuable natural habitat and causing considerable soil erosion. In the context of wartime conditions this effort was considered justifiable, but soon after the survey had been completed, peace was declared and different economic conditions for agriculture prevailed. The maps were intended to show the distribution of different land use types. Each region of Britain was accompanied by a written land use memoir, providing a statistical breakdown of the proportion of land in different use categories. Such was the rate of agricultural change during the 1950s that a second land use survey was considered necessary. The second survey made at a scale of $2\frac{1}{2}$ inches to the mile (approximately 1:25,000), undertaken for England and Wales in the early 1960s, lacked the incentive of wartime austerity and the survey made far less impact than the first.

Knowing the distribution of land use provided a useful first step in land management but added little to the measurement of land capability. It was not until 1961 that a robust land capability classification, typical of its era and based on physical and biological properties, was developed by the Soil Conservation Service of the United States Department of Agriculture (Klingebiel and Montgomery, 1961). Using this classification system, land capability was calculated from known relationships between the growth and management of crops and physical factors of soil, site and climate. Based on the severity of limitations for crop growth, land was categorised into eight capability classes. Classes 1–4 were suitable for cultivation and had a relatively wide range of potential uses whereas Classes 5–8 were unsuited for cultivation and had a restricted range of uses. Apart from Class 1 land, the remaining classes had increasingly severe limitations on use due to physical restrictions. The original classification scheme denoted the limitations by use of the subclass divisions, *w*, *s*, *e* or *c*. The letter *w* denoted restrictions on use caused by excessive wetness, the letter *s* denoted soil limitations such as shallow soils, stoniness, poor structure or inherent low fertility, *e* denoted liability to erosion and the letter *c* to climatic limitations.

Kellogg (1961) noted that Class 5 in the original land capability classification was often redundant and when subsequently adopted for use by the Canadian Land Inventory and for the UK land capability classification the scheme used only seven classes (i.e. without the original category 5). In the UK classification scheme an additional physical restriction was added, category *g*, denoting gradient and soil pattern limitations.

As with all classification schemes, an element of interpretation was involved in categorising units of land to specific categories. However, the methodology used in the US system minimised the risk of misclassification by providing clear verbal descriptions of conditions that each land use class must meet while conditions relating to the subclass categories were usually defined by precise numeric limits, such as those for gradient (*g*) and climate (*c*). Bibby and Mackney (1977) adapted the US classification for use in the UK. In Canada, considerable work has been done to extend the classification of land based on soil criteria to that of agricultural-capability mapping. The Canada Land Inventory (CLI) identifies the capability of the land to grow cereal crops (http://srmwww.gov.bc.ca/rmd/lrmp/spatial/agri.htm). Capability is determined based on soils (texture, fertility, drainage, salinity, stoniness, erosion, depth to bedrock), climate (heat units, frost-free days) and topography (slope, flooding). An agricultural-capability inventory interprets the potential or feasibility of use of areas for agricultural production, including specialty crops and rangeland. It uses CLI agricultural-capability information, as well as specific information on availability of water, access to existing agricultural areas and infrastructure such as roads and bridges and availability to markets to determine the potential of areas for agricultural use.

7.13 Desertification – an extreme case of degradation

The 1980s and 1990s witnessed major famines throughout Africa, human disasters brought into the living rooms of the wealthy developed world by means of television and the press (Phillips and Mighall, 2000: 127–34). The famines also introduced a new 'sound bite' for many people – 'desertification'. Desertification involves 'land degradation in arid, semi-arid and dry sub-humid areas resulting from various factors including climatic variations and human activities' (http://www.unep.org/unep/program/natres/land/home.htm) but despite 20 years of use, the term is still confused with a similar word – 'desertisation'. This term means the formation of new desert areas and is due to long-term (hundreds of years) natural changes in climate. While Hulme and Kelly (1993) had shown that climate change could accentuate the severity of drought, rarely was it responsible for acute famine among people who inhabited the drought-ridden countries. Unlike desertisation that occurs by *natural* processes, desertification is usually caused by an overuse of natural resources by humans. It has become the most extreme form of land degradation and represents the end point of a process that takes place over an indefinite time period. The landscape created due to desertification may take on a desert-like appearance, leading to an initial belief that desertification was the result more of climate change and due less to human impact (Wellens and Millington, 1992). Desertification is most commonly associated with areas of the world that are typified by a long dry season. The apparent changes that have occurred in climate during the 1980s and 1990s have meant that symptoms of desertification can now be associated with locations that experience approximately 500mm of precipitation a year, for example the Canterbury Plains in New Zealand, the Fenlands of England and the Argentinian Pampas, as well as much drier areas within the African continent.

Desertification is a global problem. Depending on the definition of the term, desertification occurs in 110 countries, more than 80 of which are developing nations, and covers between 33 and 40 per cent of the land area (45–60 million km^2) of the Earth (Barrow, 1991). It is in developing countries that the main impact of desertification can be seen: physical changes to the landscape are reflected in economic instability that often causes political unrest. Each year 6 million hectares of productive land becomes desertified and a further 21 million hectares become so impoverished that the land is barely useable. Desertification makes the attainment of sustainable development an

impossibility as the continuous loss of soil and biodiversity can only be offset by the use of ever-greater amounts of expensive chemical fertilisers or by opening up new land areas. The latter solution is rarely an option as all land with an agricultural potential has already been brought into use while extreme poverty does not allow the purchase of chemical fertilisers.

Poor land management practices are frequently the major cause of desertification. These include:

- overcultivation of soils that are of marginal suitability for agriculture
- overgrazing especially by sheep, goats and camels; cattle grazing is less damaging provided that cattle numbers do not become excessive
- removal of timber especially for firewood
- deforestation, particularly when this occurs in the upper reaches of the watershed
- inappropriate application of irrigation water resulting in salinisation and alkalisation of the land.

By the year 2000 desertification affected the lives of about one-sixth of the world's population (1 billion people), many of whom are considered stateless, caught up in the great tides of human migrants looking for land on which to grow food. In the past, people living in countries most prone to desertification had adapted to the fluctuating chances of survival by migrating on a seasonal basis (*transhumance*), a practice that had evolved to enable herdsmen to move their animals according to the seasonal movement of the rain belts. By doing this, the animals would be unlikely to overexploit their food supply, thus minimising the chance of land degradation. The imposition of enforced national boundaries has meant an end to most seasonal migration of humans, for example across sub-Saharan Africa (see Box 7.3).

Such is the complexity of desertification that, armed with our current knowledge, it is probably true to say that we do not have the capability to repair the most degraded land areas. The UNCED Earth Summit (Quarrie, 1992), recognising the scale of the problem, decided to focus attention on *preventing* further desertification. Land that

BOX 7.3

The problem of the Sahel

According to Batterbury (1998), about 50 million people live in the semi-arid zone that forms the southern edge of the Sahara Desert. The Sahel extends across nine north African countries and, with the possible exception of Senegal in the west, are among the poorest countries in the world with a gross domestic product per person of generally less than US$300 per annum. The abnormally low figures for Ethiopia and Somalia were due to civil war during the early 1990s. Comparison of data collected by the UN Population Division and the World Bank at the beginning and end of the 1990s reveal that all but two countries registered a decline in GDP per capita, a feature that can undoubtedly be blamed in part on desertification (see Table 7.6). The relentless downward spiral of the economy has led to desertification being named the 'creeping hazard' because of its slow, almost imperceptible stranglehold on a country.

It is difficult to calculate 'normal' rainfall figures for the Sahel region. Three major drought periods

occurred during the twentieth century: 1910–1916, 1941–45 and a long period from the late 1960s until the early 1980s although this period was punctuated by individual wetter years. Low rainfall during the 1970s was responsible for about 100,000 deaths from starvation. In theory, regions such as the Sahel should benefit from the work of the Famine Early Warning System (FEWS) developed by the USAID programme. FEWS monitors the price of food staples at local markets and a rapid price rise may indicate an impending food shortage due to drought. Reliance on 'food aid' schemes put in place by developed nations and NGOs during the 1980s as a means of offsetting the effects of desertification were largely replaced in the 1990s by schemes that improved the resilience of foodstuffs and livestock numbers production systems in times of drought. Schemes that focused on soil improvement, agro-forestry and water conservation have been given priority with emphasis being given to locally-based efforts using local indigenous knowledge-based systems (IKS) wherever possible. By the start of the twenty-first century, there were thousands of farmers' cooperatives, small-scale NGO projects and externally funded development projects directed at environmental rehabilitation schemes that focused on supporting the rural population. Hopes of a 'green revolution' in Sahel agriculture have not occurred. New varieties of sorghum and millet, both dietary staples for the region, have been introduced by the International Crops Research Institute for Semi-Arid Tropics (ICRISAT) but the new varieties have been shown to require *more* moisture, fertilisers and pesticides in order to achieve higher grain output. Elsewhere in the region, eradication of diseases such as river blindness has allowed people to move permanently into areas that were previously avoided but such examples are small scale in extent and have added little by way of newly cultivated areas.

Table 7.6 Comparison of GDP figures for Sahel countries, 1992–1997

Country	1992	1997	+/– change
Burkino Faso	258	239	–7.36
Chad	243	224	–7.82
Ethiopia	53	116	+218.86
Mali	284	262	–7.75
Mauritania	571	475	–16.81
Niger	329	206	–37.39
Senegal	812	570	–29.80
Somalia	36	no data	
Sudan	186	292	+56.99

(*Source*: based on data from World Bank and UN Population Division)

was already desertified would be managed on a year-to-year basis. Such a policy, while being pragmatic, is an admission of the failure of the management systems that have been applied in low-latitude arid, semi-arid and dry sub-humid areas.

The land management practices given in Table 7.3 can help to prevent desertification but in order to achieve the best chance of success they must be accompanied by a social action improvement plan (see Table 7.7).

Table 7.7 Social action plans most likely to minimise desertification

Accelerated programmes of tree planting using drought-resistant, fast-growing native species where possible

Planting shelterbelts between agricultural plots. Leguminous tree species should be used where possible to assist with soil nitrogen fixing

Training to be provided to farmers on correct methods of irrigating crops to prevent salinisation of soils

Monitoring of all existing agricultural and pastoral technologies to establish environmentally sound best practice

Conducting research on existing land use practices to reduce pressure on most sensitive areas and to establish environmental best practice

Introduce soil–crop management systems including crop rotation and fallow periods

Involve and educate local communities, especially young people, in the fight against desertification

Encourage the retention of young men in the community to assist with manual labour

Create the capacity of local communities and pastoral groups to take responsibility for the management of their land resources on a socially equitable and ecologically sound basis

Provide opportunities for alternative livelihoods, especially for women, as a means for reducing pressure on the land and diversifying income structure

Preparation of long-term management plans (15–30 years) for integrated development based on sound environmental science

Improve education facilities especially for girls

7.14 Rehabilitating the land

A commonly-held assumption is that an increase in the density of population in an area inevitably leads to a greater anthropogenic pressure on the land surface and, in turn, to a greater potential for degradation. This situation has undoubtedly been true for much of the industrial period but was based on the Victorian philosophy in which every effort was made to extract ever-greater quantities of resources from the land with little thought given to what would now be called 'quality of life'.

Towards the end of the twentieth century, as developed nations gradually moved into a 'post-industrial era', there emerged an awareness that consumerism was based on finite resources such as raw materials and fossil fuel. Issues such as how to safely dispose of industrial and domestic wastes or of coping with declining human health hazards resulting from life in a smog-filled city has compelled a reassessment of the standard of life we wish to enjoy in the twenty-first century. Among the many problems facing humankind is how to rehabilitate land that has been degraded due to misuse in the past. In section 7.10 the question was asked whether it was necessary to rehabilitate degraded land now that new technology has allowed similar amounts of food and timber production from a smaller area of land. What will the situation be in another 25 years? Although predicting the future is fraught with uncertainties we *can* say with every confidence that the rate of technological advance will be several orders of magnitude greater in the next 25 years than it has been in the last 25 years. There is every possibility that biochemists will discover how to replicate the process of photosynthesis, eventually freeing us from a reliance on green plants as a first stage in the food chain. When this happens, our reliance on field-based agriculture will be unnecessary. Advocates of the new technical revolution, for example von Weizsäker *et al.* (1998) and Hawken *et al.* (1999), have shown that a large, affluent and environmentally conscious population can make a smaller impact on the resource base than a small, impoverished and environmentally ignorant population.

In contrast, a strong argument against a complete reliance on new technology emerged at the UNCED Summit in Rio de Janeiro in 1992 when politicians accepted the need to adopt a precautionary approach to the management of the environment. So many uncertainties remain. For example, demographers cannot predict the dynamics of world population growth. When will our world population stabilise and, of greater importance, what will be the size of the world population? Another critical question is how, or if, ethical and moral attitudes towards the way in which we feed the world population will change. For example, will future generations accept the current situation in which approximately one-third of the population in developing countries remain undernourished (Jones and Hollier, 1997: 134–41)? Will developed countries still need every hectare of agricultural and forest land to ensure a sufficient supply of resources or will synthetic alternatives be considered sufficient to take the place of traditional agricultural production? Kloppenburg and Burrows (1996) claim that because modern agro-technology is driven by industrial and economic motives, new developments in agriculture are unlikely to take any greater account of ecologically sound and socially just production methods than the agricultural systems they replace. In contrast, Wasson (1987) has suggested that new technology is more than capable of counteracting the effects of land degradation in terms of sustaining food output (see section 7.10). However, until we have categorically shown that new agricultural technology can be relied on to deliver our future needs, it is prudent that we continue to investigate ways of rehabilitating degraded lands. In reality, only the wealthiest nations will be able to adopt the most advanced forms of new agricultural production, leaving the poorest nations that already suffer most from land degradation to struggle on with the agricultural systems that have caused land degradation to occur in the first place! Only when, at some point in the future, new technology allows the manufacture of synthetic photosynthesis (independent of the green plant growing in the soil) and with the establishment of a food production system that is totally under the control of the biochemist will it be possible to agree with Wasson's suggestion that land degradation no longer matters.

If, as was proposed in section 7.4, a common set of causes stand behind land degradation, is it also possible to assume that there is a standard set of 'solutions'? One of the recommendations that emerged from the UNCED Conference in 1992 was for the coordination and transfer of research findings between nations experiencing land degradation (Quarrie, 1992: Chapter 34). As well as each nation conducting its own research into specific problems of degradation, it is necessary to make the results as widely available as possible. In the past, results of field trials on the prevention of land degradation have usually been published as internal reports that quickly collected dust on a library shelf. A revolution in the way in which information can be disseminated has taken place using the internet. Not only are the results of research projects freely published but also 'read-only' access to the databases is sometimes possible. In this way, researchers can exchange data in order to test theories on new environmental models. It is possible for researchers to re-examine previous land management practices which are known to have caused land degradation, for example overgrazing on the South African veldt lands in the 1960s and 1970s (Acocks, 1975) or the burning of the Indonesian rainforest in the late 1990s (Barber and Schweithelm, 2000). Most exciting of all are the latest developments in which the models use real-world examples to study, for example, soil erosion or mineral cycling. Recent technological advances in internet and distributed computing, especially the wide availability of high-speed broadband connection, have proved to be particularly suitable to address the problems encountered in environmental modelling. Many decision-support systems (DSS) designed to deal with environmental management can benefit from collaborative effort and make them an ideal candidate for implementing as a distributed, component-based application. In the most sophisticated models, a host computer linked to the internet holds a specific model and sample data with which to validate the model. Remote users (the 'client') can access the computer model and study theoretical changes to an individual variable or even introduce their own data to the model (Ahmed, 2001). Collaborative research can help reduce the time needed to reach management decisions and often save valuable time that would otherwise have been spent on field evaluation studies. Application of new technology can therefore offer many positive benefits in the way in which many aspects of the environment can be studied and are not only confined to land degradation problems but also to many forms of environmental pollution, climate change and sea level rise.

7.15 Conclusion

This chapter has shown that the current interpretation of land degradation is based on a change in the *utility* of the land as assessed by prevailing human values. While the effects of degradation manifest themselves as a deterioration in one or more of the physical environment indicators (such as soil erosion, silting up of rivers, disappearance of specific plants and animals) the inconvenience and price of degradation is experienced mainly by the population living in the degraded area.

It often takes many hundreds of years for the slow-acting processes that result in degradation to be recognised, by which time, society may be disinclined to embark on costly rehabilitation programmes, the benefits of which may not be seen for many years. Alternatively, entirely new methods of agricultural production may make land rehabilitation unnecessary. Ideally, degradation should be managed by application of small-scale, appropriate technology applied as soon as the first signs of degradation are recognised. In practice, our approach to the management of land degradation shows signs of neglect and disinterest. Many similarities exist between the way in which we manage land degradation and the provision of human healthcare. Until recently, attention in medicine was focused mainly on curative surgery. High-profile surgical operations dependent on expensive technological infrastructure and specialist nursing skills have become the preferred method of treatment in western nations. Preventive medicine based on identifying an optimum lifestyle have not received an equal status by government, by the medical profession or by the public. A similar situation has existed for land degradation in which the causes of degradation have been overlooked in favour of high-profile, expensive engineering solutions. It is now recognised that through the early implementation of appropriate land use systems that match the capacity and land capability of an area it is possible to prevent serious degradation and thereby achieve a more sustained land use pattern.

Successful land management strategies to redress land degradation have been developed in many parts of the world typified by the Tennessee Valley Authority, designated in 1933, and 52 years later, by the Murray–Darling Basin Initiative in Australia (see section 5.3.2). While such schemes have been highly successful they could have been avoided if an appropriate land management policy had existed in the first place. Without the benefits of experience it is often impossible to predict the long-term result of specific land use policies. In theory, we should now be able to avoid the worst examples of degradation by insisting that all new land use practices are subject to rigorous impact assessment. In reality, we can be certain that pressure to grow more food will ensure that farmers will be required to extract every tonne of agricultural output from their land and, as a result, degradation of the land will remain a problem well into the present century.

Useful websites for this chapter

CIESIN thematic guides
http://www.ciesin.org/TG/LU/process.html

Guide to Spatial Land and Resource Information in LRMP
http://srmwww.gov.bc.ca/rmd/lrmp/spatial/agri.htm

The History of the Ancient Near East
http://ancientneareast.tripod.com/Qalat_Jarmo.html

RUSLE On Line Soil Erosion Assessment Tool
http://www.iwr.msu.edu/rusle/

United Nations Environment Programme Drylands Ecosystem and Desertification Control Programme
http://www.unep.org/unep/program/natres/land/home.htm

UNEP Land Degradation
http://www.unep.org/documents/
default.asp?documentid=74&articleid=1054

8 Why retain biodiversity?

The diversity of species is necessary for the normal functioning of ecosystems and the biosphere as a whole.... There are moral, ethical, cultural, aesthetic and purely scientific reasons for conserving wild beings.

WCED, 1987

This chapter examines the reasons why retention of the maximum profusion of plant and animal species (so-called *biodiversity*) has become of such crucial importance. Loss of biodiversity is a global problem and has many different types of consequence for developed and developing nations. No single way of halting the loss of biodiversity is likely to be acceptable to all nations and yet so important is the retention of the maximum number of species for the well-being of the planet that all countries must contribute to the task of preventing further loss of biodiversity. In this chapter, consideration is given to:

- why biologists now think biodiversity is important for the health of the biosphere
- the causes of biodiversity decline
- an investigation of ways of reversing the decline by new, comprehensive sustainable management methods of the most threatened habitats.

8.1 What is biodiversity and why is it important?

Biodiversity is a measure of the variability among all the living organisms that comprise a biolo-gical community. It is a term that incorporates all nature's diversity, and includes the number and frequency of ecosystems and species, as well as the genetic diversity that comprises the individual organisms (Dobson, 1996). The term is a contraction of *biological diversity* which, simply expressed, is taken to mean the species richness of a defined area. The US Congress Office of Technology Assessment has defined biodiversity as 'the variety and variability among living organisms and the ecological complexes in which they occur' (OTA, 1987) or as Williams *et al.* (1994) have stated, the study of 'the irreducible complexity of the totality of life'.

McNeely *et al.* (1990) state that biodiversity is an 'umbrella term' for expressing the degree of nature's variety. For the majority of ecosystems this will include a number of different constituent populations that inhabit a specific habitat and an indication of their relative abundance. In theory, it is possible to express biodiversity as the total number of species (i.e. all micro-organisms, plants and animals) per unit area of land, water or air. In practice, however, it is almost impossible to calculate a single definitive figure for the biodiversity of an area due to the following reasons:

- a majority of species will be microscopic in size and consequently will escape detection
- depending on the time of year, the populations of annual plants and newly born animals may be very much higher than outside the growing season or breeding season and biodiversity values made at this time will be overestimates

- the animal population is also difficult to assess due to the constant movement into and out of the area under study.

Apart from the cataclysmic changes (such as glaciation, major periods of volcanic activity or impact by meteorite) that periodically occur on our planet, and which have resulted in major losses of biodiversity, it is undisputedly the case that, at present, the almost exclusive cause of biodiversity loss is due to human actions. Van Kooten *et al.* (2000) claim that biodiversity loss now ranks as one of the greatest problems facing society.

For most ecosystems, we still only have a general idea of the total number of species that inhabit the geographical space. On a world scale, Dobson (1996) quotes 'around 6 million' different species of plants and animals although he admits that other writers have quoted from 1.5 million to 30 million species, depending on how many species remain to be discovered. It is more usual for calculations of biodiversity to refer to a sample of a specific group of species that inhabit an area, for example the number of trees within a tropical rainforest or the number of reptile species living in different latitudinal ranges. Palmer (1992) states that one hectare of tropical rainforest can support up to 200 different species of tree or that 7000 different flowering plants have been identified in west African forests, 40,000 throughout Brazil and 13,000 on the island of Madagascar.

Because the term *biodiversity* is in such widespread use, within the scientific literature as well as more widely in the popular press, the term has taken on an extensive variety of meanings. Gaston (1996) has classified the different interpretations into three headings. The most common interpretation is that of biodiversity as a concept that signifies 'the variety of life'. Second, biodiversity is a measurable entity involving the counting of species, habitats, biomes and possibly also monitoring the discovery of new species and the extinction rate of existing species. And, third, the term 'biodiversity' can be used in a social or political context. Because society in general recognises the idea of 'biodiversity' as being a desirable condition, it has attracted considerable social and political attention, directed mainly towards maintaining environmental conditions that will encourage optimum levels of biodiversity.

With such a wide and often intuitive range of meanings, it is inevitable that any study of the term 'biodiversity' will be complex and one of the main difficulties involved in its study will be the breaking down of the pre-existing ideas that so many people have of the term.

8.2 The main reasons for biological diversity

Two main factors have overall control of biodiversity. The first is the occurrence of a benign environment that has existed over a lengthy period (many tens or even hundreds of thousands of years), during which no major cataclysmic event such as a substantial period of rapid climate change has occurred. Second, free migration of species between regions encourages a diverse genetic composition and results in maximum hybridisation and recombination of DNA. Under these conditions, the total number of species inhabiting the Earth can be assumed to increase by means of an evermore specialised evolution that allows species to occupy a more specific *niche space* (see Box 8.1). The apparent infinite capacity of a finite biosphere to accommodate an ever-increasing number of species is a difficult concept for humans to understand as it appears to be in conflict with our own experience of human population density problems. Finding answers to the different effects of increasing human numbers compared to an increase in non-human species numbers represents a major challenge and one that has a practical significance for the human population.

In particular, answers are needed to two questions:

1 How does nature support a diverse and increasing population of species, but in the process does not cause resource exhaustion?
2 What lessons can be learnt from natural biodiversity that would enable our own population to make better use of natural resources and avoid the threat of environmental crises?

The concept of the ecological niche

Early in the twentieth century, ecologists became aware that similar or *ecologically equivalent* species had evolved in different parts of the globe where the physical environment was broadly similar. Elton (1927) explained this phenomenon by applying the term *niche* to *the functional role and position of the [animal] organism in its community*, for example, grazing animals such as the North American bison performed much the same ecological role as the kangaroo of Australia. As so often happens, further research resulted in what initially appeared a straightforward concept becoming one that is now complex and often misunderstood. Part of the confusion undoubtedly arises because of the pre-existing common meaning of the term *niche* – a comfortable or suitable position for life. The assumption is sometimes made that the ecological meaning of the term is no more than a physical location occupied by a species, providing a suitable position for life.

Ecologists soon attempted to use the principle of the niche to explain how two species can never occupy identical living spaces. Gause (1934) stated that: 'as a result of competition two species hardly ever occupy similar niches, but displace each other in such a manner that each takes possession of certain kinds of food and modes of life in which it has an advantage over its competitor.' This statement has become known as the *principle of competitive exclusion* and is a fundamental concept in explaining how species subdivide an area in such a way that the maximum number of species can inhabit a living space and at the same time minimise competition between species. In spite of later work that raised new questions about the concept of the niche (for example, Caughley and Sinclair, 1994: 145), the principle of competitive exclusion remains widely accepted as a fundamental ecological principal.

The concept was expanded further when Hutchinson (1957) described the niche as a concept that included every aspect of the environment that might influence the distribution of a species. In Hutchinson's view, the niche was a multidimensioned space, or *hypervolume*, that constantly changed over time, defined by the set of resources and environmental conditions that allow the species to persist. Hutchinson named his concept of the niche the *fundamental niche*. In yet another interpretation, Odum (1971) used the term *ecological niche* to define the *role* that the organism played within the ecosystem in which it lived, whereas Jarvis (2000) proposed replacing *ecological niche* by *biological niche* to explain how a species is confined to a specific *function* within a food web.

Despite the profusion of ideas that have emerged on the concept of the niche all workers agree that occupancy of a niche space is an essential requirement for a species in order that it may exist. Once a niche space is formed, the species occupying the niche becomes part of a complex interconnected web of material and energy linkages. The main restriction on the proliferation of niche spaces is the availability of energy; niche spaces can continue to be added to a community provided sufficient energy exists within the system to support the species in the niche. Because the main source of energy within the biosphere is solar (radiant) energy it follows that niche proliferation will be greatest in those regions of the planet that receive the greatest input of solar radiation. Latitudes that lie between the two sub-solar points (marked by the Tropic of Capricorn and Tropic of Cancer) demarcate the zone with the greatest *potential* to receive maximum energy. However, radiant energy is not the only requirement for life. A regular supply of moisture is also needed. Where maximum energy and moisture distribution coincide, the potential exists for the greatest diversification of niche spaces. It is in these places that biodiversity reaches its maximum profusion.

The answer to the first question is surprisingly straightforward. However, the mechanisms that lie behind the answer are incredibly complex (see Box 8.1). The second question is more difficult to answer and involves social, cultural and political issues. An attempt to answer this question will be given in the final chapter of this book.

Apart from favourable climatic conditions outlined in Box 8.1, attainment of optimum biodiversity requires an environment that remains relatively stable, free from events such as major volcanic activity or rapid climate change or shifts in the pattern of ocean currents. Land masses located in low latitudes appear to be the most favoured sites, having escaped the destructive impact of glaciation during the Pleistocene period that engulfed high and mid-latitudes over the last one million years. Conditions during the Pleistocene era exterminated vast numbers of plant and animal species, a sequence sometimes called the *Pleistocene overkill* (Goudie and Viles, 1997: 73–4). A new force of change emerged in the last 10,000 years, *Homo sapiens*, resulting in our becoming the major force shaping the course of evolution. The impact of people on natural ecosystems has been to modify, simplify and eliminate species considered harmful to or in competition with humankind for resources. Until recent times the major concentration of people has occurred in the mid-latitudes leaving the very low and high latitudes relatively unpopulated. It is away from the heavily modified mid-latitudes that we can still find maximum biodiversity and in particular it is in low-latitude countries where conservation strategy is now concentrated in an attempt to retain maximum biodiversity (Furley and Newey, 1983).

8.3 Indices of biodiversity richness

Within a given area, biodiversity is measured according to two indices:

- species richness, by which the frequency of a species generally increases with decreasing latitude, i.e. increases from poles to equator
- species density per unit area that decreases with increasing elevation.

However, species richness and species density values are not necessarily the best indicators of biodiversity but are often used as surrogates as this data is often the only available value. The addition of several other values can help improve our knowledge of biodiversity.

Some areas of land show a highly localised concentration of species that occur nowhere else; these species are called *endemic* species and they appear to have evolved to suit very specific geographic conditions. Isolated mountain peaks sometimes show a high proportion of endemics. In Scotland, the highest mountains in the county of Perthshire have retained a flora which is thought to be a relict of the late-glacial conditions (see Colour Plate H(a) for an example of the small, colourful purple saxifrage). Remote islands are also occupied by endemic species, notably the finches of the Galapagos Islands (Nelson, 1968). Plant species growing on many of the islands of the Mediterranean Basin also show extreme adaptation to the climate and geology of the islands. The giant squill that grows on the Mediterranean mainland has become adapted to growing in small pockets of soil on an otherwise totally degraded limestone landscape (see Colour Plate H(b)). In both these examples, the endemic species have probably been able to survive by extreme specialisation; in some cases, as in mountain environments, the species comprise the remnants of a previously more extensive flora that have retreated into a *core* area. These constitute the so-called *relict* species. By combining data on species richness and the proportion of endemic species, it becomes possible to demarcate the so-called biodiversity 'hot spots' (Reid, 1998). Based on endemic land animals, 16 countries show the highest biodiversity (see Figure 8.1). By mapping the aggregate figures obtained from this table, we discover the distribution pattern shown in Figure 8.2. These can be summarised as follows:

- Large land areas (USA, Russia, China) retain vestiges of high biodiversity. Regions with low human population density still exist in remote areas of these countries. These areas have sometimes been designated *wilderness* areas in the USA, while in Russia they are designated *zapovedniks* – areas devoid of people and in which 'pure' conservation policies can be applied to save threatened species.
- Central America and much of low- and mid-latitude South America retain high diversity often due to inhospitable climate and inaccessible terrain that make the region unattractive to modern development. However, in recent decades areas such as the

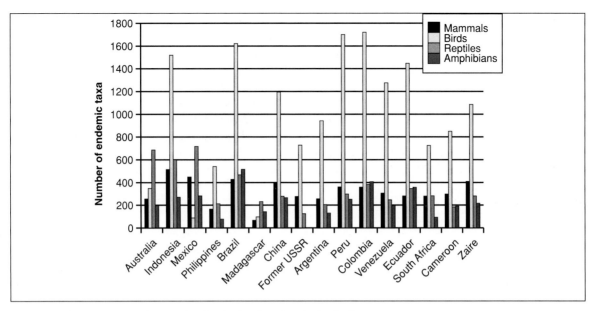

Figure 8.1 Countries with highest levels of endemism based on land animal taxa
(*Source*: based on data from McNeely *et al.*, 1990 and World Conservation Monitoring Centre, 1992)

Amazon Basin have been opened to prospectors, foresters and farmers mainly due to the construction of Trans-Amazon routeways during the 1970s and 1980s.

- The region with the greatest remaining biodiversity is the Far East. Australia possesses most mammals and reptiles, the second greatest number of bird taxa and, despite the general aridity of much of the area, contains the third most numerous amphibian taxa. Australia's pre-eminent position may be due to the highly accurate recording of species, the highly specialised nature of the Australian fauna to suit the environment and the region's isolation from newer lines of evolution that failed to reach Australia.

- Apart from Madagascar (an isolated island with a high level of endemic taxa), Africa shows a disappointing level of animal biodiversity. Cameroon and Zaire have a relatively high number of amphibians and South Africa is rich in reptiles. Two and a half centuries of colonial land use have proved extremely effective in destroying the

species richness that would inevitably have existed in this continent.

Figure 8.3 shows the distribution of the 18 most biodiverse locations as calculated by the number of endemic higher plants (the tracheophytes). All these sites are located at low latitudes and in addition are lowland or foothill locations that satisfy the optimum site conditions outlined at the beginning of this section. Some sites occupy either isolated, remote sites or locations with highly specialised climatic conditions. This is particularly so of the Cape Region of South Africa with its distinctive *fynbos* vegetation that has become adapted to frequent natural fires that burn during the tinder dry and windy summer climate (Furley and Newey, 1983).

8.4 Patterns of abundance at a single location

Irrespective of geographic location there is an amazing consistency in the structure of species data. For example, if butterflies and moths were

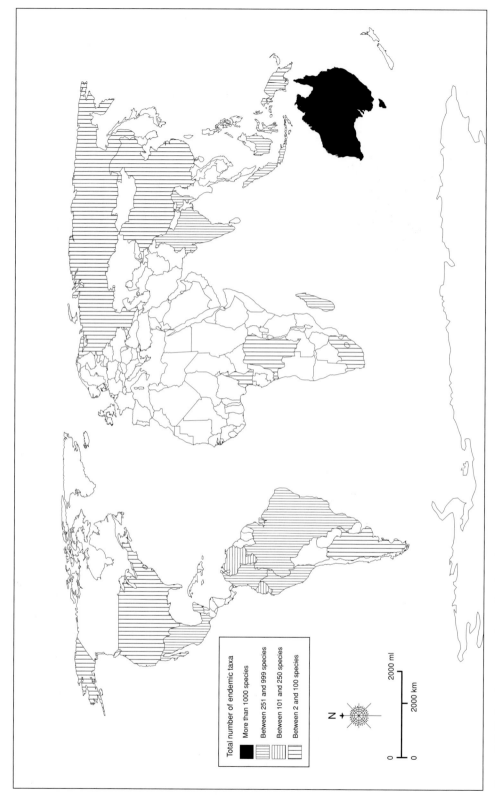

Figure 8.2 Biodiversity expressed as a function of endemic land vertebrates (mammals, birds, reptiles and amphibians) (*Source*: based on data from World Conservation Monitoring Centre, 1992)

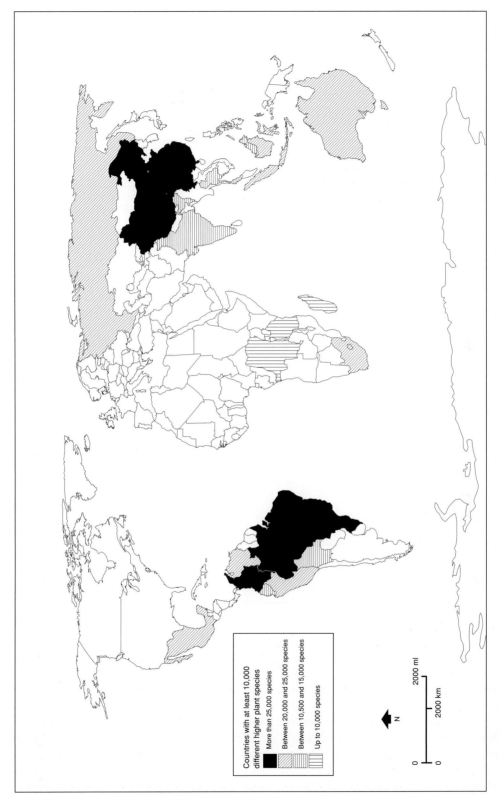

Figure 8.3 The 18 most biodiverse regions of the world, based on number of endemic higher plants (*Source*: data from McNeely et al., 1990)

Countries with at least 10,000 different higher plant species

More than 25,000 species

Between 20,000 and 25,000 species

Between 10,500 and 15,000 species

Up to 10,000 species

N

0   2000 km
0   2000 ml

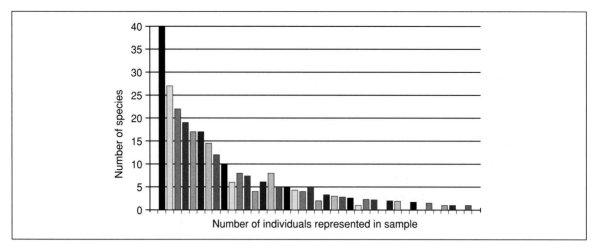

Figure 8.4 Theoretical frequency curve typical of the relative abundance pattern of species number observed at a single geographic location

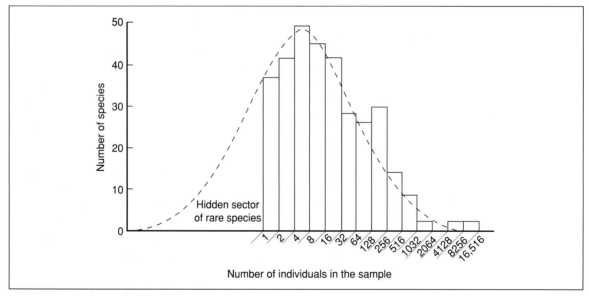

Figure 8.5 Relative abundance of butterflies and moths arranged on a logarithmic x-axis scale
(*Source*: after Dobson, 1996)

collected from a chalk downland in southern England we would typically discover that one species constituted about one-quarter of all the species collected while about 20 per cent of species would be represented by only one individual occurrence. In other words, the population at most locations is dominated by one main species and the species richness is made up of a large number of solitary individuals. This pattern of species occurrence is found throughout many different ecosystems and can be summarised by means of a species frequency curve (Figure 8.4). Dobson (1996) has reworked this data as shown in Figure 8.5. In this graph, each division on the x axis represents a doubling of the abundance of species found at a site. If a normal curve is superimposed on the graph it can be seen that an approximate fit is obtained, although the left-hand part of the curve is 'missing'. This is where the most rare species could be expected to occur and if sufficient time and effort were available for field surveys to be completed it is theorised that the 'missing' rare species would be found. That the number of individuals plotted against number of species approximates a normal distribution curve is extremely helpful to scientists involved in calculating species diversity. It becomes possible to calculate the 'expected' numbers of species and individuals at a location from the 'observed' numbers. When the observed numbers of species and individuals fall below that of expected numbers then we can say that a loss of biodiversity has occurred.

8.5 Why is biodiversity declining?

The causes of biodiversity loss are many and varied. Van Kooten *et al.* (2000) claim that the current causes of loss of biodiversity are mainly due to overexploitation, introduction of exotic species and habitat conversion. As outlined in section 8.1, at the present time, *natural* causes of biodiversity loss are probably of minimal importance, but for completeness must be included in the following list. Natural causes of biodiversity loss include:

- a failure of natural genetic evolution to retain the competitiveness of a species against a competitor, for example on a world scale, coniferous trees appear to be succumbing to the more vigorous advance of the hardwood angiosperm trees or the horse which appears to have reached the end of its natural genetic diversity
- major environmental change that causes a rapid loss of suitable habitat, for example the rapid climate change that occurred at the end of the Ice Age, 10,000–12,000 years ago. Fire, volcanic explosion, flooding and disease are other natural events that can wipe out species, while creating opportunities for new species to colonise the area.

There is little doubt that at present, the major cause for species loss are *anthropogenic* causes, that is, changes in the environment that have been brought about by humans. Swanson (1995) has claimed that human society chooses the amount of biodiversity to retain through its choice of development pathway. Consequently, the main causes of biodiversity loss can be attributed to *social*, *economic* and *political* pressures exerted on non-human species. These include:

- clearing of forests to make way for agriculture
- the use of inorganic fertilisers, pesticides, herbicides etc. in the control of the agri-system
- the spread of urban and industrial land use and the associated pollution that emanates from these sources
- 'disturbance pressure' exerted by an increasingly extensive and numerous human population on natural fauna and flora and the consequent loss of undisturbed habitats
- climate change due to the burning of fossil fuels.

Care is needed not to confuse a decline in the level of biodiversity with that of the disappearance of species due to the natural process of species extinction. Theories in support of the latter abound, for example see Stanley (1987). Events such as catastrophic landslides that cause giant tsunamis (Marshal, 2000; Walker, 2000) or meteors that crash into the planet are used to support the *catastrophe theory* that cause significant alteration to the environment and result in major species loss to the plant and animal life of the Earth. By contrast, the majority of biodiversity loss usually involves small, often unidentifiable environmental changes that take place in the background, causing the loss of one or two species. The reduction in number of individuals gradually registers on the right-hand side of the x axis in a graph of the type shown in Figure 8.4 and also as 'lost species' on the left-hand side of Figure 8.5. It is important to stress that the majority of species have become extinct during the geological time period when humankind played an insignificant role. However, recent research on the impact of the rapid spread of humans across Australia around 50,000 years ago suggests that the arrival of *Homo sapiens* supports the 'Blitzkrieg' theory of the extinction of the continent's exotic megafauna soon after humans arrived (Bowler *et al.*, 2003). Other researchers have suggested that climate change could be to blame. Only in the last 4000 years or so can we confidently state that roles have been reversed and anthropogenic causes are now almost certainly responsible for the majority of species extinctions.

8.6 Anthropogenic factors that threaten biodiversity

Throughout most of the twentieth century, the term biodiversity has largely been confined to the vocabulary of the biologist and the environmental scientist. Harper and Hawksworth (1995) claim that the term was 'discovered' by the popular press in 1989 and is now widely used by politicians and is especially common in the burgeoning number of intergovernmental reports on the environment. As with so many scientific terms that have passed into general use, the real understanding of the term may be far poorer than the widespread use of the word suggests. A poll conducted in 1998 on 1000 members of the public visiting the American Museum of Natural History concluded that members of the general public were relatively unaware of the loss of species and the threats it posed to the long-term stability of the planet (http://www.enn.com/features/1998/09/091698/fea0916.asp). This lack of appreciation of the ecological importance of biodiversity fits well with the assumption often made by environmental economists, for example Pearce and Moran (1994) and Field and Field (2002), when they propose that if society can perceive no economic gain from a natural resource (in this case, biodiversity) then there will be no incentive to look after that resource.

As with so many environmental issues, the antecedence of biodiversity began during the 1970s in California. A perception developed among a small sector of society that all species had a right to exist and that a richness of species number

was synonymous with a healthy environment. Conversely, a decline in species number was associated with environmental deterioration. A diverse fauna and flora became a criterion that, simply put, provided an index to the general state of the environment. Loss of species became associated with issues such as logging the old Californian redwood forest, removal of the Amazonian rainforest or destruction of the world's coral reefs due to pollution from land. This laudable concern for the retention of species at all costs reached its extreme in California when environmental groups began campaigns of civil disobedience and even threatened the lives of people involved in felling redwood trees, all in the name of conserving the existence of a threatened woodland bird, the Spotted Owl. Similar controversy occurred over the fishing for tuna and the inadvertent killing of dolphins (McKinney and Schoch, 1998: 563). Such action became known as the 'environmental guerrilla movement' and these more extremist views are usually associated with the deep environmental beliefs introduced in section 2.4 (see also Phillips and Mighall, 2000: 295–6). These attitudes place the well-being of non-human species above the needs of ourselves. Such views inevitably bring conflict as they run counter to the equally deep-seated human psyche of technocentrism in which we are driven towards a total control of the environment in which we live.

The contest between humans and nature has become one of the most fundamental behavioural characteristics of our species. Until comparatively recently, we have been on the losing end of the competition with nature and even today, a small let-up in human endeavour would allow nature to reassert its control over humans. By the final decade of the twentieth century the balance began to swing firmly in our favour and we have achieved 'control' through the use of a wide range of technology – chemicals, knowledge of genetic material and DNA code – and through the prodigious use of fossil fuels to supplement our own limited physical capabilities.

As with all conflicts, the victors assume an obligation to take care of the vanquished. This requirement has become easier to implement because of changes that have occurred in the motives surrounding our concern for maximising biodiversity. The deep environmental view of biodiversity that first emerged in the 1970s was replaced at the beginning of the twenty-first century by a more utilitarian, pragmatic approach. This change has largely come about through the revolution in our understanding of the genetic code of plants and animals and the consequent capability to transfer desirable genetic material between species. In as little as five years, a sound utilitarian reason for maintaining biodiversity has become a reality. We now argue that maximum biodiversity must be retained to guarantee the most diverse reservoir of genetic material for future use. The rapid advances in genetic science at the start of the twenty-first century has done more to change our attitude towards retaining biodiversity than the previous 50 years of the environmental movement. The claim by Wood *et al.* (2000a) that an appreciation of the need for maintaining maximum biodiversity 'must begin with an understanding of that conflict [between people and nature] . . . and that the roots of that conflict must be addressed if [a new approach to biodiversity] is to succeed in the long run' has been largely overturned. Although genetic material is not yet traded on the international commodity markets, DNA has become as valuable a commodity as oil, iron ore or grain. Ferraro and Kiss (2002) have proposed that the time may have come to use conservation funds to 'purchase biodiversity' through leasing land with diverse plant and animal populations. They show that in Guyana, forested land can be leased for US$1.25 per hectare per annum. Instead of using the leasing rights to clear fell the forest, they propose letting the forest remain as a biodiversity reserve.

8.7 Agricultural impacts on biodiversity

While natural ecosystems exist best without human intervention, agricultural systems require the constant attention and protection that only humans can provide (Tudge, 1988). Agricultural systems come into direct conflict with natural ecosystems for living space, soil nutrients and water. Agricultural crop species generally will not tolerate shade created by the presence of other plant species, so competitor plants must be constantly removed. Unlike wild species, agricultural crops thrive when grown in single-species plots, thus encouraging the development of monoculture practice. Over time, small agricultural clearings became progressively larger, coalescing to form fields surrounded by scrub, hedgerows and small areas of woodland. Eventually, an agri-system was created comprising the planting and harvesting of crops. Between these two events in the agricultural calendar, an endless succession of weeding and scaring away of birds and predatory animals was undertaken by all members of the human community. Apart from a gradual increase in the amount of land devoted to agriculture, relatively few other improvements occurred until the advantages of crop rotation were discovered in the late 1700s. The first use of chemical fertilisers began in the 1940s as well as the use of 'total weedkillers' to eliminate the agricultural weeds. From the 1960s, increasing reliance was placed on the use of 'selective' chemical pesticides, herbicides and insecticides to eliminate pests.

Apart from technical advances, another invisible pressure has caused changes to agriculture: the inexorable increase in the human population. An increasing population served both as a stimulus for bringing more land into cultivation and also provided more people to tend the crops and animals. One by one, competitive species were either deliberately or inadvertently removed. Plants and animals were forced out of suitable habitats by the constant hunting, weeding and general disturbance caused by people, most of whom were dedicated to the production of food. By the 1750s much of the European land surface below 400m above sea level had been converted to agriculture. From the 1850s a new pressure was exerted on the environment, as intensification of production became a feature of the European agricultural system. At about the same time, the great drive to open the central and western plains of North America had started. A sequence involving the clearance of native vegetation that had taken thousands of years in the Old World, took only 100 years in the New World and square plots of land were leased out, checkerboard fashion, alongside the railway lines that gradually spread westwards. In Africa it took even less time: between 1945 and the mid-1980s the destruction of native habitat reached 89 per cent in Gambia, 87 per cent in Liberia and Rwanda, while further east, 97 per cent of native habitats were destroyed throughout Hong Kong province and 94 per cent across the much larger area of Bangladesh (IUCN/UNEP, 1986a, b). While western countries now try to persuade those countries that still retain a diverse fauna and flora to restrict clearance of forests for agriculture it must be remembered that European countries, followed later by the USA, Canada, New Zealand and South Africa have been responsible for a massive and irreplaceable loss of biodiversity.

As shown in sections 7.10 and 7.11, the best land has already been used for agriculture, leaving the ill-drained or the most exposed, steeply sloping or poorest soils for natural habitats. In addition, non-agricultural land became confined to ever-smaller areas of land, often located in remote areas and characterised by inhospitable climates surrounded by a sea of agricultural land that made it difficult for species to move from one suitable site to another. The taking of new land for agriculture has now largely come to a stop simply because most land with an agricultural potential has been used. Some new grazing land is still being formed particularly as a result of illegal forest burning in the tropics. These poor quality grasslands become degraded after as little as five years, leading to farmers burning more forest to create more poor grassland in a spiral of desperation to grow food. This is especially the case in Indonesia and, nowadays, to a lesser extent in Amazonia, where national governments have condoned illegal deforestation in the knowledge that legally to enforce a strict no-burning policy would result in starvation of the rural population. The burnt-over areas of the tropics now suffer the greatest human-induced rates both of biodiversity loss and of species extinction and are the focus of much of the work of international bodies such as the International Union for Conservation of Nature (IUCN) and Worldwide Fund for Nature (WWF).

8.7.1 Impact of agricultural chemicals on the biosphere

Until the end of the nineteenth century comparatively little use had been made of chemicals to destroy unwanted weeds or insect pests. Human labour was cheap and women and children were used to hoe out weeds or pick caterpillars from crops in a never-ending, back-breaking job (see Plate 8.1). Chemicals have been used in limited quantities in vineyards since Roman times, when solutions containing copper sulphate had been effective in preventing fungus disease on vine leaves (Mellanby, 1967). Little was known of the way in which the chemicals achieved their goal, less still on the way the chemical elements reacted with the soil, the water and the biota to which they were added. Dramatic visual evidence of the effectiveness of non-metaliferrous ores such as lead, copper and zinc on vegetation can still be seen from the locations of old mines (see Plate 8.2). The toxicity of these ores still prevent all but the most tolerant of plants from growing on these sites, often several hundreds of years after mining has ceased.

The signs of chemical damage are often difficult to trace and throughout the 1960s, when suspicions first emerged that agricultural chemicals could be responsible for inadvertent damage to plants and animals, the causal links especially between persistent organochloride pesticides (such as DDT) could not be proved with absolute certainty. However, new analytical techniques enabled the detection of minute quantities of organochlorides in the eggshells of birds of prey (Ratcliffe, 1980) and helped prove that these substances caused the breaking of the eggs during incubation, resulting in a dramatic decline in the numbers of birds of prey, particularly the peregrine falcon. A ban on the sale of these substances led to a rapid recovery in numbers.

By the early 1900s 'total' weedkillers such as sodium chlorate and arsenic compounds were used to kill all vegetation at a site. So effective were these substances that nothing would grow on the sprayed areas for up to six months following application. A major breakthrough in the use of chemicals was achieved in the 1930s with the

Plate 8.1 Four women using hand hoes to cultivate a vegetable patch, Zanskar, Indian Himalayas

discovery of the first selective herbicides – the dinitro compounds, marketed as DNOC. Their use has caused great harm to thousands of species, being poisonous to mammals either by inhalation or by absorption through the skin. In 1942 the first 'hormonal' weedkillers were available. These are synthetically produced organic compounds, the most common of which is MCPA (see http://ace.ace.orst.edu/info/extoxnet/pips/MCPA.htm for details of this product). When applied to weeds, it causes grotesque growth of cells leading to rapid death, yet making no impact on narrow-leafed grasses or cereal crops.

While the use of synthetic chemicals enabled a great increase in the *control* of unwanted plants and animals and led to major increases in agricultural productivity, their use was also responsible for much unwanted and unintended damage to ecosystems. Cunningham and Saigo (2001: 266–71) claim that 90 per cent of all pesticides never reach their intended target due to misapplication or drifting spray and, consequently, unintentional destruction of non-agricultural species occurs. For example, the same authors state that each year, 20 per cent of all honeybee colonies in the USA are destroyed through indiscriminate chemical contamination. The bees can only rebuild their numbers against this onslaught by relying on their ability to breed at a far higher level than would be necessary to maintain a stable population – an example of the considerable *biotic potential* that most species possess but only rarely need to use, see Section 1.4.1.

For a number of reasons, chemicals rarely kill all their intended target. Not least, more and more individuals appear to develop a resistance to specific chemical sprays, a condition known as *pest resurgence* or *rebound*. Alternatives to the

Plate 8.2 Disused Ystumtuen lead mine in the Rheidol Valley, Wales, showing destruction of vegetation in the vicinity of the waste material dumped from the mine. Mining operations ceased early in the twentieth century

ever-increasing spiral of using higher concentrations of chemicals include:

- reintroduction of crop rotation in place of monoculture
- use of biological control to reduce the number of pests to an acceptable level
- reversing the trend towards intensive agricultural production
- organic farming methods
- better husbandry to ensure a minimisation of wastes on which pests can live from one season to the next
- reintroduction of hedgerows.

8.8 Effect of urbanisation on biodiversity

So attractive is the urban lifestyle that almost 60 million people worldwide leave the countryside and move into the cities each year (United Nations Environment Programme, 1999). An urban lifestyle represents an extremely efficient way of organising and making available services to a population – provided that the economy of the country can support the enhanced level of services required by an urban population. To accommodate the ceaseless demand for urban expansion, cities constantly try to expand onto surrounding land. In the developed world, strict planning policy has effectively prevented unbridled expansion but in the developing world, an effective restriction on outward expansion is impossible to enforce. Of the 20 largest cities of the world, New York, Los Angeles, London and Moscow all effectively ceased to expand during the 1990s whereas Mexico City, São Paulo, Calcutta, Mumbai, Jakarta and Karachi have all continued to show massive expansion. The largest of all the urban areas, Mexico City, extended over 1250km² by 1990 (Fellmann *et al.*, 1995).

The environmental problems created by these burgeoning megacities are incalculable. Their impact extends well beyond the formal boundaries of the urban area. Many people work in the city but live outside; the daily journey by road to and from the metropolis generates air pollution that degrades the natural environment, destroying most native vegetation and with it the animal life too. Vermin thrive in the frequently found unsanitary conditions and not only bring considerable health risks for humans but also kill and eat many of the other small rodents and bird life.

The natural hydrological resources are also changed. Sewers and drains discharge their often untreated wastes directly into rivers that, in turn, become biologically dead. The rivers flow beyond the city boundary carrying with them the pollution and extend like poisonous arteries into the countryside. Worse still, infiltration of wastes into the soil eventually reaches the groundwater table where it contaminates huge volumes of water. Often, groundwater is tapped by means of uncontrolled wells and the water is used for irrigation and consumed by humans even though it is contaminated with heavy metals, organochlorides, and herbicides such as 2,4-D and 2,4,5-T. These contaminants become incorporated into plant tissue and eventually pass into the food chain where they accumulate in the biotic components. The chemicals become concentrated in the fatty tissues of the secondary consumers and often result in genetic defects, growth abnormalities or premature death of the animals concerned.

Wilcove and Bean (1994) have examined the impact of polluted rivers and changes to the natural flow of river water on ecosystems and species in North America. In 1994 only 42 free-flowing rivers of over 125 miles existed in the USA. These authors stated that of the 3.2 million miles of streams in the lower 48 states of the USA only 2 per cent remained free-flowing and were relatively unspoilt, the other 98 per cent of the US rivers being subjected to development or pollution. The consequences for aquatic species have been considerable, one-third of the 800 North American fish species are classified as 'at risk' or possibly extinct (Master, 1990). Comparable figures for mussels and crayfish were 73 per cent and 65 per cent respectively. The species extinction rate of aquatic life has doubled during the twentieth century and some species now find themselves isolated in upstream reaches, cut off from the genetic diversity of the previous geographically more extensive population. These isolated species are destined to become 'relict' species (see section 8.3).

Pringle (1997) has suggested that while the loss of habitat and biodiversity in North America has been serious, conditions in the former Soviet Union and in central and eastern Europe are far worse. The River Danube receives three-quarters of the sewage load generated from the urban area of Belgrade in an untreated form and the river transports this effluent loaded with detergent residues, chemicals and industrial wastes as well as organic wastes into the Black Sea. Of the 26 economically valuable fish species in the Black Sea, only five species remain. A similar situation exists in the River Vistula and its tributary, the Bug, Poland's main river system, and the source of

drinking water for a large proportion of the population. Pesticides and heavy metal contamination is high and concentrations in the human food chain frequently reach danger level, posing a serious threat not only to species comprising the food chain but also to human health and economic development. One condition of acquiring EU membership by the former eastern European countries is that mechanisms must be put in place to bring environmental management up to existing EU criteria (Environmental Development Consultants, 1997). The economic costs of achieving this are high. Hall (1990) has quoted an estimated cost of environmental compliance throughout eastern and central Europe as $200 billion. (Note that this figure is a 1990 valuation.)

8.9 Habitat destruction

A study of the palaeontological (fossil) record shows that only a fraction of all the species that have ever lived are still present on the Earth today. Species have constantly passed into extinction while others have evolved to form new lines of development (Stanley, 1987). We can confidently deduce that the disappearance of species has been a recurrent event throughout geological history. However, new species evolve at a rate that is at least equal to the rate of species loss. This raises the question whether the current concern over biodiversity loss is merely another example of a natural trend in the long-term history of the planet or whether the current rate of biodiversity loss really is significantly greater than in the past. The assumption is commonly made that the present rate of species loss is inevitably greater now than in the past because of the current level of human impact on the biosphere (see Cohen, 1997 for a full discussion of the linkages between the growth of human population and the impact on habitat and species number).

In the geological past, the loss of species was often accompanied by, or may have been due to, changes in the circumstances by which new rocks were formed and within which fossils were preserved. These changes in the geological conditions are often identified as 'geological unconformities'

– occasions when new environmental conditions were caused by catastrophic events such as the inundation of land by the sea or the deposition of thick layers of meteoric dust or volcanic ash (Stanley, 1987). Six major periods of species extinction have occurred (Goudie and Viles, 1997), for example at the end of the Permian geological period about 250 million years ago when sea floor protozoans, marine invertebrates and mammal-like reptiles disappeared (May *et al.*, 1995). The best known of the mass extinctions was the disappearance of the large dinosaurs that occurred at the end of the Cretaceous period 65 million years ago, an event that also saw the demise of many plankton, marine invertebrates and marine reptiles.

The current concern is that species loss is due not to geological reasons but, rather, results from human activity. Events such as deforestation, draining of wetlands, ploughing of grassland and to the increasing use of chemical sprays to control insect pests and disease transmission are the main reasons for a decline in biodiversity. Although we are quite certain that species are being made extinct we cannot say with certainty how many species are being lost and neither do we know if the rate of species extinction exceeds the natural evolution of new species (see Box 8.2).

BOX 8.2

How many species are we losing and how many are we gaining?

According to a Worldwatch Institute Report (1998), we are living through the fastest mass extinction of living things in the 4.5 billion-year history of our planet. There have been other occasions when *natural* extinction occurred at very high rates. For example, 65 million years ago when the dinosaurs largely disappeared and, more recently, between 10,000 and 12,000 years ago at the end of the Ice Age when *Homo sapiens* contributed to the extermination of up to 75% of all large mammals, such as mammoths, cave bears and sabre tooth tigers (Martin and Wright, 1967; *New Scientist*, 1993). As we move towards the present day, the causes of species extinction are increasingly the result of human activity.

Great care has to be taken to interpret the published figures on species loss correctly. Some of the figures are based on studies in small areas that are 'rounded up' to apply to whole continents or even the whole planet. Other statistics relate to specific plant or animal types – most commonly to flowering plants, trees, birds and primates. Our knowledge of losses among aquatic organisms and of the non-vertebrate animals is exceptionally poor and it is almost certain that tens of thousands of small and seemingly insignificant organisms have disappeared before they have even been discovered and classified. The Worldwide Fund for Nature has reported that 30% of species diversity has been lost between 1970 and 1995 (WWF, 1999) but this figure does not take account of the massive species loss that occurred earlier in the twentieth century. Because we have no idea of how many species we had worldwide at, for example 1 January 1900, we cannot calculate the percentage losses that have occurred between that date and 1 January 2000. Crude estimates for extinction have been provided by Myers (1988) for higher plants. Based on fossil evidence one species has become extinct every 27 years over the past 400 million years, while Raup (1986) has calculated that 90 species of vertebrates have become extinct every century for the past 200 million years.

Some highly accurate figures do exist and data provided by Lawton and May (1995) represent some of the most accurate calculations of species loss. One habitat type that has suffered more species loss than most others is the Mediterranean habitat. Due to the very high rates of endemism, species number is higher than most other areas of the world but as human intrusion into these areas has increased, so extinction rates have soared. Work by Greuter (1995) has shown 37 species of higher plants have been lost from Mediterranean regions since the 1940s. Although this is thought to represent the highest rate of higher plant extinction in the world, it only represents about one species per thousand in the Mediterranean biome as a

BOX 8.2 CONTINUED

whole. Greuter asks whether this rate is significantly more than the rate of natural extinction and, when set into the context that the Mediterranean Basin region has been 'home' to mankind for up to 10,000 years, is this rate of extinction something we should be worried about?

Bibby (1995) presents evidence to show that about 100 bird species worldwide, equivalent to 1% of all birds, are thought to have become extinct since 1600. Ninety of these were confined to islands and were probably rare endemic species. A further 200 extinct birds have been identified from fossil remains.

Very detailed information exists on the state of the insect population of Britain. Two factors help calculate species loss. First, Britain has a relatively small and hence manageable number of insect species at about 22,000 and several thousand other invertebrates. Second, knowledge of these groups is probably greater than for any other area in the world, studies of insects dating back 300 years (Thomas and Morris, 1995). Of the 59 species of British butterflies, 8% have become extinct over the last 250 years and 7% of the 41 species of British dragonflies. In the Netherlands, another country with good data, 21% of butterflies have become extinct in historical times and 46% are in long-term decline.

According to theoretical calculations based on the number of fossil species, both the plant and animal kingdoms should still be *expanding* in terms of species numbers. According to Niklas and Tiffney (1995) diversification among vascular plants has occurred through the geological record and no evidence exists to suggest that this process has changed during the current period of time. It is probably the case that scientific interest in biodiversity is so recent that we have not yet found evidence to support the theory of species diversification. In a few infrequent instances new species do come to light. For example, on 22 January 2001, at a meeting of the International Primatology Society held in Adelaide, Australia, zoologists announced the remarkable discovery of nine new lemur and two new marmoset species (http://www.conservation.org/xp/CIWEB/home). Lemurs are unique to Madagascar while the marmosets were discovered in south central Amazonia. The discovery of the lemurs is highly remarkable given that their habitat is shrinking rapidly. The discovery of new primate species is highly unusual, ten new monkey species having been added to the list since 1990.

8.9.1 Fragmentation of habitats

Fragmentation occurs when a homogeneous area of vegetation becomes subdivided into a number of smaller, desegregated patches and separated from each other by habitats different from the original (Wilcove *et al.*, 1986). All species require a minimum space in which to live. For plants, this space is shown by the spread of the roots below ground and the span of leaves, branches and stalks above ground. A large oak tree may occupy a volume of 1700 cubic metres above ground and 1000 cubic metres of root space. Plants remain rooted in the same space for the duration of their life and have a predictable space requirement depending on their ultimate size, whereas animals, being mobile, occupy a variable space that undergoes constant change depending on the

stage of the life history and season of year. Many animals are 'territorial', that is, they mark out a living space within which they can obtain all the requirements necessary for completing their life history. For example, species as diverse as robins, badgers and wolves establish territories of approximately 1 hectare (Lack, 1972), 4 hectares (Neal, 1958) and 100–500km² respectively (http://large-carnivores-lcie.org/wolf.htm).

All territories must provide a certain 'minimal area' that can provide sufficient space within which the individual organism can complete its life history (Shimwell, 1971). In more and more cases, the increasing number of humans has resulted in an expansion of the human 'territory' and a consequent fragmentation of the space available for other species. On some occasions, the remaining area comprises the wrong shape, for example narrow, linear corridors that follow the verges of motorways or river banks or strips of unused land at the edges of agricultural land that are periodically sprayed with herbicides or insecticides. Other units of land exist in a temporary state, for example derelict industrial land on which species colonisation may begin, only to be destroyed by a new phase of development after a few years.

Fragmentation increases the ratio of habitat edge to total area, exposing the remnant area to more disturbance from surrounding land use. Although fragmentation was thought to affect larger animals more than smaller ones, Klein (1989) has shown that increasing fragmentation of tropical forest has a significant effect on small beetles with higher fragmentation leading to fewer and smaller dung beetles.

Detecting fragmentation of habitats has relied on detailed research work, mainly by historical geographers and biogeographers working from old maps and estate records. Work by Webb and Haskins (1980) on the decline of lowland heath in the county of Dorset, southern England, has shown a fall from 30,000ha in 1800 to less than 5000ha in 1980.

Since the mid-1970s satellite technology combined with photogrammetric and GIS analysis has enabled a more accurate recording of the fragmentation of tropical rainforest. In 1990 the first comprehensive data on forest fragmentation in the Amazon Basin based on 167 Landsat images showed that 404,000km^2 of forest had been cleared compared to land-based estimates of 343,975km^2 (*Nature*, 1990).

8.10 Managing biodiversity

Success in the global economy requires countries to operate in a 'competitive' and 'efficient' manner. The same principles are applied to the ways in which we relate to our natural resources and the environment. Unfortunately, we do not have a clearly defined understanding of what this means in a practical sense. Bean (1997) suggests that achieving successful management of our environment involves far more than developing a 'cookbook of "how-to" steps for practical managers'.

Successful management of our environment needs to ensure a sufficient supply of resources both at the present time and in the future. In addition, all non-human species must be provided with every opportunity for their long-term survival.

Plants, animals, soil, air and water provide the essential resources on which human society is founded. Although the precise mechanisms for obtaining food have changed over time we still mainly rely on fisheries and agricultural systems for the majority of our food. Different types of plant materials, such as cotton, timber, sunflower seeds and groundnuts provide essential raw materials for industrial processing. A quarter of modern medicines and drugs contain at least one vital plant extract (Myers, 1984), while products as diverse as ice cream, pharmaceutical items and our clothes contain gums, colouring or flavourings derived from plants. In short, the diversity of our modern consumer society relies greatly on the existence of natural raw materials. Myers estimated the value of these components to be worth $20 billion per annum to the US economy in the early 1980s!

In addition to the economic value of natural components, we must also consider the aesthetic value of land- and seascapes. One of the poorly understood requirements of the human animal appears to be the need to spend some time 'communicating with nature'. Each individual has a very different level of need. For some, watching a spectacular sunset through a car windscreen will be sufficient while for others the need will be to stand alone on a remote mountain-top or walk beneath the canopy of an old forest. It is difficult to place an economic value on solitude or magnificent scenery, yet without a recognisable value the resource becomes a 'free good', open to all, owned by no one and consumed by all. The new discipline of environmental evaluation and environmental economics is struggling to develop methodology that can 'price' environmental intangibles (Garrod, 2000; O'Connor, 2000).

In the past, when extensive areas of empty land existed, a policy of allowing free use of the land and its resources was an acceptable, if somewhat unconstructive, land management strategy. As the world population increased so planning how we used the land and its natural resources became an essential part of modern life. Preventing shortages of natural resources, providing sufficient agricultural land to feed a growing population and minimising harmful pollution of land, air and water became key objectives for modern society. In the west, the land use model employed to deliver such a system has been based on a free market economy, in which the use placed on the land was determined by whoever was willing to pay most for the land. However, from an early date some nations recognised the need to designate some land as free public space, owned by the state and retained mainly as wilderness areas. The first formal example of wilderness or recreation space was created in 1872 when Yellowstone National Park covering parts of Montana, Wyoming and Idaho in the USA was designated (Rettie, 1995). Since 1872, national parks have been designated in all countries of the world apart from Iraq.

Another major step forward in the management of biodiversity occurred in 1980 with the publication of the *World Conservation Strategy* (WCS) (IUCN/UNEP/WWF, 1980). The WCS was supported by UNESCO, FAO, WWF, IUCN and UNEP and proposed a single integrated approach to the management of global problems. The WCS was based on three propositions:

1 Such is the impact of humankind on the biosphere that species and populations (both plant and animal) cannot successfully compete with humans, and must, therefore, be helped to retain their capability for self-regulation.
2 A basic requirement for self-regulation is the retention of the physical resources of the biosphere (climate, soils, water) in as an unpolluted state as can possibly be achieved within the limits of modern scientific knowledge.
3 Concern was expressed that the disappearance of many individual species of plants and animals was leading to:
   ● a simplified trophic structure
   ● extinction of species that would result in a serious loss of genetic diversity.

If left unchecked, the loss of natural areas would result in a reduction of biodiversity and, eventually, would lead to reduction in the number of opportunities for evolutionary diversification. This trend is the opposite of that which is necessary if maximum flexibility is to be retained in the organic resource base.

The launch of the WCS heralded the beginning of a new era in species management. Conservation of nature was no longer seen as the sole preserve of natural scientists. Political, economic, ethical and moral reasons were to assume equal status to that of scientific arguments in support of biodiversity. For the first time the argument in support of the 'right to exist' was applied to non-human species. The deep green ecological movement (see section 3.8) advocated that all species had an equal right to live and that humankind, as the dominant species, had an ethical and moral responsibility to ensure suitable environmental conditions for all species.

The ecological importance now placed on retention of maximum biodiversity has stimulated a new approach to be taken to conservation of habitats and the species that occur within them. Natural parks (concerned primarily with the conservation of nature) as distinct from national parks (concerned with designating land of high scenic quality in which members of the public can observe nature) have provided a safe future for many threatened habitats. The work of the International Union for the Conservation of Nature and Natural Resources (IUCN) has enabled the designation of 851 million hectares of land intended specifically for conservation – some 6.4 per cent of the total world land area (World Resources Institute, 2000).

In recent years, a new incentive for the retention of maximum biodiversity has emerged from an unusual direction. An increasing number of wealthy and environmentally aware tourists from the developed world are prepared to spend large sums of money and often suffer considerable discomfort visiting the last remaining wilderness areas of the world. Many conservationists argue that by opening up remote areas, the very uniqueness of the area is reduced. For example, Yellowstone National Park attracts 3 million tourists a year and suffers traffic jams as bad as many urban areas, while the combined annual visitor figures for all US national parks amounts to 267 million people (Cutter and Renwick, 1999). Even the most sensitive and well-managed eco-tourism developments will inevitably cause some degradation of the habitat but as compensation, eco-tourism can provide a stimulus for alternative wealth generation in some of the poorest countries of the world. Establishing a balance between a limited growth of sustainable eco-tourism and nature conservation is difficult to achieve but, if successful, the money generated from eco-tourism can contribute to the retention of biodiversity.

8.11 Ways of preventing biodiversity loss

Despite much conjecture on behalf of the popular press that scientists will soon be able to recreate a woolly mammoth or even a dinosaur from a fragment of DNA taken from a fossilised remain, it is not possible to recreate a species exactly in its original form once it has become extinct. The only alternative is to ensure that we manage the biosphere in such a way that we retain the maximum number of species. The most common approach has been the creation of protected land areas such as scientific nature reserves, national parks and *NATURA 2000* sites. More complex methods involve 'debt-for-nature swaps' (see section 8.14).

The IUCN target figure intended specifically for land and coastal conservation in each country is 10 per cent of a nation's territory. However, it is impractical to consider *all* land as suitable for conservation as other essential land uses such as water catchment areas and land used for agriculture must be given priority. Conserved land must be selected with care, to ensure that it represents a representative sample of habitat types found within a nation. Wherever possible, land designated for conservation must be the *best* examples of specific habitat types. The successful designation of land for conservation purposes must be accompanied by appropriate legislation to safeguard its long-term future and to prevent other competing land interests from intruding into the conservation zone.

It may also be necessary to ensure that at certain times, such as during the breeding season or until rehabilitation and stabilisation of the habitat has been achieved, access to the site may have to be restricted. Successful implementation of such a policy must ensure that the general public understands the need for restricted access and that long-term conservation benefits must outweigh the public's short-term requirements.

McNeely *et al.* (1900) have suggested that an effective retention of biodiversity will require increased resources, including personnel, finance and political commitment. They have identified six steps required to promote biodiversity. These are:

1 **Policy restructuring**. The existence of separate government departments (e.g. transport, agriculture, employment) has ensured that a sector-based structure is unhelpful in achieving a coherent, integrated policy on conservation management. Instead, government policy should be based on integrated management (the so-called 'joined-up government' advocated by the UK Labour government of 1997) and should have as one of its aims, the interweaving of conservation (in its broadest sense) into the fabric of government.

2 **Integrated land use management**. A policy of land use coordination can be achieved through a national planning policy supporting a sustainable land use plan. Integral to the planning policy should be a national conservation strategy (NCS) providing a broad national environmental management plan. The exploitative colonial-style government whereby industrialised northern hemisphere countries took primary resources from low-latitude, undeveloped nations has been largely replaced by ideas of sustainable development, although the actions of some multinational companies show some similarities to former colonial practices. A national conservation strategy can assist governments in the preparation of an integrated land use plan by setting a sustainable level of development based on the long-term availability of biological resources.

3 **Species protection.** Individual plant and animal species make up complex ecosystems. Ecologists associate a strong and vigorous ecosystem with a maximum species number and diversity. A wide range of long-established legislation exists to protect species at both national and international level (e.g. International Convention on Whaling, 1946; Convention on Conservation of Migratory Species of Wild Animals, 1979). While species-specific action has often been necessary to safeguard the existence of species severely exploited in the past, species are best conserved as part of a programme of total ecosystem protection.

4 **Habitat protection.** Examples of specially protected habitats date from very early in human history. In the Arabian peninsula, traditional forms of resource reserves known as *hema* had their origins over 2000 years ago in the pre-Islamic period and may have been developed in response to the scarcity of renewable resources in the difficult natural environment that prevailed throughout Arabia (Draz, 1985). In recent times most national governments have provided legal means for protecting habitats that support large numbers of species or unique biological resources. The IUCN has designated five categories of conservation (see Box 8.3). Soulé and Wilcox (1986) have shown that by themselves, categories I, II and III in the IUCN list will not be able to conserve all species, genetic resources and ecological processes. It now appears certain that in order to retain existing levels of biodiversity far greater areas of land must be devoted to conservation than modern societies appear willing to consider. McNeely *et al.* (1990) have argued that one solution would be the designation of different types of protected areas that allow compatible human use of

---

BOX 8.3

IUCN categories of protected land

IUCN protected areas management categories
from 1993 United Nations List of National Parks and Protected Areas

I. *Strict nature reserve/scientific reserve.* To protect nature and maintain natural processes in an undisturbed state in order to have ecologically representative examples of the natural environment available for scientific study, environmental monitoring, education and for the maintenance of genetic resources in a dynamic and evolutionary state.

II. *National park.* To protect outstanding natural and scenic areas of national or international significance for scientific, educational and recreational use. These are relatively large natural areas not materially altered by human activity where extractive resource uses are not allowed.

III. *Natural monument/natural landmark.* To protect and preserve nationally significant natural features because of their special interest or unique characteristics. These are relatively small areas focused on protection of specific features.

IV. *Managed nature reserve/wildlife sanctuary.* To assure the natural conditions necessary to protect nationally significant species, groups of species, biotic communities or physical features of the environment where these may require specific human manipulation for their perpetuation. Controlled harvesting of some resources can be permitted.

V. *Protected landscapes and seascapes.* To maintain nationally significant natural landscapes which are characteristic of the harmonious interaction of man and land while providing opportunities for public enjoyment through recreation and tourism within the normal lifestyle and economic activity of these areas. These are mixed cultural/natural landscapes of high scenic value where traditional land uses are maintained.

natural resources. However, even when the new and improved knowledge of habitat management is taken into account, the level of use will inevitably have to be set at far lower levels than currently adopted if species and their habitats are to be managed at a sustainable level (Pressey, 1996).

5 *Ex situ* conservation. Where it is impossible to designate sufficient areas of conservation land, resort may have to be made to offsite (*ex situ*) conservation facilities (see Box 8.4). The main examples include: botanic gardens, game parks, zoos, seed and gene banks. Other less-managed examples include: hedgerows, wood lots, waste and derelict ground. Most formal *ex situ* facilities are involved with the protection of wild species that are of known economic value.

6 **Measures to protect the biosphere.** Kennet (1974) claims that the first conservation laws to be democratically made were a series drawn up in the first half of the seventeenth century that controlled the capture of turtles from the shores of Bermuda. Modern legislative action intended to prevent environmental damage to the environment by humans, and all other life forms, is considered to have taken place in 1872 when the US Congress passed an Act to set apart a 'certain Tract of Land lying near the Headwaters of the Yellowstone River as a public Park'. Most of the earliest legislation linked conservation laws with species or habitats that had economic or symbolic value to society (O'Connell, 1996). The British government passed its first Clean Air Act in 1956 and was a legislative milestone in establishing cleaner air in the major UK cities. Since that time almost every country has enacted a host of legislation to ensure that we

---

BOX 8.4

Ex situ *conservation programmes*

For a small number of highly threatened species, relying on natural reproduction to propagate the species is no longer a viable means of assuring the future of the species. Small numbers of scattered, aging or inbred individuals cannot be relied on to renew the species and in these cases it may be necessary to turn to so-called 'offsite' or *ex situ* methods of reproduction. In this approach, 'captive' breeding methods can be used, in which suitable breeding partners are brought together in locations such as zoos, game parks and botanic gardens. The success of captive breeding programmes is highly varied, some species such as pandas have a very low success rate whereas Arabian oryx reproduction has been good. Conway (1988) claims that over 3000 vertebrate species have been bred in zoos with over half a million individuals raised in this way. By far the greatest effort at *ex situ* conservation has involved crop plants (food resources) and some 500 species have been involved in breeding programmes (McNeely *et al.*, 1990).

All captive breeding programmes involve the eventual release of the progeny back into the wild and it is vital that a suitable habitat exists to receive and sustain the species. It is obvious, therefore, that captive breeding programmes form only a small part of a total conservation strategy in which the retention of a suitable habitat forms the most essential component. If sufficient good-quality habitats exist and within which hunting, poaching and collection of vulnerable species is strictly controlled, then *ex situ* programmes should only be necessary for the most vulnerable of species.

The future of *ex situ* species conservation currently faces its biggest ethical challenge. The ability of scientists to identify the genetic code of all plants and animals combined with the major advances that have occurred in cloning of species means that it is possible to replicate a population from one single male individual. Even the limited genetic resource that such a procedure would represent can be minimised by the transfer of genetic material from closely related varieties. But the implications this has for the future evolution of life on Earth are immense! Humans now have the scientific ability to control evolution.

---

receive safe supplies of drinking water, clean air to breathe and safe levels of chemical substances in our food. Incidental to these human safeguards, many plants and animals have also benefited from the improved environmental conditions. In the latter part of the twentieth century it became apparent that many environmental problems could only be solved by international collaboration and thus there began the trend towards teams of scientists working in tandem with world governments to set safe standards and limits for a wide range of environmental problems. The Convention for the Protection of the Ozone Layer (Vienna, 1985), the Montreal Protocol on Substances that Deplete the Ozone Layer (Montreal, 1987) and the Climate Conference of Parties (Kyoto, 1997) have all contributed in major ways to ensuring a safer biosphere for all life forms. Regularly scheduled meetings of groups such as the International Conference on Climate Change ensure that environmental issues are foremost in the minds of politicians.

8.12 Role of eco-tourism in retaining biodiversity

Great hopes have been pinned on eco-tourism as a means of providing a financial resource to help support conservation of natural resources especially in the poorer countries of the world (World Resources Institute, 2000). Eco-tourism, also known as 'safari' or 'green' tourism comprises 'travel to natural areas that conserves the environment and sustains the well-being of local people' (http://www.ecotourism.org). Its spectacular growth since the 1980s has been a mixed blessing, offering economic opportunities for some of the poorest peoples of the world but also bringing exploitation and environmental degradation. In countries such as Kenya, Belize, Ecuador and Costa Rica, eco-tourism has provided a major boost for their income, but in many other instances, eco-tourist income is a 'paper' figure only. Most of the value of an eco-trip remains in the country from which the tourists originate, for example international

air fares, some hotel charges and trek leader and administration costs may account for 90 per cent of the total cost and usually do not contribute to the economy of the country in which eco-tourism occurs. The income derived from hiring local guides and porters, local transport and entrance fees to national parks is all that remains in the host country (see Plate 8.3). By western standards the level of remuneration for local services is minimal.

Most countries now impose a departure tax on tourists leaving the country but the revenue raised from this tax is not specifically allocated for environmental support. One interesting development has been the decision of the regional government of the Balearic Islands to introduce an additional eco-tax directed specifically on visitors (see Box 3.1 for details). Both the tourist industry and conservation movement will anxiously watch the effect this tax has on visitor numbers. If its impact on visitor numbers is small then we might expect similar taxation to be applied to tourists visiting countries such as Kenya, New Zealand and Belize all of which rely greatly on the quality of the environment to 'sell' tourism. Already, tourists leaving Belize pay a departure tax that amounts to an income of about three-quarters of a million US dollars a year (Sweeting et al., 1999). Generating an equivalent amount from an eco-tax imposed on tourists would radically change the financial structure of conservation and land rehabilitation in the country (http://www.belizetourism.org/revenues.html).

At present it is usually impossible to calculate how much tax revenue is used to support conservation projects in a country due to the unwillingness of finance ministers to make use of 'hypothecation' of funds for specific purposes. In Belize, despite the importance of eco-tourism, no government minister is in sole charge of conservation. Funding takes place mainly through a contorted link involving the Minister of Resources and the Forestry Department. Other eco-tourist developments are funded through the Minister of Agriculture and Fisheries and from the Fisheries Department, while ancient monuments and antiquities are funded through the Minister of Tourism and Environment via the Department of Archaeology. It comes as a disappointment to find that

Plate 8.3 Practical eco-tourism in the Ecuadorian rainforest. An indigenous guide explains the potential medicinal uses of rainforest species to eco-tourists

proactive eco-tourism in Belize is in the hands not of the government but of private individuals and organisations. The results of the excellent conservation work published in the Rio Bravo Conservation and Management Area (RBCMA) (Programme for Belize, 1996) provide no indication of how funding is to be forthcoming. The equally useful Directory of Belizian Protected Areas and Sites of Nature Conservation Interest (NARMAP, 1996) is slightly more revealing, acknowledging 'support' from the Belize government, the US Agency for International Development and technical assistance from the Worldwide Fund for Nature and Winrock International for Agricultural Development.

The eco-tourism and safari holidays industry has been most successfully developed on the African continent especially in Kenya and South Africa where very large areas of land have been given over to a network of game reserves. Countries such as

Botswana, Namibia and Rwanda are also intent on developing smaller, specialist eco-tourist industries. In South Africa, an extensive network of government-organised national parks and game reserves cater for the eco-tourist and also provide employment for wardens, trackers, veterinarians, drivers, mechanics and hotel and camp staff. In Kenya, as well as government-run national parks, private ventures have been developed that provide high-quality eco-tourist resorts that generate sufficient revenue to finance conservation projects. One example, the Sarara Lodge located in the Namunyak-Mathews mountain range, supports large populations of elephants, buffalo, rhino, lion, leopard, cheetah and kudu. The objective of the organisation is to provide high-quality, low-impact tourism that generates significant income for the local Samburu community and encourages the local population to protect the indigenous plant and animal life so that they may continue with their traditional lifestyle.

In northern Botswana, the Khwai Development Trust (KDT) provides employment for about 400 people of the Babukakhwae or 'river bushmen' ethnic group. Located just outside the Moremi Game Reserve, the KDT is working to build revenue from eco-tourism that in turn will support sustainable development of the traditional Babukakhwae lifestyle. Other organisations operate at a trans-African level, for example the 'Tusk Trust' that generates funds outside Africa for support of threatened species in a number of African countries. In a similar manner, the African Conservation Foundation (ACF) (http://africanconservation.com) provides guidance and financial support for conservation projects within Africa.

Many of the achievements made between conservation and eco-tourism have been shown to be exceedingly fragile, especially when local unrest in countries that are often politically unstable disturbs the security of the tourist industry. The bombing of the Sri Lanka International Airport in Colombo on 23 July 2001 by Tamil Tiger guerrillas destroyed a booming tourist industry worth US$224 million a year. Sri Lanka had been designated a 'green globe' destination in 1999, a travel trade award for eco-tourism but the unrest destroyed the many years of work that had justified the award. In a similar vein, the civil war in Rwanda in 1994 between Tutsis and Hutu tribes eliminated a tourist revenue in that country of US$1.02 million. In this instance, tourist revenue had been used to train and support the local population to manage a highly successful mountain gorilla-breeding programme in Volcanoes National Park (Gossling, 1999). More than a decade of support for an endangered species was lost in the space of a few months and, worse, the gorilla population was decimated as it became used as a much-prized source of 'bush meat'.

8.13 Can science help maintain biodiversity?

Modern society takes for granted the benefits of an environment that is free of contagious diseases, has low levels of all forms of pollution, provides access to a proficient medical care system, a nutritious and reliable food supply, and a clean, diverse and stimulating environment. In our quest to provide a 'safe' environment in which to live, we have undoubtedly eliminated many plants and, especially, animals that were considered hazardous to our well-being. Other species have been eliminated, not because of a deliberate extermination policy on behalf of humankind, but due to subtle and often unintentional changes, such as disturbance of food webs or destruction of breeding territory. As the World Conservation Strategy made clear, such has been the level of disturbance caused by human development that non-human species and populations must be provided with help in order to retain their capacity for self-renewal. It is the role of biological and environmental scientists to establish *how* we can create conditions suitable for this self-renewal process.

However, before scientists can help find ways to re-establish threatened species we need to know which species are in decline as well as those that may be increasing. Only by means of careful monitoring of species type and number over time can we be certain as to which species need our attention. The monitoring of species, and the habitats in which they live, involves ecologists skilled in locating and identifying species that are frequently microscopic in size. Data from many thousands of sites must be collated and stored in archive data banks so that statistically significant trends in species numbers can be detected. Two international groups are responsible for the collection of data on biodiversity. The main international database on threatened species is organised by the World Conservation Monitoring Centre (WCMC), a partnership set up in the mid-1960s between the International Union for the Conservation of Nature (IUCN), the United Nations Environment Programme (UNEP) and the Worldwide Fund for Nature (WWF) with a headquarters in Cambridge, England. To date, information has been collected on what are considered to be the most threatened species, amounting to about 60,000 plants and 2000 animals worldwide. Information is analysed according to geographical area and main species group and is published in the form of 'red data books'. (The comprehensive IUCN website provides full details: http://www.iucn.org/themes/ssc/redlists/categor.htm.) McNeely *et al.* (1990) suggest that the data in the red data books should be taken as indicating only a part of the problem and in reality the situation regarding threatened species is likely to be far worse than that shown by the data. Despite this drawback, the red data books allow scientists to determine a management strategy towards those species most in need of attention.

The second organisation, the Convention on International Trade in Endangered Species of Wild Fauna and Flora (CITIES) monitors the movement and trade of rare species (except for fish and whales). (See the CITES website for details of the work of the organisation: http://www.cites.org/.)

The monitoring of species over time is an expensive process and is sometimes difficult to justify on a cost-benefit basis. Data becomes more valuable when a continuous long run of information exists so that it can be used to predict changes in number and forecast future trends (Fitter and Fitter, 1987). Ideally, a data set extending over at least 30 years is required before sound statistical information can be drawn from it. In addition, data from as many different sampling areas as possible is required before a regional picture can be formed of the changes in fortune of species.

Threatened species are placed into one of eight main categories, a summary of which is as follows:

1 EXTINCT (EX). A taxon is extinct when there is no reasonable doubt that the last individual has died.
2 EXTINCT IN THE WILD (EW). A taxon is extinct in the wild when it is known to survive only in cultivation, in captivity or as a naturalised population (or populations) well outside the past range.
3 CRITICALLY ENDANGERED (CR). A taxon is critically endangered when it faces an extremely high risk of extinction in the wild in the immediate future.
4 ENDANGERED (EN). A taxon is endangered when it is not critically endangered but is facing a very high risk of extinction in the wild in the near future.

5 VULNERABLE (VU). A taxon is vulnerable when it is not critically endangered or endangered but is facing a high risk of extinction in the wild in the medium-term future.
6 LOWER RISK (LR). A taxon is lower risk when it does not satisfy the criteria for any of the categories critically endangered, endangered or vulnerable. Taxa included in the lower risk category can be separated into three subcategories:
   conservation dependent (CD). Taxa which are the focus of a continuing taxon-specific or habitat-specific conservation programme targeted towards the taxon in question, the cessation of which would result in the taxon qualifying for one of the threatened categories within a period of five years.
   Near threatened (NT). Taxa which do not qualify for conservation dependent, but which are close to qualifying for vulnerable.
   Least concern (LC). Taxa which do not qualify for conservation dependent or near threatened.
7 DATA DEFICIENT (DD). A taxon is data deficient when there is inadequate information to make a direct, or indirect, assessment of its risk of extinction based on its distribution and/or population status. A taxon in this category may be well studied and its biology well known, but appropriate data on abundance and/or distribution is lacking. Data deficient is therefore not a category of threat or lower risk. Listing of taxa in this category indicates that more information is required and acknowledges the possibility that future research will show that threatened classification is appropriate.
8 NOT EVALUATED (NE). A taxon is not evaluated when it is has not yet been assessed against the criteria.

The IUCN has set up a special Species Survival Unit (SSC) to study the decline in species numbers. Table 8.1 provides a summary of the species in each threat category and Figure 8.6 shows the geographical areas under greatest cumulative threat

Table 8.1 Summary statistics for globally threatened species, 2000

	EX	EW	CR	EN	VU	LRcd	LRnt	DD	Total
Mammals	83	4	180	340	610	74	602	240	2,133
Birds	128	3	131	182	680	3	727	79	2,123
Reptiles	21	1	56	79	161	3	74	59	454
Amphibians	5	0	25	38	83	2	25	53	231
Invertebrates	375	14	326	431	1928	31	321	608	3,277
Fish	80	11	156	143	450	16	131	168	1,250
Plants	73	17	1014	1266	3331	244	707	370	7,022
Totals	764	50	1888	2479	7223	373	2587	1577	16,490

(*Source*: based on data from IUCN Species Survival Commission, 2000)

of species loss. The size of each divided circle is proportional to the number of species under threat within that area. No account is taken of the size or variety of habitats. A small circle in Antarctica representing ten threatened species is almost invisible.

8.14 Debt-for-nature swaps

Many European countries still carry a legacy of environmental damage caused by exploitation of natural resources from the early decades of the Industrial Revolution. The impact of industrialisation extended well beyond the boundaries of the rapidly growing towns and cities, mainly through the effects of pollution of land, water and air. Pollution often reached levels that were damaging to both human health and to plants and animals. Industrialisation and urban development in Europe and North America took precedence over nature conservation until about mid-way through the twentieth century. The extent of the species and habitat losses during this time is difficult to establish as the losses occurred before extensive scientific recording of the habitats was possible (Evans, 1992: Chapter 2).

As industrialisation of national economies gradually moved away from the old developed countries to the new developing nations, so the threat of habitat disturbance and species loss has been transferred to those countries that had, until the late twentieth century, retained maximum levels of biodiversity. Deforestation, uncontrolled access to minerals, burning of rainforest and rapid growth of towns and cities are the universal hallmark of 'development' and are still responsible for habitat and biodiversity loss.

In an effort to ensure that developing countries do not repeat these mistakes, countries retaining high levels of biodiversity are being encouraged to consider new ways of retaining their remaining natural habitat. For countries such as Indonesia, selling the rights to clear away rainforest generates vital hard currency from First World logging companies as well as providing a source of land that is quickly colonised by land-starved farmers. Such actions can only be considered to provide a short-term expediency, but in reality the need to repay the interest on massive foreign debts (amounting to a total of $1.3 trillion for all the developing nations by the early 1990s (McKinney and Schoch, 1998)) prevents almost every developing nation from taking a long-term view of its natural resources. Forests are cleared to make way for unsustainable agricultural systems that are often designed to produce crops for export to mid-latitude markets rather than to produce food for local consumers. The sale of tropical timber on the international market is often the easiest way for a country to obtain foreign currency and the consequences of exploitive deforestation on habitat and species loss is given little consideration.

In an attempt to disentangle the entrapment of developing countries from a financial spiral created and encouraged by massive loans provided

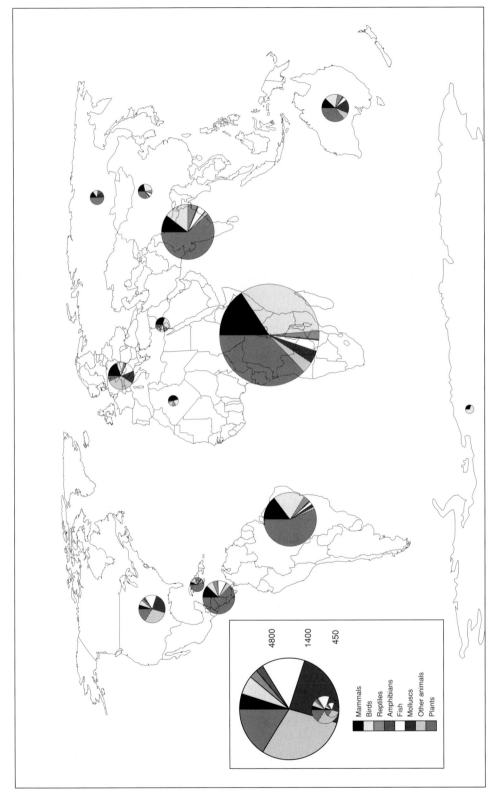

Figure 8.6 Global distribution of threatened species, 2000
(*Source:* based on data from IUCN Species Survival Commission, 2000)

by the World Bank, some of the old industrialised governments have searched for new ways of providing sustainable economic development for the less developed nations of the world. If financial aid from the west could be linked to a locally based organisation, responsible for sustainable eco-development, it might enable the use of natural resources based on traditional indigenous knowledge systems (IKS), for example the indigenous soil and water conservation (ISWC) practices described by Critchley *et al.* (1994). Instead of cutting down a forest on a once-for-all profit basis, local people could be given financial assistance to bridge the gap between the immediate loss in income and the time it takes to develop a sustainable forest management programme. An example of sustainable forest use comes from an area of Indonesia. Cutting the forest and selling the timber yielded $3600 per hectare. Calculating the sustainable yield (i.e. not cutting the trees down) yielded $4800 per hectare. In addition, the forest remains intact and generates 'invisible income' calculated to contribute 70 per cent of the income for the local population (Bryant *et al.*, 1997).

Schemes that provide financial assistance in which the safeguarding of natural resources has been clearly identified and managed are known as 'debt-for-nature' swaps (see Box 8.5). Under this

---

BOX 8.5

Debt-for-nature 'swaps'

For the governments of the Third World countries, environmental protection is not always a top priority: in order to promote social and economic development and raise currency for the external debt, the forests are logged and mining, tourism, agriculture and timber projects proceed without regard to their natural sustainability.

The history of the debt-for-nature swaps

When in the early 1980s several highly indebted Latin American countries indicated they could no longer make payments on their foreign debts, a secondary market for Third World debt obligations was established. Some banks heavily engaged in this business, preferred selling these risky debt notes at a discount (Schmid-Schonbein, 1992). This discount could reach up to 85% of the face value in extreme cases and enabled the acquisition of high nominal values at a very low price. Soon, a debt exchange programme was developed to assist the borrower nations to reduce their burden. In 'debt-for-equity swaps', the creditors were offered bonds or local currency to invest in domestic, industrial or commercial projects in lieu of repayment in hard currency. This idea of offering other assets in exchange for the debt inspired environmentalists to propose debt-for-nature swaps, in which Third World countries offer to conserve their natural resources in exchange for a reduction of debt (Fuller, 1988).

Many participants are involved in a debt-for-nature swap: for example, an international environmental organisation, in most cases a non-governmental (NGO) like the Worldwide Fund for Nature (WWF), an international bank, the governmental agencies of the country concerned, its central bank and a local environmental organisation responsible for the conservation project funded. The involvement of so many participants does not simplify negotiation of the debt-for-nature swap.

Debt-for-nature swaps were successfully put into operation in Bolivia, Costa Rica, Ecuador and Mexico, as well as in several African and Asian states. In 1989 the Dutch government swapped $33 million in debt for £10 million worth of forestry and soil conservation programmes and water resource management in Costa Rica. In the 1990s, a new field for nature swaps opened up in eastern European countries and in 1991 the USA agreed to cancel 10% of the Polish debt if money was spent on clean-up programmes for toxic wastes associated with old industrial sites. In total it is estimated that swaps managed to cover about US$100 million of debt and generated nearly US$60 million in funds for conservation.

---

scheme debt-ridden countries set a monetary value on their ecological capital (usually their forest resources) and use these as collateral for continued international trading.

Unfortunately, debt-for-nature agreements have not been a successful method for ensuring the management of natural resources and at the same time assuring an improved set of living conditions for the local population. A number of defects in debt-for-nature agreements have been identified:

- There has been a mismatch in timescales. Debt-for-nature agreements have been used as short-term solutions to a very long-term problem.
- The valuation of natural resources has presented difficulties, leading to an undervaluation of resources and, in turn, an underpayment to the participating developing countries.
- Debt-for-nature deals compromise national sovereignty and do little to reduce the total debt burden or to change the causes by which debts are incurred.
- Policing the agreements has been difficult. For example, gangs of poachers have raided protected wildlife reserves especially for elephant tusks (ivory) and illegal logging of tropical hardwood forests is commonplace.
- Most of all, it remains difficult to convince politicians and the population of the developed world that the maintenance of their lifestyle can be assured by making payment for *not* developing a resource in developing countries. This is contrary to all rationale we have grown accustomed to in the developed world. Our new environmental awareness informs us that the stability of the biosphere can be assisted through events such as the retention of the remaining areas of tropical forests for the purpose of absorbing CO_2 and helping to prevent global warming. Retaining natural ecosystems helps maximise biodiversity and by so doing increases the possible chance of discovering new and useful genetic material or medicinal substances. We have assumed many of these features as free 'God-given'

resources but, as stated by Nicholson (1970): 'Man has made it inevitable, not only that he should manage nature, but also henceforth learn to manage himself as part of nature.'

Perhaps the time has now arrived to accept the need to pay for the retention of items such as clean air, wilderness areas and species diversity. Payment to countries that still possess these natural resources should not been seen as *aid* but as an essential insurance premium, as necessary a part of our contemporary lifestyle as personal or property insurance.

8.15 Establishing a priority for conservation of species

Reference to Figure 8.6 shows that the geographical areas where species are under greatest threat are sub-Saharan Africa, south and southeast Asia, and South America. To provide a context in which to interpret this pattern it is necessary to take account of the following:

- Low-latitude regions of the world have a greater species richness (a greater number of species per unit of living space) than mid- and high-latitude countries. For example, tropical rainforest covers only 6 per cent of the land surface yet contains an estimated 70 per cent of all non-microbial land species. This is thought to be due to the relative abundance of heat and moisture in the current climatic regime and enables a profusion of life forms to exist. In addition, low latitudes escaped the most recent glaciation that in northern and mid-latitudes was responsible for wiping out many species and disrupting the landforms. Furley and Newey (1983: 166–7) provide examples of the number of species occurring in different geographical areas. For example, the number of ant species declines from 222 species in Brazil to three in the Arctic and the number of bird species declines from 1500 in Ecuador to 56 in Greenland. Table 8.2, taken from Jarvis (2000: 191), provides information on

Table 8.2 Species richness for trees, mammals, amphibians and reptiles in North America

	Trees	Mammals	Amphibians	Reptiles
North	10	20	0	0
Central	40	40	10	10
South	100	50	40	80

(*Source*: data from Jarvis, 2000)

the number of tree, mammal, amphibian and reptile species that occur throughout North America. The number of bird species increases from 30 in northern Canada to 660 in Central America.

- In the event of human activity taking place in low latitudes the *potential* for biodiversity loss is greater than elsewhere because of the impact of human interference on the profusion of species.
- Compared to northern hemisphere mid-latitude countries, development in low-latitude countries has been comparatively recent, occurring mainly since political independence was achieved in the mid-1960s. Modern development has taken advantage of new, highly efficient technology that has allowed land to be cleared in a fraction of the time it took previously.
- A substantial but largely unknown reduction in the number of species has already taken place in northern hemisphere mid-latitudes. In Europe, these losses occurred in two main phases: from the Middle Ages onwards when deforestation accelerated and during the Industrial Revolution when pollution reached extremely high levels. Compared with low-latitude countries, only small numbers of indigenous species remain and many of these have been given 'protected' status. In Europe, considerable effort has been made to minimise further reduction in biodiversity by establishing a representative network of conservation areas under the *Natura 2000* programme (see Box 8.6).

Unfortunately, many low-latitude countries are among the poorest in the world and record some of the fastest human population growth rates. The World Resources Institute (2000) provides poverty indices for some of the developing nations. In 1992 the proportion of the population in Gambia classified as 'in poverty' was 64 per cent; the corresponding figure for Haiti in 1987 was 65 per cent. Diverting scarce economic resources into conservation management has not been a priority for governments of countries experiencing such poverty. Remarkably, at the United Nations Conference on Environment and Development held in Rio de Janeiro in 1992, low-latitude countries showed an enthusiasm and a willingness to work towards sustainable conservation management that put many northern industrialised countries to shame. Countries such as Costa Rica, Ecuador, Kenya and the Maldives recognise the value to their economies that a sound conservation management policy can have on the development of sustainable eco-tourism. The highest quality environments containing pristine natural habitats and a maximum number of species are able to attract affluent eco-tourists. Entry fees to nature reserves can help support the management of statutory protected areas.

A clear conflict arises, however, between the designation, for example of National Parks which have as a priority the attraction of tourists and areas that can be termed ecologically sensitive areas (ESAs) from which tourists are discouraged from visiting. In terms of benefit to threatened species, ESAs are of greater value as they provide protected habitats for threatened species yet offer little in the way of economic return to the economy of a nation. Conversely, national parks can generate a substantial tourist income but may be unable to contribute much to scientific conservation of species. The existence of a specific natural landform or a species that attracts human interest can also dictate a nation's approach to conservation. In Belize, the existence of the world's second longest offshore reef system figures prominently in that country's tourist literature, while the existence of the Giant Panda (China), Asian tiger (India) and African elephant (Kenya) have all dictated the conservation programmes of the countries concerned.

BOX 8.6

The EU Natura 2000 scheme

The European Union (EU) took its first steps towards habitat and species conservation in 1979 with the Birds Directive (Directive 79/409/EEC) which required Member States to designate special protection areas (SPAs) specifically for rare and migratory birds. Following the signing of the Convention on Biological Diversity at the Earth Summit in Rio de Janeiro by all EU Member States in 1992, legislation was quickly adopted for the conservation of natural habitats for wild flora and fauna. This legislation, called the Habitats Directive (Directive 92/43/EEC), forms the backbone of protected sites throughout EU nations. The Habitats Directive subdivides Europe into five biogeographical regions: continental, Mediterranean, alpine, macaronesian (comprising Madeira, the Canaries and the Azores) and Atlantic. A total of 168 distinct natural habitat types have been identified from within the five regions and national governments are required to identify representatives of all types occurring within national boundaries. The designated sites make up the **Natura 2000** network of protected sites. These comprise special areas of conservation (SACs) and also incorporate the SPA sites. The Habitats Directive also requires Member States to establish an effective system to prevent the capture, killing, injuring or damaging disturbance of specific endangered species.

The intention of the Habitats Directive has been to 'maintain or restore the extent and quality of rare habitat types and to ensure that rare species can survive and maintain their populations and natural range on a long-term basis and taking account of economic, social, cultural and regional requirements' (DoE, 1995). Each Member State has been required to compile a list containing the areas of habitat types and species in the Directive. The Commission will then use the lists to designate a European network of SACs by 2004. Significantly, the Directive requires marine SACs to be designated as well as land sites. A total of 75 habitat types occur in Britain and of these 22 have been given *priority* status due to the threat of disappearance.

Within the 168 habitats requiring protection, 632 animal and plants have been identified 'at risk' with some again receiving *priority* status. A total of 40 listed species will be included from Great Britain. In addition to Britain's contribution of sites and species to the Habitats Directive, the UK established (in 1994) a biodiversity action plan (UK government Command No. 2428, 1994) designed specifically to assist with the conservation of species and habitats over the following 20 years.

8.16 Conclusion

The diversity of habitats that remain on this planet means that it is impossible to recommend a simple, 'one-size-fits-all' policy to prevent further loss of biodiversity. Different countries will experience different social and economic conditions that make some options to retain biodiversity impossible to implement. An internationally accepted 'best policy' guideline would enable all countries to work towards the highest possible standard of biodiversity, as follows:

1   All countries should be encouraged to identify habitats of local, national and international conservation importance and provide appropriate levels of management to ensure their long-term survival.

2   Countries should recognise their obligation to the world community and maintain the maximum possible number of habitats in pristine condition. By this means maximum biodiversity can be achieved and the future health of the biosphere assured.

3   In recognition that land and organic resources have an economic value, each country should establish a means of using their conserved areas in a sustainable manner. This could involve a mix of eco-tourist developments, sustainable cutting, collecting and prospecting for biomaterials and designation of environmentally sensitive areas in which all exploitation was banned.

4 Harvesting biomaterials by multinational companies (such as logging, food, pharmaceutical and medical companies) should incur a financial surcharge, used to pay the local population who would become the custodians of natural resources in their area.

5 'Clear felling' of forests or 'once-and-for-all "mining" of organic resources' should be prevented at all costs. Interest-free loans from the international community should be made available to help the poorest nations achieve this objective.

6 A proportion of patent rights and royalties derived from commercially valuable biomaterials should be returned to the country from which the base material was obtained.

Many impediments currently exist to prevent such an idealistic set of guidelines from being implemented. The poverty trap into which many of the developing countries fall has already been referred to. Others include: undervaluing unique organic resources; ignorance of the biological systems in tropical environments; the inertia of exploitative development inherited from a previous economic era; economic and political dominance of northern hemisphere developed countries; economic power of multinational companies; intertribal conflicts in developing countries and corruption of key personnel responsible for selling off organic resources. Implementing an ecologically sound and sustainable management policy for the organic resources that survive on the planet would represent a clear indication that humankind had finally recognised its responsibility to itself and to all other species.

Useful websites for this chapter

African Conservation Foundation (ACF)
http://africanconservation.com

Association for Biodiversity Information
http://www.abi.org

Biodiversity: an overview
http://www.wcmc.org.uk/infoserv/biogen/biogen.html

Biodiversity and world map
http://www.nhm.ac.uk/science/projects/worldmap
(an older site but provides a good introduction to the topic)

CITES
http://www.cites.org/

Convention on Biological Diversity
http://www.biodiv.org
(this site has a good Biodiversity Search Engine)

Conservation International
http://www.conservation.org
http://www.conservation.org/xp/CIWEB/home

PRESSRELConservation of Arctic Flora and Fauna
http://www.grida.no/caff/

Convention on International Trade in Endangered Species of Wild Fauna and Flora
http://www.wcmc.org.uk/CITES/

Eco-tourism Information Centre
http://www.life.csu.edu.au/ecotour/EcoTrHme.html

Eco-tourism Society
http://www.ecotourism.org

Ecology and biodiversity
http://conbio.rice.edu/vl/

Ecotravel centre
http://www.ecotour.org

International Eco-tourism Society
http://www.ecotourism.org/textfiles/stats.text

IUCN red list categories
http://www.iucn.org/themes/ssc/redlists/categor.htm

IUCN Species Survival Commission
http://www.redlist.org/info/tables.html

Large Carnivore Initiative for Europe (LCIE)
http://large-carnivores-lcie.org/wolf.htm

National Biodiversity Network
http://www.nbn.org.uk

National biodiversity profiles
http://www.wcmc.org.uk/nbp/

Pesticide information profiles
http://ace.ace.orst.edu/info/extoxnet/pips/MCPA.htm

Species Survival Commission
http://www.rbgkew.org.uk/conservation/cpdu/ssc.html

UK Biodiversity
http://ibs.uel.ac.uk/ibs/other/ukbiodiv/ukbiodiv.htm

Worldwatch Institute
http://www.enn.com/features/1998/09/091698/fea0916.asp

9 Looking forward to the next 100 years

As our knowledge of environmental devastation and the global consequences of this devastation grow, so do questions about how we ought to think about and act towards the natural world.

Gruen, 1997

One of the most momentous achievements of the twentieth century, space travel, has allowed humankind to gain a new perspective of our planet. Set in the enormity of space, the finite nature of the speck of dust and drop of water that is our home, has finally become visible to the 6 billion or so human inhabitants of planet Earth. For the first time we are able to recognise the true dimensions of our planet and the finite resources it contains. The consequences of exhausting the resources or of polluting the atmosphere, oceans and land suddenly become all too obvious. Viewed from the perspective of space, our planet *is* insignificant, but as the opening quotation to Chapter 2 recognised, *it is all we have*. This final chapter attempts to bring together the separate strands, to take stock of the current state of the environment and to identify some of the environmental challenges that may face us in the twenty-first century.

9.1 Putting the environment into context

Promising signs emerged in the 1990s that we were beginning to recognise the necessity of managing our planet as a *global resource*. Environmental science – for so many years considered a 'soft' science by the establishment – finally began to pro-duce the hard evidence that showed environmental deterioration really was taking place as a result of human actions. Global warming, critical shortages of fresh water, loss of biodiversity and overfishing of the oceans were just a few of the problems that assumed serious proportions during the 1990s. Spurred on initially by green pressure groups, many of the established political parties began to incorporate 'green issues' in their manifestos. World political leaders began to respond to the advice being given by the scientific community that unless changes were made in the management of the resources of the planet there would be very serious repercussions for the continued economic and social stability of the entire planet.

In the UK, government set out the *scientific* basis for policy in response to a report on Environmental Research Programmes prepared by the then Advisory Council on Science and Technology (ACOST) in 1992 which stated that:

> Sound science should underpin all environmental policy. The environment cannot speak for itself and we require a clear understanding of its present and future condition to guide its stewardship. Research to improve our understanding for future action is still one of the best precautionary measures.

Also in 1992 the United Nations Conference on Environment and Development (UNCED) held in Rio de Janeiro marked a major international step forward in the way in which scientific knowledge and political willingness to respond to that knowledge came together. In retrospect, the long-term achievements of the Rio Conference have been disappointing, but at least it stimulated a global

Biodiversity Action Plan and, of even greater significance, led to the introduction of local plans that considered regional environmental issues.

Many of the environmental problems we now face are the result of deliberate attempts made by our ancestors to alter the natural environment. These changes were almost always made with good intent, but because we knew insufficient about how the environment and all its component parts were interrelated, our technology was insufficient to cope with the adverse changes that our use of the biosphere caused. It is also the case that in the past, society placed less emphasis on the quality of the natural environment than it does today. Concern over issues such as pollution and of the well-being of non-human species simply did not register with the majority of people in the rapidly industrialising world.

Two reasons dictated the way in which we responded to environmental problems in the past: ignorance of the way in which the environment worked and the lack of appropriate technology to detect and repair degraded environments. The development of both ecological and environmental knowledge during the twentieth century has largely eliminated our ignorance of the natural world, although there are undoubtedly many areas where we still only possess rudimentary knowledge both of the components and the interrelationships between components of the biosphere. The second problem, a lack of appropriate technology, has undergone a radical change in the late twentieth century. In theory, our society is now in a better position to interact intelligently with the environment and, by so doing, to make neutral or only slightly negative impact on its components.

In addition to better knowledge and more appropriate technology, there is also a willingness to set environmental standards to which society must comply. The standards are implemented by means of legal statute. The setting of environmental standards for water pollution provides a clear example of how scientific knowledge, technology and legislation have worked together to achieve a safer management of disposal of polluted water and for the supply of fresh water. Although the link between illness and disease in humans and the consumption of water polluted with sewage would not be understood until the late eighteenth century, the unhygienic practice of disposing sewage onto the streets of sixteenth-century England led to the earliest attempt to legislate against environmental pollution. The Bill of Sewers 1531 was one of the earliest attempts to protect public health by making it the duty of people responsible for sewers to 'cleanse and purge the trenches, sewers and ditches'. Much later, the River Pollution Prevention Act 1867 made it a criminal offence to pollute water with solid matter, sewerage or material from manufacturing and mining, while the River Boards Act 1948 established the river boards as the regulators of water pollution. Under the Water Resources Act 1963, the river boards became river authorities, then regional water authorities (Water Act 1973) and, eventually, the National Rivers Authority (Water Act 1989).

The sequence of legislation relating to the control of water pollution illustrates how the increasing complexity of pollutants released by industry and the increasing demands for fresh water by a rapidly growing population required a constant updating of legislation to ensure that a satisfactory state of public environmental health was maintained. To ensure that our elected politicians continue to update environmental legislation requires continuous monitoring and research of the environment, work that requires environmental and social scientists to record existing pollutants and to investigate potential new areas of environmental damage (Haslam, 1994).

Second only to water pollution as the main threat to public health is air pollution. In the UK, the Alkali Act 1863 was among the very first pieces of legislation based on scientific standards enacted anywhere, to safeguard the health of the population. It set a specific reduction of 95 per cent in the discharge of all noxious and offensive gases compared to existing levels. However, as it was impossible to determine the *exact* level of existing pollution (because of inadequate technology to monitor pollution levels), the amount by which pollution was to be reduced was fixed at an arbitrary figure. The Alkali Act also established the first environmental regulator, known as the Alkali Inspectorate, to enforce the legislation. Such was the rate of industrialisation that the legislation soon required amendment and, 11 years later, the second Alkali Act (1874) introduced the principle of the best practical means (BPM) to prevent the escape of noxious and offensive gases from alkali works. It also introduced the first statutory emission limit, for hydrogen chloride, the uncontrolled release of which caused severe respiratory illness among workers, see p.34.

Introduction of the principle of BPM was an attempt to overcome lack of precise knowledge of pollution levels and provides an early example of how legislators were forced to compromise the purpose of an act because of lack of knowledge pertaining to a specific environmental problem. As a result, a compromise situation was reached in which industrial premises were allowed to emit higher than permitted pollution levels due to exceptional local circumstances. The circumstances included equipment that could not meet the new pollution standards but which in itself was not life expired or a factory that was the sole source of employment in an area of already high unemployment and which would be forced to close or make expensive alterations if it were required to meet new anti-pollution measures.

BPM has been the cause of much confusion ever since it was introduced. No absolute definition for the term exists; it relies instead on local conditions and the interpretation of the term by the local health inspectorate. As a result, emission standards for similar industries in different regions of the same country can show a wide range of permitted values. Only when improved methods of monitoring pollution emissions became available in the final decade of the twentieth century, combined with the more stringent application of pollution legislation, was it possible to replace the concept of BPM.

By the mid-1990s, an understanding had emerged in government and industry, that for environmental management to be successful it was necessary to obtain global participation in the setting and achieving of permitted standards regarding a wide range of environmental variables. No longer was 'best practical means' an acceptable management strategy. To set internationally acceptable standards for environmental management, the International Organization for Stand-ardization (ISO) based in Geneva has established a voluntary code of best practice. The ISO has existing links to 145 countries and works in partnership with international organisations, governments, industry, businesses and consumer representatives. The environmental 'Standards' are documented agreements containing technical specifications or other precise criteria to be used consistently by organisations and industry as rules, guidelines or definitions of characteristics, to ensure that materials, products, processes and services are fit for their purpose. The ISO 14001 Environmental Management Standard (EMS) requires companies to 'establish a procedure to identify and have access to legal requirements'. This can be readily achieved by compiling an environmental legislation register, which can stand alone or be integrated into an environmental aspects register.

The ISO 14000 'family' of International Standards apply to environmental standards (see http://www.iso.ch/iso/en/prods-services/otherpubs/iso14000/index.html for details of the 2002 environmental management criteria). Table 9.1 provides a summary of the standards. Although collectively called ISO 14000, the specific certifiable environmental standard is ISO 14001 to which organisations can gain certification. By the end of 1999 Japan had most companies with ISO 14001 certification with 1174 organisations. The UK and Germany were placed joint second each with 650 accredited companies.

Within the European Union (EU) and the European Economic Area, an additional voluntary environment management scheme for organisations has been operating since 1995. The Eco-Management Audit Scheme (EMAS) aims to promote continuous environmental performance improvements of activities by committing organisations to evaluate and improve their own environmental performance. In February 2001 EMAS was revised to EMAS2 to include all sectors of economic activity including local authorities and integrates ISO 14001 as the environmental management system required by EMAS2 so that progress from ISO 14001 to EMAS2 will not entail duplication. Participation in EMAS requires that an organisation operates within total compliance with all relevant environmental legislation,

Table 9.1 The ISO 14000 'family' of International Environmental Standards

ISO 14001/04	1996	Environmental Management Systems – specification with guidance for use. General guidelines on principles, systems and supporting techniques
ISO 14010	1996	Guidelines for Environmental Auditing – general principles
ISO 14011	1996	Guidelines for Environmental Auditing – audit procedures – auditing of environmental management systems
ISO 14012	1996	Guidelines for Environmental Auditing – qualification criteria for environmental auditors
ISO 14015	2001	Environmental Assessment of Site and Entities (EASO)
ISO 14020	2000	Environmental Labels and Declarations – general principles
ISO 14021/24/25	1999	Environmental Labels and Declarations
ISO 14031		Environmental Performance Evaluation
ISO 14041/42/43		Lifecycle Assessment – lifecycle inventory analysis, interpretation and impact assessment
ISO 14050		Environmental Management – terms and definitions

works to minimise the production of pollution and aims for a continuous improvement in environmental performance. (See the following website for details: http://europa.eu.int/comm/environment/emas/about/summary_en.htm.)

The setting of voluntary environmental standards and allowing industry and commerce to police its own compliance with the standards is, of course, open to abuse. In a survey completed by the UK Environment Agency in 2003 on the level of environmental awareness and legislation compliance on behalf of small and medium-sized (SMS) businesses, 86 per cent of these businesses believed they were not harming the environment. The same survey revealed that 60 per cent of UK waste production and 80 per cent of pollution was produced by SMS businesses. Only 18 per cent could name environmental legislation that applied to them, fewer than 17 per cent had an environmental management system in place and only a further 11 per cent planned to introduce one (see http://www.environment-agency.gov.uk/netregs/legislation/?lang=_e). It is evident that the voluntary standards set by ISO 14001 and EMAS2 are not being implemented by a large proportion of the business community and that, in the main, it is large, international companies that are complying with environmental legislation.

9.2 Supplying energy in the twenty-first century

'Future gazing' is a perilous business for we have no way of knowing what problems await us or how, or indeed if, technology will be able to solve them. At worst, the future of all the life forms that currently exist on our planet may suffer total destruction, for example if a large meteorite made impact with our planet or a new disease suddenly evolved to strike down the entire human population. Evidence exists that meteorites have previously hit our planet, killing off almost all the prevailing life forms (Spray *et al.*, 1998). With our present level of technology, we might be able to predict exactly when and where the next meteorite would hit our planet but could not alter the course of the meteorite! All available evidence currently suggests that such a disaster is many thousands of years away and by which time technology may allow us to alter the trajectory of the incoming meteorite or even to transfer some sectors of society to the relative safety of an orbiting space station.

Long before we face annihilation from a meteorite impact it is certain that we will have caused many new, unexpected environmental crises caused by the introduction of new technology. For example, throughout history we have continually sought new ways of harnessing different sources of energy (Pickering and Owen, 1994). Energy generated in ways other than human or animal muscle power or from wind or water have all involved technological breakthroughs but each has eventually created new environmental problems. Steam power derived from the burning of coal dominated the period from the 1800s until the 1940s, only to be replaced by oil until the energy crisis of 1973, when electricity became the energy source most in demand. The advantages of electrical power are well known: it can be generated from a number of different fuel sources, it is easily transported by power line, it is 'clean' at the point of use and it is adaptable for heating, lighting, transport and for powering industry and commerce (Blunden and Reddish, 1996). The use of electricity has soared during the last decade of the twentieth century especially in the most technically advanced countries. Figures from the California Energy Commission (http://www.energy.ca.gov/electricity/silicon_valley_consumption.html) show that overall consumption throughout the state rose by 15.9 per cent between 1990 and 2000. For Silicon Valley, overall consumption rose by 16.47 per cent and residential consumption rose by 19.58 per cent. Figure 9.1 shows the predicted increase in generating capacity between 1995 and 2010 for EU countries and worldwide. For the EU Member States, an increase of 28 per cent is forecast, while for the world as a whole the increase is an incredible 56 per cent (http://europa.eu.int/comm/energy_transport/atlas/htmlu/mardep1.html). Care is needed in using this data as it is based on generating company predictions, which, in recent years, have proved over-optimistic.

Although we already possess many different ways in which we can generate electricity, for example from coal-fired thermal power stations, oil- or gas-fired generation plant, from nuclear fission, by hydro generation and, most recently, by wind turbines or generators using wave or tidal power, this range of options does not satisfy our insatiable demand for electricity. If classical economic theory applied to the generation of electricity, we could anticipate that society would use

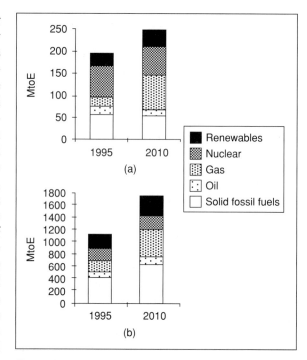

Figure 9.1 Electricity-generating capacity, 1995 and 2010 (predicted) (a) EU Member States and (b) worldwide. MtoE = Million tonnes of oil Equivalent

the cheapest means of generation, but, in practice, this is no longer the case. Producing electricity by burning fossil fuels (coal or oil) releases large amounts of air pollution, mainly sulphur dioxide, carbon dioxide and hydrogen sulphide. For example, the New Brunswick Power plant at Coleson Cove, 15km southwest of Saint John in Canada, emitted 40,500 tonnes of SO_2 to the atmosphere in 1994 (for further details see http://www.gnb.ca/cnb/promos/NB-Power/Refurbishment-e.asp). This oil-fired power plant, of 1050MW output, was constructed in the 1970s with environmental protection equipment appropriate for that period. At this time, the technological solution used to reduce pollution concentration in the vicinity of the plant was simply to build higher chimneys at the power stations in order to disperse the exhaust gases over a greater area. By the 1980s this method proved insufficient to disperse pollutants and gas-cleaning devices such as electrostatic precipitators and 'wet scrubbers' were retro-fitted into the chimney stack to remove the pollutants physically or chemically before they escaped to the atmosphere.

To meet the forecast demand for electricity in the final years of the twentieth century, governments in the developed world turned to nuclear power plants as the best way of providing 'clean' energy. At first, the technology allowed only inefficient Magnox nuclear reactors to be built, then advanced gas-cooled reactors (AGRs) and finally pressurised-water reactors (PWRs). While the generation of electricity from nuclear reactors does not generate gaseous pollution, the spent fuel rods, comprising mainly uranium-238 and plutonium-239 remain highly radioactive for thousands of years. The thorny problem of the safe disposal of nuclear waste has proven controversial with the general public (Middleton, 1999; Phillips and Mighall, 2000). McKinney and Schoch (1998: 199–224) have presented an extensive discussion on the role of nuclear power, claiming that nuclear plants have an excellent safety record. The impact of generating electricity from a nuclear power plant is almost certainly less than the pollution impact created by a power plant that burns fossil fuel with its consequent release of pollutants to the atmosphere. Generation of power from nuclear plants has three main problems, two real and the other of a more intangible nature: the entrapment of the spent fuel rods in a leakproof location presents a major technical problem while the rise of international terrorism has raised real concerns about radioactive waste being used for the manufacture of the so-called 'dirty' bomb. Finally, there remains the unwillingness of the general public to accept nuclear power plants as safe, despite the outstanding safety record of the industry and the infinitesimal impact that generating power from plutonium or uranium has on the environment. Lobbying of politicians by pressure groups has meant that for the time being, nuclear power stations are no longer considered a practical means of generating electricity. However, the acceptance by European governments that current levels of atmospheric carbon dioxide need to be cut by 60 per cent by 2050 has meant that generating electricity by means other than the burning of fossil fuels is required. Renewable energy sources may, in the future, meet a substantial proportion of our needs, but some governments are now hinting that a new generation of nuclear power stations may be needed to provide a base load electricity production. The UK government's white paper, entitled *Our Energy Future – Creating a Low-Carbon Economy*, published in February 2003, stated (page 12) that: 'we do not rule out the possibility that at some point in the future new nuclear build might be necessary if we are to meet our carbon targets.' (The white paper can be accessed via http://www.dti.gov.uk/energy/whitepaper/ourenergyfuture.pdf.)

One of the most illusive technological achievements of the late twentieth century was how to generate electricity by means of fusion power. In this process, which attempts to replicate events within the Sun, four hydrogen molecules are fused under conditions of immense pressure and heated to form one molecule of helium. Nuclear physicists and engineers have been working on ways to achieve fusion power for about 50 years and despite several promising breakthroughs remain some way from success (Key, 2001). In March 1989 the scientific world had been shocked into thinking that a simple 'cold fusion' experiment had unlocked the process of fusion power, but the experiment was invalid (see http://www.alternativescience.com/cold_fusion.htm for a full discussion). It is unlikely that fusion power will be a commercial proposition before mid-way through the present century. The popular press have identified fusion energy as providing almost unlimited, cheap, 'clean' electricity that would enable our civilisation to embark on a new and unimaginable phase of development. Von Weizsäcker *et al.* (1998) argue strongly against the development of fusion power, claiming that it requires vast quantities of radioactive hydrogen as a base material for fuel generation and that the containment of this material is highly problematical. The fusion of hydrogen creates high intensities of neutron flows that defy virtually any attempt to contain them until they join unpredictably in many different types of atomic reactions by which time they have contaminated the surfaces they have come into contact with intense radioactivity.

Generating electricity by means of fusion power would not, therefore, provide the 'clean' energy that modern society yearns for, neither would it be available for all nations. It would, however, enable almost unlimited amounts of electricity to be generated. Only the most technologically advanced, wealthy nations would be able to afford the huge first economic costs of nuclear fusion power stations. It is unlikely that the power generated in the new generating stations would be cheap. Private companies and governments would seek to recoup the huge development costs associated with fusion power. Countries that could afford to generate electricity via fusion would have a super-abundant source of power that would enable them to embark on a new phase of technological growth. For example, abundant electricity would make the recycling of previously used materials an economic proposition, but it would also make the processing of newly mined, low-grade minerals a practical proposition thereby heralding a new phase of opencast (strip) mining. Accessing low-grade minerals is usually accompanied by vast amounts of mining wastes, and disposal of these wastes, once a common feature of the coal-mining industry in the nineteenth century, could once again become a major problem (see Box 9.1). Abundant electrical energy would also allow a major change in our agricultural systems. At present, the cost of irrigation is largely a function of the cost of energy used to pump the water from beneath the ground. The availability of abundant energy from fusion power would enable deep aquifers to be tapped and would inevitably lead to the overdrawing of these ancient reserves, the consequences of which are largely unknown (see section 5.3.3 and Box 5.3). Would the release of large amounts of previously unavailable fresh water into the hydrological system bring about changes to the climate of the planet? In a world that will undoubtedly be several degrees warmer by 2100, the availability of more moisture in the hydrological cycle may be sufficient to cause a significant increase in global precipitation and possibly add to an already higher sea level caused by global warming (UKCIP02, 2002).

9.3 Forecasting future technological impact on the environment

While it is impossible to predict with certainty specific events in the future, by looking at the pattern of technological advance over the previous ten to 25 years, it is possible to make an educated guess at the technology that will be available in the next 25 years. Of one thing we can be certain: the rate of technological achievements in the future will rise exponentially, superseding even the most spectacular scientific discoveries and technological achievements of the previous 50 years. As we look further into the future it becomes increasingly difficult to avoid becoming caught up in the more extreme and, at the current time, unsupportable ideas of new technology. The only certainty is that technology will continue to provide the means of energising society and generating vitality and creativity (Trewavas, 1999).

One of the main benefits of technology is that it should permit society to achieve a specific task more safely or without expending as much time and energy than previously necessary. Technology should also allow us to make use of resources more efficiently making it possible to carry out the same amount of work for the expenditure of less time and effort or using the efficiency of new technology to perform more work in the same timespan. We usually opt for the latter option as a greater work output usually results in a greater economic gain. By performing more work we consume more natural resources (raw materials), use more energy and generate more pollution. Commoner (1971: 52) was among the first to recognise that the use of technology 'amplifies the effect human beings have on the biosphere'. Because the use of new technology is usually accompanied by economic gain or improvements in the quality of life for society, it is not in the interest of the individual to be overcritical of the technology. Aware of the unwanted damage caused to the environment by inappropriate technology, governments now support the existence of organisations responsible for assessing the likely impact of new technology. Changes to existing land use almost always require a compulsory environmental impact assessment (EIA) to be carried out at the expense of the developer, in which the likely impact of development on the environment must be assessed.

Many aspects of new technology prove to be beneficial to the environment. Air quality in developed world cities is immeasurably better today than in the 1970s as a result of fuel efficiency

BOX 9.1

How disasters can change our environmental awareness

Towards the end of the twentieth century it became apparent that a serious imbalance was emerging between the timescales on which human events occurred compared with the timescales of natural events. Improvements in technology allowed major changes to the built environment to take place very quickly and without opportunity for the natural environment to readjust. Computer-aided design of developments cut preparation time, more efficient ground preparation equipment enables construction times to be shortened and pre-assembled building components allows buildings to rise quickly on greenfield sites. In the past, the impacts of changes to the built environment occurred on a smaller scale and often took many years to manifest themselves. For example, mining of coal from shallow shaft mines generated only small amounts of waste, but gradually as mining technology allowed shafts to go deeper, then so proportionately more waste rock and shale was generated. Disposal of mining waste became a problem. In the absence of planning laws, mine owners deposited the waste in the cheapest way possible by dumping the spoil near the mine (see Plate 9.1). Often, the spoil began to undergo spontaneous combustion, releasing toxic gases that killed the vegetation and were harmful to animals and humans (see Plate 9.2). Rainwater flowing through the waste can carry away high concentrations of chemical elements causing water pollution and transfers fine particles of coal shale far from the spoil heap. Worse still, spoil heaps

Plate 9.1 A large, newly created coal spoil tip in the Rhondda Valley, South Wales, in 1975. An old spoil tip, now largely vegetated, appears on the left of the photograph

BOX 9.1 CONTINUED

Plate 9.2 Old coal waste often undergoes spontaneous heating and smoulders for years, releasing noxious fumes into the atmosphere. In this photograph, taken near Aberdare in South Wales, the coal waste was about 50 years old before it began to combust. All the vegetation has been cleared, possibly to prevent the local population becoming alarmed at the death of the vegetation due to the fumes and the heat

can become supercharged with water and where the waste had been placed on steep slopes, the material could begin to move down slope, often quite rapidly after periods of heavy rain. In the mining village of Aberfan, in South Wales, disaster struck on the 21 October 1966, when 144 people, 116 of them children, were killed when a tip of coal waste slid onto the village. The accident tribunal attributed the cause of the tragedy mainly to poor management on behalf of the nationalised coal industry responsible for the disposing of the waste but some blame can be attributed to the ignorance attached to a practice of disposing of coal waste that had been common for more than 100 years. The inhabitants of Aberfan were tragically involved in a disaster that was destined to occur, somewhere and at some point in time. In hindsight, the tragedy could have been avoided had higher standards of environmental management been in operation.

As the built environment continues to expand into the natural world, it becomes increasingly necessary that we become proactive in identifying at the earliest possible stage the nature and the physical location of potential environmental conflict. Technological solutions to many impending problems already exist.

improvements in vehicle engines. The replacement of carburettors by electronically controlled fuel injection systems coupled with catalytic converters to clean exhaust gases and the use of lead-free petroleum or the use of low-sulphur diesel fuel in commercial vehicles has helped reduce air pollution levels in cities. Unfortunately, the potential gains offered by such technology have been offset by the increased use of the motorised transport. For example, in Scotland, the volume of traffic on major roads increased by 55 per cent between 1984 and 2000. Despite this increase, carbon dioxide emissions from transport in Scotland fell from 8.6 million tonnes (mt) to 8.3mt between 1990 and 1999. The ban on the sale of leaded petrol from 1 January 2000 resulted in the concentration of atmospheric lead falling from about 250 nanograms per cubic metre (ng/m^3) in 1985 to about $10ng/m^3$ in 2000. The figures for nitrogen dioxide (NO_2) concentration changed less and in spite of almost 90 per cent of all petrol vehicles being equipped with catalytic converters, annual concentration of NO_2 fell from 51 micrograms per cubic metre (mg/m^3) in 1992 to $45mg/m^3$ in 2001 (Scottish Executive, 2002). Clearly, a trade-off has operated between what society is prepared to accept and pay for in terms of air quality and the perceived need to allow road traffic levels to rise.

If society wished to return to an air quality that was last achieved in pre-industrial times we would have two options: one, to de-industrialise (an impossibility as industry provides the means of generating wealth) or, two, to develop technology that was either *completely* non-polluting or that could *totally* retain the pollutants for recycling. The second option is theoretically possible and its attainment is restricted only by cost. Society must assess the level of atmospheric cleanliness it desires; technologists must then devise a means of obtaining that standard and economists will be able to calculate the cost of attaining that cleanliness. Finally, if society accepts the cost, legislators will be required to provide governments with the powers of achieving the air quality.

Apart from the problem of petroleum-driven vehicles causing air pollution, their use also causes traffic congestion and traffic accidents resulting in injury and death to road users. Technology can offset both these problems. In-car traffic routing displays, using global positioning satellite technology (GPS) can direct drivers to less congested routes while experiments with radar distance-finding equipment to control the minimum distance between moving vehicles appears able to regulate the flow of traffic and to maintain movement. A model devised by Treiber and Helbing (2001) showed that fitting 10 per cent of cars with driver assistance distance regulation systems could ease most congestion problems while fitting 20 per cent of cars could make congestion disappear altogether.

The threat of global warming, brought about to a great extent by the build-up of greenhouse gases emanating from the burning of fossil fuels in thermal power stations, has resulted in new technology being developed to support renewable sources of energy, notably wind power and solar energy and, to a lesser extent, biofuels. Renewable energy is recognised as having none of the traditional polluting problems of fossil fuels or nuclear energy. For this reason, renewable energy is described as 'clean' energy. Unfortunately, electricity generated from renewable sources is, at present, substantially more expensive (at best, between two and four times) to generate than from traditional means and is less reliable due to its dependence on the availability of solar energy or wind. However, the implications of continuing to rely on the burning of coal as the main source of generating electricity has such severe consequences for the biosphere, and ultimately for society, that governments worldwide are prepared to offer substantial incentives to private companies to develop more efficient technology for renewable energy generation. According to data provided by the Worldwatch Institute (http://www.riverdeep.net/current/2000/11/112200_worldwatch.jhtml), electricity generated by harnessing the wind rose by 39 per cent in 2000 (see Figure 9.2) and the production of photovoltaic (solar) cells increased by more than 30 per cent. In both cases the increases occurred from a very low base level.

Germany is one of the leading nations in the development of renewable energy sources. It has become the world's largest producer of wind-generated energy, accounting for one-third of the world output in 2002. It is also a major producer

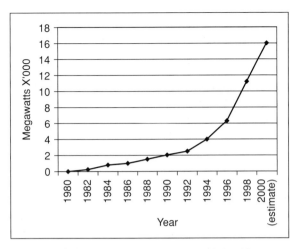

Figure 9.2 Growth in energy generated by wind turbines (*Source*: based on data from Worldwatch Institute, 2000)

of solar energy and biogas energy (Common Ground, 2002) (see Figure 9.3).

9.4 Incentives for using new technology

Nineteenth-century philosophers, economists and political revolutionaries were strong advocators of technology as a means of improving social conditions for the working classes. The most prominent was Karl Marx who, along with Friedrich Engels, can be credited with co-founding scientific socialism that has since become one of the most influential tenets of modern society. Phillips and Mighall (2000) state that technology provides the following benefits to society:

- It improves the 'efficiency' with which work is completed.
- It improves the cost effectiveness and, in turn, the profitability of industry.
- It increases the competitiveness of a company, or a country, over its competitors.
- It reduces the expenditure of human effort and, in turn, releases time for other non-work-related activities. This enables an improvement in the quality of life for the workforce.

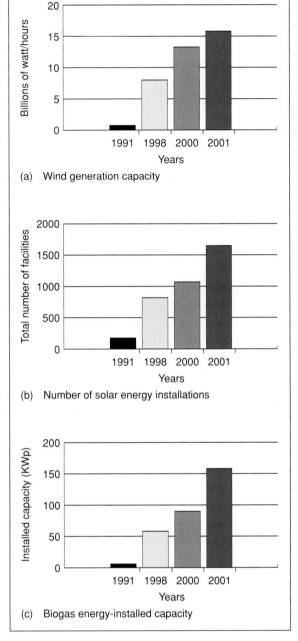

(a) Wind generation capacity

(b) Number of solar energy installations

(c) Biogas energy-installed capacity

Figure 9.3 Growth in renewable energy sources, Germany 1991–2001 (*Source*: data from Common Ground, 2002)

- Companies specialising in the development of high-tech solutions have opportunities for exporting technological solutions to other, less developed nations. This enables the generation of export income and helps transfer technology to less developed countries.

It is notable that the list of opportunities created by the use of technology does not include any specific benefits for the environment, plants or animals. Karl Marx, in his work *Das Kapital*, published in three volumes between 1867 and 1894, considered labour to be at the centre of the society–nature relationship. Human labour permits both the use and exploitation of nature, the benefits of which are improvements in wealth and social conditions for humankind. Provided the use and exploitation of the environment did not create conditions unfavourable for society then Marx gave little consideration to the decline of nature. The benefits to society from the use of technology were seen as entirely positive: for example, an increase in personal wealth, a healthier and longer lifespan, better housing, education and transportation.

However, the continued application of the principles of scientific socialism brings considerable consequences for the biosphere and its non-human components. Marx and Engels established their principles in an era when knowledge of the environment was minimal. When Marx wrote *Das Kapital* the working class was wracked by disease and malnourishment. Conditions were not helped by the terrible working conditions imposed by the factory, mill and mine owners. Air and water pollution levels reached record levels and contributed greatly to the short life expectancy of the workers who lived and worked in the industrial squalor. Out of this morass emerged the principles of humanitarianism, intent on improving the conditions of the working class. Judged from today's standards, the work of men such as David Dale and Robert Owen in their 'model towns' may appear patronising but their work led the way in establishing conditions that allowed their employees a means of self-improvement. Free education for children and night-school classes for adults eventually helped to produce a reservoir of self-educated and fiercely independent people. The introduction of new technology became an established part of industrial life. Technology became the panacea by which the constraints exerted by the environment on the population could be reduced, thereby improving the quality of life for society. However, society wanted more! Instead of opting solely for an improved lifestyle, technology allowed the population to increase exponentially (see Figure 1.3), taking advantage of the improved conditions offered by the technological revolution.

The advantage of new technology has always been to provide some sectors of society with a significant advantage over other human and non-human species that did not have access to technology. The onset of the Industrial Revolution made available the advantages of technology to a larger proportion of the human population and, consequently, the natural world came under a correspondingly greater threat of change. The impact of industrialisation and urbanisation extended to adjacent and previously unaltered habitats. In particular, new agricultural practices resulted in a new phase of deforestation, drainage and introduction of new agricultural species. In particular, the drift of air and water pollution and the use of chemicals in agriculture extended the influence of technology well beyond the formal boundaries of towns and cities. For at least 150 years, between 1790 and 1950, the introduction of new industrial technology took total precedence over all else. Disappearance of plants and animals and a decline in environmental quality were considered the inevitable price to pay for the benefits society gained from new technology.

Technology has undoubtedly enabled the fortunes of *Homo sapiens* to be raised beyond the imagination of people alive in 1750 and, looking forward another 250 years, it is impossible for us to assess how technology will have benefited our descendents alive in 2250. Park (2001) has questioned whether we can be certain that technology will continue for all time to bring improvements to our lives. The question mark surrounding the ability of technology to go on solving problems far into the future has existed in the minds of ecocentrists for at least 50 years and expresses a fundamental human concern that there is something intrinsically wrong with the strategy of pinning our future well-being on technology that still has to be discovered! Commoner (1971) had argued from the ecocentric camp that far from new technology *improving* the quality of life for humans, it was deluding us into believing we were becoming more efficient and less damaging to the environment whereas, in reality, technology was leading society into a cul-de-sac. This line of reasoning goes on to suggest that because technology has, in the past, been able to raise the level of human development is no reason to accept that it will always be able to do so in the future. Park (2001) has suggested that a more appropriate argument may be that the use of technology has been responsible for *postponing* the time when society will face a problem of such magnitude that technology will not be able to solve it. The analogy can be made of a society increasingly dependent on technology travelling along a one-way street that becomes increasingly narrow the further we proceed. The further we travel, the fewer the exit points and eventually we discover that the road is a dead end and by which time it is impossible to turn the juggernaut society around and retrace our steps.

In practice, there have been many occasions when our ancestors faced seemingly insuperable problems, most commonly, those that involved the supply of food, water or natural resources. On each occasion, a strategy was devised to overcome the problem. The solution often lay in a combination of changes, including a migration of 'surplus' population to new lands, the discovery of an 'alternative' resource or the employment of a new technological solution. The solving of problems by means of a range of strategies represents a unique element of the human character. Our destiny is not *determined* by a set of immutable controls but lies in a range of *possibilities* from which we have relative freedom to choose. Sometimes we make the wrong choice, but again, we have flexibility to recognise our mistake and to reselect an alternative strategy. This apparent freedom to choose seems at odds with the way in which other species behave. The rigidity imposed on most plants and other animals by, for example, seasonal climatic events can be seen in the great patterns of species migration, in dormancy or in distinct patterns of breeding linked to climatic events. The question in the minds of many people is, 'why should we be *so* different from all other species?' The corollary is, 'Are we deluding ourselves into believing that technology allows us to behave differently?'

If, through the use of technology, we have reached a false conclusion on the way in which we can use the planet's resources, then the collective wisdom of all past and present scientists, technologists and philosophers can be cast into doubt! If there had been unsolvable crises in the past that could not have been solved by technology, it is likely that the human population would have experienced a population 'crash' to a level that was sustainable by a pre-existing level of technical skills. If such an event had occurred this would indeed represent an example of the 'catastrophe theory' proposed by the 'ecocentric lobby' (see Table 2.1). The nearest that our population has approached a population crash was in the mid-fourteenth century when the Black Death (the plague) ravaged the European population, reducing it by up to 75 per cent. Overcrowded living conditions and poor sanitation contributed to the rapid spread of the plague and technocentrists would argue that it was a *lack* of suitable technology in the form of appropriate understanding and treatment that led to the rapid spread of the disease.

In contrast to the assumption of impending disaster brought about through overreliance on technology, supporters of the 'cornucopian' theory argue that there is no evidence whatsoever to link the use of technology with the catastrophe theory. On the contrary, technology has allowed an ever-increasing human population to enjoy a constant rise in the quality of life. Supporters of technology argue that to exist in a state of anticipation of an insolvable critical problem would stultify the development of society and would provide an undesirable feedback on the development of technology. By pinning our belief that technology will *always* be able to solve whatever problem we face, then both as individuals and society at large, we sustain our unique human characteristic – we are strategic opportunists and any deviation from this belief would herald the end of *Homo sapien*'s reign as the dominant species. It can be further argued that by slowing down the rate of research and development into new technology we could

The cornucopian solution to drought in Australia

In 2002 Australia experienced a six-month drought that was the most severe since 1982. Agricultural output slumped and the Australian gross domestic product fell by $3.8 billion. In a scheme that recalled a plan to reverse the flow of the northerly flowing rivers of the former Soviet Union in the 1960s, a combination of industrialists and entrepreneurs have proposed that the flow of four major rivers – the Clarence, Pioneer, Burdekin and Ord – be reversed so that the dry interior agricultural land can benefit from irrigation.

The plan has been universally condemned by environmentalists on the grounds that reversal of flow would cause more problems than it would solve. For example, coastal waters devoid of fresh river water would become more saline resulting in unpredictable ecological changes. They claim that irrigation carried out in the Murray–Darling catchment (see section 5.3.2) shows the danger of changing the hydrological balance of the naturally arid Australian environment. Application of irrigation water has increased the evapo-transpiration from the soil, bringing toxic salts from deep within the soil profile to the surface.

However, the developers have funded the Farmhand Foundation to make possible a feasibility study of the river reversal. The plan has attracted the attention of many Australians who seem willing to use the latest technological possibilities to sustain their economy. As one journalist reported: 'Maybe turning rivers around is a dopey idea. So what? Mankind would never have landed on the moon if pragmatists had prevailed' (*New Scientist*, 2002).

jeopardise a still-to-be-discovered line of technology that might prove essential to solve a future problem!

The evidence, therefore, appears to lie in favour of those who support the use of technology to help humankind prosper. But does the evidence support the use of technology above all else and to the complete exclusion of sympathy for the natural processes at work within the biosphere? Box 9.2 provides a case study from Australia of the way in which a cornucopian attitude strikes such fear into the minds of ecocentrists. Finding a compromise between the two extremes, the eco-centric assertion that only low-technology solutions are acceptable, and the 'blue sky' maximum use of advanced technology strategy of the techno-centrists, is extremely difficult to rationalise. In many ways, the concerns outlined earlier by Commoner and Park reflect the same fears for the future as that of Thomas Malthus in 1798. No one can foresee the technological advances that await discovery, but neither has anyone been able to forecast the level of environment quality that will be considered desirable to fulfil a specific quality of life. Bronowski (1973: 436) has encapsulated the human concern of relying totally on techno-

logy, suggesting that: 'The ascent of man is always teetering on the balance. There is always a sense of uncertainty, whether when man lifts his foot for the next step is it really going to come down pointing ahead.'

We have ascended to our present position in the biosphere because of our ability to make technology deliver the type of lifestyle that we (subconsciously) aspire to. Technology has become akin to an extra gene within our reproductive code. We allow the technological gene to mutate and evolve and, as for our natural genes, it has allowed us to colonise a different position in the biosphere from that which we occupied when humans made their first tentative steps towards distinct independence. Wood (2002) has suggested that a freak mutation in our natural genetic code has allowed our species to become the technocrats we are today. Whatever the nature of the freak mutation, it occurred sometime between 7 and 10 million years ago and allowed hominoids to evolve away from an ancestral stock. The mutation was small. Human DNA differs, for example, from that of our nearest biological neighbours, the present-day chimpanzees by, at most, 2 per cent. This has been sufficient to allow *Homo sapiens* to achieve total domination over the other members of the biosphere and of many of the natural physical processes that take place between the atmosphere, lithosphere and hydrosphere. Our rise to super-species status has involved more than a technical competence, for there were many times when our ancestors were faced with social and religious turmoil as well as competition from natural forces within the biosphere. Our survival has been helped by what has been called the 'selfish gene' – a trait that makes us prone to compete more than co-operate, both with our fellows and with other species. Singer (1979) and Trusted (1992) have examined whether humans differ more in their attitudes to other species today than in the past. The situation is confused. Whereas our attitudes towards fellow humans stricken by natural hazards is usually to provide charitable aid, especially when the disaster is severe, in contrast, our attitudes towards asylum seekers attempting to find sanctuary in the country in which we live is usually unsympathetic, except when the humanitarian conditions from which they are fleeing are life-threatening. A similar contrast exists between our willingness to support attempts to retain biodiversity, yet at the same time sanction the work of international logging companies to remove tropical rainforest, or release genetically modified organisms into the wild. Our schizophrenic relationship with the biosphere certainly represents a dilemma for modern society – we find it difficult to believe that we occupy such a position of strength that we can now show compassion to other members of the human race or to non-human species with which we share the planet.

9.5 Harmonising modern technology with the needs of the biosphere

Our industrialised world has reached its current state of development despite working in apparent conflict for much of the time with the natural world. Examples of the ways in which the human system conflicts with nature's way are legion – the felling of forests without a replanting programme, the depletion of fish stocks through industrial fishing and our industrial systems that fail to recycle or reuse components that have reached the end of their lifecycle. The sequence of using materials once before disposing of the product and extracting new resources contrasts to the processes at work in the natural world where reuse of materials appears to be the norm. In nature, complex finite resources are recycled through the decomposer organisms and although most natural systems are not completely 'tight' systems (that is, some of the materials 'leak' away), the system only needs to supplement a relatively small proportion of its total material needs. By contrast, in most of our industrial systems, 'leakage' of material and energy is high because we use energy only once and send our waste materials back to landfill sites. To remedy the loss of energy and materials we are constantly devising ways of extracting more of these commodities from the biosphere. There have been many occasions in the past when overuse of a key commodity has resulted in an increasing scarcity of the material, accompanied by a rise in price, which, in turn, has led to the search for a substitute material. Either the alternative resources may be used directly by existing technology or, more commonly, requires the development of a new technology. The bringing of alternative resources to the market place is discussed more fully by Jones and Hollier (1997: Chapter 2).

Provided the bringing of new materials to the marketplace costs less than recycling previously used materials, the case for recycling is difficult to justify on economic grounds alone. However, there are other considerations besides cost and since the mid-1970s the argument for recycling based on environmental principles have become stronger. The main reasons are:

1 The increasing rate of raw material consumption has placed such a great demand on reserves that the price of many raw materials has increased (but notably not oil).

2 The resources of the highest quality or easiest accessibility have been mined first, leaving only materials of lower quality or in sites that are difficult and expensive to access and refine.

3 More stringent pollution standards have meant that processing costs have risen as the

cost of installing pollution control measures have been added to the stages of processing and manufacturing.

4 Energy costs have risen, making the transport and refining costs of low-grade raw materials more expensive.

5 A major shortage of suitable landfill sites has resulted in difficulties in disposing of waste.

6 As a consequence of governments signing up to international environmental treaties and agreements to minimise pollution, to maximise biodiversity and to reduce energy consumption based on the burning of fossil fuels, materials that have traditionally been sent to landfill must now be recycled. As an incentive, a tax has been placed on all materials sent to landfill (and by so doing treasury departments have gained a valuable source of additional revenue).

7 Lobbying by environmental organisations has persuaded governments and the public in developed countries of the benefits of recycling.

8 Stimulated by points 1–6, new technology has been developed to allow the recycling of materials that were previously discarded, creating new job opportunities and wealth generation.

A strong incentive for recycling was made in the Brundtland Report (World Commission on Environment and Development, 1987). The report forecast the escalation of costs (for the reasons outlined in points 1–4 in the list just presented) of bringing new reserves to the marketplace and argued that to assure a continuation of economic growth, future development should be based on a minimisation of waste, control of all forms of pollution associated with development and a move towards greater energy efficiency. Future industrial growth, the report argued, should be centred on 'sustainable development' defined in the now classic statement as 'development that meets the needs of the present without compromising the ability of future generations to meet their own needs'. New technology was identified as providing the means of unlocking the door to major efficiency gains provided by sustainable development.

In practice, it has proved to be far more difficult to achieve even a modicum of sustainable development. Despite grand statements from world leaders meeting at the UNCED summit in Rio de Janeiro in 1992 and at the Earth Summit in Johannesburg in 2002 it has not been possible to devise an economic or industrial system that operates entirely on a sustainable basis.

The attainment of a fully sustainable society based on the best available technology remains some way in the future. It is a goal to which all nations can aspire but one that is unlikely ever to be totally reached. Radically new technology will undoubtedly enable major new advances in 'clean' use of materials. However, based on our past experience, it is inevitable that completely new environmental problems will continue to confront society in the future, making it unlikely that human society will ever be able to exist in a utopian, benign existence with the biosphere. Technological developments will need to run hard to keep up with the demands made on the biosphere by a society driven by the desire to constantly expand.

9.6 Forecasting technological change

Attempts to forecast what the future holds for humankind and for our planet is a difficult task and, some would say, a worthless exercise. In the past, many scientists have felt it their duty to warn the public of what they see as dangerous trends in the way in which we use (or misuse) technology and have attempted to warn us of the consequences if current trends were continued (the so-called 'business-as-usual' scenario). Throughout the 1960s and 1970s, a succession of forecasts were made, each of which proposed a bleak outlook for the developed world unless a move towards sustainable development was made. Well-known examples included: *The Waste Makers* (Packard, 1960), *Earth as a Spaceship* (Boulding, 1965), *Limits to Growth* (Meadows *et al.*, 1972), *Blueprint for Survival* (Goldsmith *et al.*, 1972) and *Battle for the Environment* (Aldous, 1972). Some predictions forecast wonderful new opportunities for humankind if the change to sustainable development could be made, while others saw only signs of impending disaster. The basis for many of these arguments was founded on incomplete data or too short a time period over which trends could be established. Many of the predictions for society such as those contained in *The Population Bomb* (Ehrlich, 1965) were subsequently used to make speculative television programmes that drew conclusions far beyond those contained in the original documents. These programmes did much, however, to undermine the confidence of the public in the predictions of the environmental scientist. Lessons were quickly learned from the ways in which television could be used to influence society of the problems of adopting a sustainable infrastructure.

The arguments in support of sustainable development have matured considerably since the 1970s. Some sceptical viewpoints remain about the need to change our pattern of consumerism. The work of Simon (1996) and Lomborg (2001) provide a counter-argument to the concerns of the ecocentric lobby. None of the extreme viewpoints proposed by either ecocentrists or technocentrists provides a certain indicator of what the future holds for society because the future is, itself, always changing.

Historically, our pre-industrial societies were highly vulnerable to the seemingly haphazard occurrence of natural disasters. A better understanding of these hazards combined with engineered solutions and the efforts of relief agencies have reduced the impact of natural hazards, especially in the developed world. Improved levels of technology among highly developed nations has brought a new threat in the form of highly destructive warfare. At the height of the Cold War in the late 1950s, when the USA and the USSR began development of inter-continental ballistic missiles (ICBMs), the most likely form of destruction appeared to come from nuclear war. Eventually in 1985, under the leadership of President Mikhail Gorbachev, reformist policies based on *glasnost* and *perestroika* eased ideological tension and the threat of a superpower nuclear war subsided. Faced with the first real opportunity for world peace since the end of the Second World War, it might have been reasonable to assume that society could look forward to a 'golden age' in which scientific and technological improvements in agriculture and in healthcare would mark an end to malnourishment

and ill health. By a cruel irony, society began to face a new series of threats that jeopardized the attainment of world stability. To the traditional natural threats examined in earlier chapters of the book such as land degradation, loss of biodiversity, scarcity of fresh water, misuse of marine resources, pollution, global warming and, most recently of all, introduction of genetically modified organisms (GMOs) we now face new biological hazards such as HIV-AIDS, the Ebola virus, a resurgence of tuberculosis, foot and mouth disease and bovine spongiform encephalitis (BSE), all of which threaten our well-being. If this list were not sufficiently daunting, the new century brought a new threat to world stability – international terrorism based on the use of biological 'weapons' such anthrax, smallpox and the plague.

Our species has always lived alongside environmental threats. To what extent are the 'new' threats of the twenty-first century 'real' or 'perceived'? Have we become so accustomed to living alongside threats that we need an ever bigger problem against which to pit ourselves? Is global warming, for example, the new problem and does it really change the way the planet operates or is it an inherent human fear of change itself that we most fear?

To counteract the invisible threat of change and of the way in which multinational companies working alongside the World Trade Organization and the G8 group of countries (comprising the eight most affluent nations) bring about change, a new generation of anti-capitalist and environmental lobbyists has emerged. It is difficult to propose an acceptable name for this newly established group or to identify the primary aims, as its membership is catholic and lacks a formal organisation. At the beginning of the twenty-first century, this diverse group became united through a dislike of multinational technology companies, of the World Trade Organization (WTO) and of the G8 group of nations (see Box 9.3). Some of the arguments used by the current generation of campaigners bear a strong link to the ideas of the environmental pressure group activists of the 1970s and 1980s (see section 3.6). However, the activists of the twenty-first century have the benefit of a far better environmental education and conduct their research and

BOX 9.3

The new environmental activist movement

Environmental protest groups first became widely known during the 1970s when anti-logging protesters drew attention to the activities of major logging companies intent on clear felling old forests in the western United States of America. Section 3.6 of this book has shown how the actions of people on the fringe of established political parties were eventually responsible for substantial changes to the environmental attitudes of political parties in the developed world. By the beginning of the 1990s the 'greening' of world politics appeared to be complete and the action of protest groups had subsided.

The re-emergence of protest groups at the end of the 1990s therefore came as a surprise. At the World Trade Organization (WTO) meeting in Seattle, held in the first week of December 1999, some 50,000 protesters from all over the world, and representing a diverse number of causes, laid siege to the WTO talks. The WTO is a relatively new organisation, founded in 1995 to replace the General Agreement on Tariffs and Trade (GATT) and given a legal standing equivalent to the United Nations. The WTO is a largely secretive organisation whose brief is not only to encourage free trade, but also to examine non-economic trade barriers such as food safety laws, workers' rights and environmental protection standards. One of its most controversial tasks is to consider trade disputes between nations. Members of the WTO mainly comprise lawyers with direct links to the industries being regulated. There are no rules against appointing members with conflicting interests and no appeal against WTO findings is allowed. The WTO has been criticised as a cabal that serves the interests of multinational organisations and the world's richest trading nations.

The Seattle demonstration argued, among other things, that the WTO was a threat to democracy, to world peace, to environmental health and justice and to quality of life. More fundamentally, the protest was against globalisation and the growth in power of the transnational corporation. For example, a WTO agreement prevents individual countries from departing from a world agreement that allows some electrical equipment to contain lead, mercury and cadmium all of which are hazardous to humans and the environment. Also, Europe must remain an open market to beef sales from the USA even though US beef contains growth hormones that are banned from use in the EU.

After Seattle it seemed inevitable that future international meetings of bodies such as the International Monetary Fund (IMF), the World Bank, the World Economic Forum, Summit of the Americas and the European Union summit meetings would attract the protestations of thousands of activists, intent on highlighting the damage being done to societies and the environment, especially those of developing countries.

The IMF meeting in Prague became the next protest target in 2000. By 2001 the protesting groups had become more active and disrupted the IMF meeting in Davos, Switerland in January 2001. Later that year, in June 2001, the summit of the European Union (EU) leaders held in Gothenburg, Sweden's second city, was reduced to tatters. A small group of protestors smashed capitalist symbols in the city during the worst riots witnessed in Sweden for many years and overshadowed more than 80 peaceful protest groups participating in the Non-Violence Network who campaigned under the slogan 'People Not Profit'. The damage to the city hardened the already negative feelings many Swedes held towards the EU and forced all future EU summits to be held in Brussels where security of EU officials is tighter. However, such a departure from the traditional pattern of holding the summits in each of the 15 Member States in turn, has been described by some as admitting to the need to create a 'fortress Europe' image of a democratic union of European nations intent on dismantling international borders and enabling free movement within Member States.

Worse was to follow when the G8 meeting (comprising politicians and their advisers from the eight most industrialised nations of the world) met in Genoa, Italy. An estimated 100,000 protestors 'hijacked' the occasion and completely overshadowed the purpose of the meeting. The civil disobedience associated with the protesters resulted in damage to property, looting, personal injury to the protesters and police and ultimately, death.

BOX 9.3 CONTINUED

What has the new phase of civil unrest achieved? Such action raised many issues concerning civil responsibility and the right to peaceful protest as a means of bringing about change. That such violent protest should be carried out on behalf of the environment, of its resources and its inhabitants illustrates the level of concern felt by the protesters. At present, there is little political agreement how to respond to such action, apart from a determination not to give in to the protesters. Future meetings are likely to be held in inaccessible locations – while one suggestion is that the venue will be a warship on the high seas! Despite an almost complete lack of focus due to the large number of different protesting groups and the disparity of their issues and causes, the protestors have gained strength from their numbers and sympathy from the general public as a result of the heavy-handed intervention of police and military forces. Based on past experience, it is likely that a few central themes will emerge from the multitude of different issues, and as the 'issue–attention' cycle takes effect, politicians will, for reasons of electoral expediency, be forced to give ground to the demands of the protesters.

As a postscript to this section, in light of the political turmoil resulting from the terrorist attack on the World Trade Center in New York on 11 September 2001, the removal of the Taliban from power in Afghanistan, and unrest in the Middle East, it appears that protest groups have redirected their attention to anti-war and humanitarian causes.

disseminate their knowledge of the biotech companies by means of the internet. Some of their arguments are based on newly emerging concepts contained within environmental ethics.

9.7 Environmental ethics

The way in which we relate to the environment changed very considerably during the closing decades of the twentieth century. This was due, in part, to the greater understanding of the way in which the biosphere worked and also to the way in which humans interacted with the biosphere. As always, spectacular incidents in which the biosphere was damaged shaped the public viewpoint and, in turn, influenced the way environmental managers and politicians responded with new methods of environmental monitoring and control.

More significantly, the way in which individual human beings and society as a whole thought about the environment also began to change. Helped by the medium of television and its ability to bring pictures from space exploration into our homes, it has been possible to comprehend the finite dimensions of our planet and its resources. Environmental education from kindergarten to university

has ensured that almost all people below the age of 25 now understand the complex issues surrounding the use of the biosphere and the environment. A large proportion of the population also understand the need to develop a social responsibility to the environment and, as a consequence, the need to work towards keeping the environment in as healthy a state as possible. In short, our way of thinking about the environment has undergone a profound change in the space of about 25 years. We have adopted an approach to the environment that reflects the knowledge-based society in which we live.

Simply knowing about the components of the biosphere and the environment cannot ensure that we develop a healthy relationship with the environment. For that to be achieved we must understand the moral relationship that exists between human beings and the rest of the world. This requires that we develop an environmental ethic. The way in which each individual human being relates to our surroundings is governed by an immensely complex set of factors that includes our childhood upbringing, level of education, economic wealth, political background and exposure to advertising and pressure groups (Botzler and Armstrong, 1993). The way we treat the environment and its components is governed by our individual ethical belief, which determines the moral relationships between the natural world and ourselves. As many of the great philosophical thinkers across the ages have discovered, it is impossible to categorically prove that any one moral attitude towards the environment, or any other issue, is any more correct than that of a different moral attitude.

Whenever we have to make decisions about using any part of the environment, for example a natural resource such as fresh water or soil or a tree, we are involved in making an ethical decision. For example, when we need to use water for washing or food preparation, we use water with little thought about our 'rights' to use that resource. However, if our use of water deprived a neighbour of water then it is possible that our 'right' to use water would be altered. Under these changed circumstances, our modified use of a resource would represent an environmental ethic at work.

We are constantly required to apply ethical decision making in a much broader sense to the way in which we live our daily lives. For the majority of people, the decision not to steal, lie or cheat is a simple one; it becomes an ethical decision to live in a morally correct fashion. Many western religions believe that ethical values can be gained through study and reasoning and consider that 'good ethics' are learned through a study of God's teaching (McKinney and Schoch, 1998). However, in the contemporary secular world, if the traditional concept of what constitutes 'good ethics' is extended to other aspects of modern life, the identification of ethically correct behaviour becomes more difficult. Do we drive to work in a fuel-inefficient car or do we use a combination of walking and public transport to make less impact on the environment? As individuals, how far is it our responsibility to recycle unwanted materials? Should it be the responsibility of the municipal refuse department to recycle our domestic rubbish? More controversially, do plants and animals have ethical rights? Should they have an equal right to clean air, water and living space as do humans? If plants and animals have rights, do humans, as the dominant organism, have an obligation to ensure that the rights of other living organisms as well as the existence of unpolluted rivers, oceans and landscapes are protected? All of these questions require

a specific environmental ethic between ourselves and our relationship to the environment.

Cunningham and Saigo (2001) identify a number of different ethical attitudes based on the reasoning of the great philosophers (see Table 9.2). It is significant to note that the evolution of philosophical ideology has been paralleled by the changes in the way we relate to our social and natural environments. It may be no coincidence, that the emergence of a *utilitarian* philosophy, which considers it desirable for actions to produce the greatest benefit for the largest number of people, coincided with the Industrial Revolution in Europe. At this time (post-1750), the growth of urban areas throughout Britain and Europe was proceeding rapidly. During this time, little attention was given to the social and working conditions of society. Gradually, an emerging group of humanitarians such as William Wilberforce (1759–1833), campaigner for the abolition of slavery, and Robert Owen (1771–1858), founder of the cooperative movement and of the model industrial settlement at New Lanark (1799) and later (1825) of the New Harmony settlement in Illinois and Indiana, were to establish new ethical standards for the treatment of people. However, the utilitarian attitude did not originally extend to the environment and in Europe during the period 1750 to about 1900, more species became extinct and more habitats destroyed than at any time before or since that period (Ziswiler, 1971).

The so-called *modernist* ethical philosophy held brief power at the beginning of the twentieth century when it was thought possible to categorise the environment into precise behavioural modes. The theories of William Morris Davis (1850–1934) and Penck (1888–1923) attempted to categorise landscapes into distinct phases. Davis' theory of 'normal' landscape evolution involved three stages: 'youthfulness', 'maturity' and 'old age' and led to a compartmentalising of geomorphological ideas on landscape evolution for more than 50 years leading to a rigidity of landscape interpretation. A similar rigid theory was applied to vegetation succession (Tansley, 1923) in which vegetation was deemed to pass through a *seral* development that saw a vegetation community develop from pioneer to climax community. It did not concern

Table 9.2 Main ethical attitudes based on the reasoning of the great philosophers and the links to modern environmentalism

Ethical 'brand' name	Main proponents	Main features	Environmental application
Universalists	Plato, Kant	Fundamental principles of ethics are universal. Rules of right and wrong are valid regardless of interests, attitudes, desires or preferences	'Deep green' beliefs that all plants and animals, as well as inanimate objects, have equal rights as humans to exist
Relativists	The Sophists	Moral principles are relative to a particular person, society or situation. No absolute ethical principle applies to any situation. There are no facts, only interpretations	Each person or society is free to develop their own relationship with the environment, all of which are valid. There is no 'right' or 'wrong' way of treating the environment
Nihilists	Schopenhauer	Existence is arbitrary and the only meaning to life is the constant struggle for survival. There is no such thing as 'moral' behaviour. Only the *ability* to survive is of relevance, not the *means of survival*	Survival of the fittest
Utilitarians	Original proponent was Jeremy Bentham. Concept modified by J. S. Mill and later still by Gifford Pinchot	Pleasure is the only thing worth having in its own right. Selecting an ethical approach that achieves the greatest good for the greatest number of people is the best act society can make. When modified, this concept proposed that pleasures of the intellect were superior to pleasures of the body	Pinchot argued that the purpose of conservation was to protect resources for the 'greatest good for the greatest number *for the longest time*'. Utilitarianism remains popular in western society although its application to environmental matters can be used to justify antisocial behaviour
Modernism	Descartes, Bacon, Newton	By adopting a 'positive' approach, the modernists believed they could find universal laws of morality through the objective application of science	Application of a modernist approach achieves moral progress and universal justice resulting in an even share of resources and achievement of greatest happiness
Postmodernism (relativism)	Derrida, Lyotard, Foucault	Society comprises many diverse, marginalised and disempowered groups. The classical meaning of truth, reality and meaning do not exist. We can never know anything in its totality	The environment is an arbitrary and constantly changing social construction, with infinite interpretations. This approach can lead to 'stalemate' as every opinion must be considered valid

(*Source:* based on data from Cunningham and Saigo, 2001)

the modernist geographers if real plant communities did not show evidence of perfect seral development; a theoretical development for a vegetation community had been established and little attempt was made to validate the theory with real-world evidence.

It became evident that modernist philosophical theories did not apply to the natural world and in particular to the biological kingdom. Based on field evidence, geomorphologists and biogeographers gradually recognised that instead of following predictable stages of development, landscapes and vegetation progressed through an interminable development process in which a number of equally valid 'end points' could be recognised. These ideas neatly meshed into the *postmodern* school of thought that shows scepticism of previous rigid theory and instead argues that many different theories, processes, groupings and conclusions are possible. In this approach, 'end points' rarely remain defined for long as environmental circumstances constantly undergo change. Reality consists of complex, ever changing, chaotic conditions. Instead of precise rules, development takes place by means of 'fuzzy logic' that is used to explain the indeterminate end point for most natural development.

9.8 People's attitudes towards nature

The set of attitudes each person holds towards the environment is termed our 'worldview'. For many of us, our worldview may be set by the age of five years. Once fixed, it proves very difficult to change our own particular worldview even if clear evidence exists to the contrary. Furthermore, worldviews may not be arrived at rationally, they may be based on outdated or politically biased information but despite all their imperfections, our worldview can strongly influence our day-to-day behaviour.

Our worldview affects all our personal relationships including our attitude towards the environment. In our day-to-day lives most of our contact is with fellow human beings, for example parents, friends, partners, people we meet in cafés, shops and at work. How we relate to the different components in our lives will be determined by our

worldview and one of the most fundamental determinants is how we think of ourselves, as an individual, in relation to all other components of our environment.

One of the most fundamental considerations that determines our environmental worldview is deciding where *Homo sapiens* fits into the overall structure of the relationship between the non-living world (rocks, soil, water, air and climate) and the living components (plants and animals). In the most simple case, the relationship between all living and non-living components of the biosphere can be shown as the ecosystem model introduced earlier in this book (see Figure 2.1). In this worldview, plants and animals comprise the living biota and draw their nutrients, energy and living space requirements from the non-living, abiotic components comprising rocks, soil, water and climate. Humans are located alongside all the other animals and make the same requirements as other members of the biota. A different worldview, however, recognises that humankind controls its destiny more than any other plant or animal species by means of scientific knowledge and technological capability and consequently, *Homo sapiens* is clearly separate from other members of the biota. In Figure 2.2 the human component has been separated from the animal component and because of our ability to *control* our development we can place our species in the centre of the ecosystem diagram. Taken to its extreme, this scenario recognises that *Homo sapiens* dominates the rest of the ecosystem.

The ecosystem models shown in Figures 2.1 and 2.2 represent two quite common worldviews between which are a number of other widely held worldviews. The way in which the main worldviews broadly relate to one another is shown in Figure 9.4. At one end of the viewpoint spectrum is the *ecocentric* view that equates to Figure 2.1. Ecocentrism, as the name suggests, bases its viewpoint on ecological principles in which the individual species is subordinate to much larger processes at work in the world biome. Killing an individual plant or animal causes few problems for the ecosystem. Exterminating an entire species population is far more serious as an entire line of evolution has been destroyed. In the ecocentric

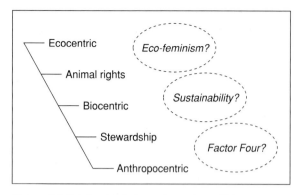

Figure 9.4 Relationship between the main worldviews

worldview, the individual is subordinate to both large-scale and long-term processes. Consequently, ensuring that conditions are suitable, for example for photosynthesis and biogeochemical cycles, take precedence over the survival of the individual – including humans.

The *animal rights* worldview has similarities to the ecocentric view apart from giving greater priority to the importance of individual animals, claiming that it is the way in which an individual animal functions within the ecosystem that provides the uniqueness to the system. Midway between the two extreme viewpoints is the *biocentric* worldview. Emphasis in this worldview remains centred on the belief that all living organisms have intrinsic values and rights regardless of whether they have economic value for humankind. Attention is concentrated on the need to maintain the greatest possible biodiversity as this is believed to allow maximum stability within the ecosystem.

Ecocentric and animal rights worldviews give little recognition to the different impacts made by humans on the environment compared to those of other animals. The biocentric worldview supports only a partial acceptance of the unique *impact* of humans on the environment, preferring instead to argue that the *demands* made by humans are no different from those of other animals. While this is a *theoretically* acceptable stance, in practice, even the most non-technical group of humans will make an impact on the environment different from that of the most advanced group of non-human animals, for example chimpanzees. An alternative worldview is therefore needed that recognises that,

due to the powers of the human intellect, our ability to apply reasoning to problem solving and decision making enable us to separate ourselves from other animals and by so doing, occupy a separate position within the ecosystem diagram as shown in Figure 2.2.

The problem becomes one of deciding how far humankind has moved away from a single animal grouping. The worldviews of *stewardship* and *anthropogenic domination* represent two possible steps that humans have followed. The first of these, stewardship, has been associated with previous stages of human development in which groups of people (equivalent to tribes or clans) have occupied specific areas of land (a territorial space) in which they have developed a strong sense of *stewardship* or responsibility for the ecosystem within which they lived. In particular, people with only a simple level of technical capability are sometimes considered to practise a land stewardship in which the amount of material harvested (in terms of numbers of animals hunted or trees cut down) is roughly equal to the capability of the ecosystem to regenerate the amount of material harvested. Supporters of the principles of stewardship argue that humility and reverence of nature are two essential human qualities necessary to make this type of relationship work. The writings of Leopold (1949) in which he argues for land owners to show responsibility in the manner in which they use the land, strongly supports the principles of stewardship. In primitive stewardship systems, hunters asked for the approval of the spirit world before embarking on a hunting expedition and before felling a tree, it would be necessary to appease its guardian spirit. Stewardship of the land recognises the need for the natural resources of the ecosystem to support all the animals that inhabit the ecosystem but accepts that the evolutionary development of *Homo sapiens* makes it inevitable that resources will be directed predominantly towards one species but not to the exclusion of others. Consequently, this worldview supports the *anthropogenic* perspective, but with the important proviso that stewardship must also apply to the non-human components of the ecosystem to ensure sustainability over a long time period. Present-day examples of a stewardship worldview include organic farming systems and societies in which recycling of waste and generation of power from renewable resources is practised.

The stewardship worldview contains a number of intrinsic problems. Because the system uses low levels of technical inputs (for example, little or no inorganic fertiliser and biological control of pests) the level of control exerted by humans on the ecosystem makes it difficult for the system to survive the natural annual variations in climate. The amount of food harvested under such a system is prone to considerable variation resulting in times of surplus food and at other, of famine. In order to minimise the effect of natural fluctuations, farmers would use simple technology to reduce some of the most easily avoided vagaries of nature. For example, ditches were cut to remove surplus water, animal manure was spread on fields to act as a fertiliser, the best animals were used as breeding stock and inedible plants were removed from crops. The success of such actions would be reflected in both an improved reliability and greater output from agricultural systems. A reciprocal relationship quickly became established in which more use of technology enabled greater amounts of food production to be achieved. Over time, fewer people were required as agricultural labour, allowing them to work in the newly industrialised societies. Unless a conscious decision is taken to retain a stewardship outlook on land use policy there will be an inevitable drift towards the final worldview, anthropocentric domination.

From the time of the Industrial Revolution in Europe, humankind has established an unchallenged position as the pre-eminent species on this planet. An ability to reshape the world to suit our needs has encouraged a belief that we are indeed superior to all other living organisms. Our position has subsequently been used to *dominate* nature. The prevailing ethical perspective condones our domination of natural systems. For ethical reasons, many people express a wish that they would prefer to support a stewardship worldview, but recognise the inability of such a system to support the current size of the human population and its associated demand for resources. Human society has become so wedded to the use of technology that there appears to be no alternative

perspective as a move from an anthropogenic worldview would inevitably mean a deterioration in the quality of life we currently enjoy.

The success of the anthropogenic domination worldview has been extremely successful in furthering the position of humankind, but what are the true costs of pursuing such worldview? Earlier chapters in this book have highlighted some of the problems that have resulted from implementing a worldview based on the total domination of one species, for example mismanaging fishery resources, the crisis facing many nations over the availability of fresh water and the concern over the loss of biodiversity. Have we passed too far down the pathway of human domination of the biosphere to turn around and look for an alternative and better direction? Would a change be possible without a massive loss in food production that would, in turn, cause famine, population decline and social unrest? Or is it inevitable that our exploitative system will grind to an inexorable halt as our agricultural systems fail through lack of chemical fertilisers or resistance to pesticides and insecticides allows agricultural pests and diseases to decimate our agricultural systems? Supporters of the anthropogenic argument claim that such a gloomy scenario can never occur, as past evidence shows that our ability to overcome adversity *always* wins the day. We can always find a solution to our problems.

There is an increasing concern that relying on a 'blue sky' research policy in which the 'technological fix' always wins the day may not always be correct. Was Commoner (1971) correct in claiming that the more we rely on technology the more vulnerable we become to even a partial failure? Are there viable alternative strategies that would allow modern society to reduce our dependence on technology? The final section examines three possible alternative worldviews.

9.9 Alternative worldviews

Our knowledge of the biosphere has benefited greatly from the overall improvement in natural and earth sciences. Advances in our technical capabilities have enabled our engineers to design and build more efficient machines, and to produce less pollution, consume less energy and make less noise. From the science of cybernetics (coined from the Greek *kybernetes* meaning 'a steersman'), we have gained an insight into the ways that complex ecosystems comprising plants, animals and the physical components communicate and control their systems. Through knowledge has come the ability to use parts of the biosphere more efficiently and hopefully to cause less harm in the form of degradation and exploitation. Technology, therefore, allows us to use the biosphere and its resources more efficiently, but at the same time has also allowed us to raise our demands made on the biosphere in the belief that our greater scientific and technical skills provides greater control over our actions. This approach typifies the anthropomorphic domination of the biosphere as discussed in section 9.8.

While at the most sub-microscopic level of the biosphere, the geneticist has been able to unravel the incredibly complex workings of DNA within individual cells, we have nothing like a comparable understanding of how plants, animals and the physical components of the biosphere operate together. The emerging knowledge of the so-called noosphere – the space occupied by the totality of information and human knowledge collectively available to humankind combined with all the processes associated with life operating in this space – is slowly allowing us an insight into the mechanisms that control the biosphere.

There is little doubt that humans are becoming increasingly separate from the rest of the natural world and it is to the social sciences that we must turn for new ideas on the ways in which humans can interact with their environment. In addition to the worldviews discussed in section 9.8 there are also a number of quite radical new ways through which we can view our environment. These are shown in Figure 9.4 as ecofeminism, sustainability and *Factor Four*, that sit alongside specific worldviews. The boundaries of these new viewpoints are diffuse and are open to provide information on the environmental debate and also to receive information from the traditional sciences. The viewpoints are dynamic and can respond to the rapidly changing world around us.

9.9.1 Eco-feminism

The role of women in the grassroots management of the biosphere has been both substantially underestimated and undervalued. In as many as two-thirds of all countries, especially developing countries, women hold predominant responsibility for managing the biosphere and are responsible for cultivating crops, collecting water and firewood and making food to support their families. In short, women must ensure that they can provide the basic commodities of life. If they cannot, their children will suffer. Men usually hold responsibility only for the livestock and are responsible for cultivating the land.

In recent years, the rise of the feminist viewpoint in society has highlighted the shortcomings in the male-dominated value system that usually applies to management of the biosphere. In developing countries, such a value system places greater prestige on features such as a wealth system based on the number of cattle held by the males of the family compared to the lower social value of crop production that is a female responsibility, yet the latter provides the food necessary for the family. In a patriarchal worldview, men are superior to women which results in an exploitation of women and female-dominated activities. NGOs and aid organisations have gradually learned that women (especially African women) possess a deep local knowledge that is passed from mother to daughter and has probably developed over long periods of time. The development of these skills is based on very local ecological circumstances and expressed in local languages, often through storytelling or song (Brokensha *et al.*, 1980). The development of an intimate knowledge of the local environmental condition has been called an *indigenous knowledge system* (IKS) and is built around a firm cultural basis and shared cognitive system. Agricultural systems using IKS typically show a wide ecological tolerance, maximise the number of crops grown, show maximum diversification during the poorest climatic periods and, most importantly, are capable of adaptation to suit changing local conditions (Gregory-Knight, 1980).

Eco-feminism has been proposed by workers such as Rocheleau *et al.* (1996) and Warren (1997) as an alternative to patriarchal systems. They suggest that any environmentalism or environmental ethic that does not make the connections between the treatment of non-human nature and the treatment of women, people of colour and other issues that involve the underclasses is 'grossly inadequate'. Their worldview seeks to replace the male-dominated value system based on rights, prestige, ownership and responsibility by a pluralistic, non-hierarchical, relationship-oriented philosophy (Cunningham and Saigo, 2001). The eco-feminist worldview suggests that care, reciprocity and values of kinship promote a richly textured understanding of the relationship between humans and the surroundings in which they live and allows cooperation between individuals to replace the competitive forces that have resulted in exploitation both of people and non-human components of the biosphere.

It is clear that the eco-feminist viewpoint differs fundamentally from all the other worldviews and therein occurs a potential weakness. Adoption of an eco-feminist approach would require a mindset change on behalf of all members of society. To achieve such a change would require a spiritual change akin to a religious revival, but one based not on religious fervour, but on rational scientific and ecological knowledge. It is difficult to see such a change occurring in the immediate future, and the emergence of a new environmental crisis might be required to galvanise the world society into action.

9.9.2 Sustainability

Throughout the period of human history, the level of use that we have exerted on the biosphere has always been close to the maximum extent of our technical capabilities. However, until the final years of the twentieth century, technology always appeared to lag behind what seemed to be necessary in order to provide humans with the desired control over their environment. By the final decade of the twentieth century new technological developments, particularly those in the fields of genetic engineering and microelectronics, began to provide humans with a hitherto unimaginable capability of modifying the biosphere and its components. Through the achievements of genetic transfer between species we have acquired the capability of changing the course of evolution and have altered the role of humans in determining the future of our environment.

However, for the time being we still do not possess complete control of the biosphere and the environment. We need the natural life-giving resources that only the biosphere can provide: clean air and water, soil in which to grow our crops, resources to maintain our industrial lifestyle that in turn generates wealth to support the development of yet more new technology. Society has always faced the problem that to feed the voracious appetite of technology we must use more and more of the planet's resources; we become trapped in a cycle in which we need new technology to enable society to progress and yet by using the planet's finite resources we threaten the long-term continuation of our existence.

An alternative worldview has emerged in which new technology is seen as providing a solution to the consumption of finite resources, but at the same time allowing our quality of life to go on improving. The worldview is founded on the concept of *sustainable development*, more recently, abbreviated to *sustainability*. The process by which the idea of sustainability came about has been explained in section 3.12. The Brundtland Commission (United Nations, 1987) made it clear that its findings marked a *pathway* and not a *blueprint*, in effect, a worldview that identified a series of ideas which, if followed, should result in a less destructive use of resources. The most central concepts of the sustainability worldview are based on natural processes, for example avoiding waste and reusing materials, cascading energy in order to capture as much of it as possible before releasing it as waste heat and ensuring that the ways in which we use the biosphere today do not prejudice the opportunities for resource use in the future. Mitchell (2002) observed that the Brundtland Report contained two other key concepts less frequently noted than that of the high-profile idea of sustainability. These are, the priority that must be given to the *needs* of the poor people of the world and the *limitations* created by using technology on the capacity of the biosphere to satisfy current and future requirements.

Ever since its inception in the mid-1980s, attaining even a modicum of sustainable development has proved elusive. Cynics have argued that putting the words 'sustainable' and 'development' together is a contradiction of meaning, for after all, 'sustainable' means an action that can continue uninterrupted over time, whereas 'development', as we recognise it, implies consumption in order to achieve growth. All our experience suggests that growth cannot be continuous because ultimately, it will be forced to stop due to a shortage of a critical component. On a technicality, therefore, the term 'sustainability' is preferred to 'sustainable development' as, used on its own, 'sustainability' can be interpreted more broadly. The Brundland Commission, it will be recalled, advocated that nations followed a 'pathway' towards sustainability and the signposts along the pathway are determined by the attainment of the *well-being of the nation*. Numerous attempts to construct an index of environmental well-being have been attempted, for example Thomas (1972) and Inhaber (1976). The most recent attempt by Prescott-Allen (1999, 2001), allows a measure of the extent to which a nation's environmental goals have been achieved and are determined according to series of relationships between ecosystem and human conditions. The calculation of an index of well-being is very complex but the results can show how the development of a nation is moving towards or away from sustainability. We are accustomed to monitoring the progress of society by means of economic indicators such as the gross domestic product and the index of leading economic indicators. By tradition, we fail to take account of sustainability issues. Prescott-Allen (*op. cit*) has developed an *indicator of human well-being*, based on statistics that combine measures on health, population and wealth, with those of environmental sustainability, such as water quality, species diversity and energy use, from which a comprehensive picture of the state of our world can be calculated. A well-being assessment value for 180 countries has been calculated. (The methodology used in the assessment along with a world map showing the assessment values can be found at the following website: http://www.idrc.ca/en/ev-41564-201-1-DO_TOPIC.html.) The assessment initially calculates the human well-being index (HWI) based on 36 indicators of socio-economic conditions and the ecosystem well-being index (EWI) synthesises 51 indicators of the state of the environment. A well-being/stress index (WSI) measures how much harm each country does to the environment for the level of development it achieves. Finally, the well-being index (WI) is computed based on the HWI and EWI values and allows a barometer of sustainability to be drawn which shows how far each country is from the goal of attaining high levels of human and ecosystem well-being.

The results of the well-being assessment show that two-thirds of the world population live in countries with a poor or bad human well-being index and only 15 per cent live in countries with a fair or good HWI. The disparity between the best off (countries such as Sweden, Finland, Norway, Iceland and Austria) and worst-off (Afghanistan, Syria and Iraq) is huge: the median HWI of the top 10 per cent of countries is almost eight times that of the bottom 10 per cent.

One major problem exists with any attempt to devise an index of well-being. No one knows precisely what a twenty-first century society based on indices of total sustainability should look like and therefore any index has to be an 'approximation' that will be refined and modified as our knowledge and understanding of sustainability changes. It is unlikely that governments and industrialists, so used to measuring 'performance' by GDP figures, will be keen to see it displaced as *the* indicator for the general condition of society. The well-being index provides a very different way of measuring the condition of society and unlike GDP makes a real attempt to include issues of sustainability within its calculation. The well-being index is unique in that it shows the impact our society makes on the environment. If governments are committed to moving towards an infrastructure based on indices of sustainability then it is imperative that a measure such as the well-being index is used alongside more traditional measures of economic condition.

9.9.3 Factor Four

Managing the environment is expensive. Research scientists working on environmental issues for long periods of time in university laboratories and at research institutes consume vast amounts of the world's wealth. When the results of these research projects are applied to our everyday lives it is usually the case that the consumer has to pay the cost of the research leading to the development of new technology. For example, despite considerable reductions in the cost of energy produced by renewable means it remains more expensive than energy generated from fossil fuels, therefore a consumer who chooses not to contribute to the atmospheric pollution load pays a premium for 'green' energy (see Table 9.3). The Club of Rome, an organisation committed to radical ways of solving the plethora of environmental problems facing the world, has examined how the efficiency of society could be improved through the use of the best technology. The report entitled *Factor Four* (von Weizsäker *et al.*, 1998) shows that the ways in which we use resources could be made much more efficient, that resource productivity could grow by 400 per cent, thus enabling us to live twice as well yet use only half the current level of resources. Such a hypothesis is alluring, seeming to promise 'something for nothing', yet how can it be possible?

Factor Four argues that by applying the best science and technology to everyday situations, by reducing waste and recycling, it is possible to use resources at least four times as efficiently compared to present use patterns. Conversely, it would be possible to maintain the current standard of living but use one-quarter of the resources currently used. A fourfold improvement is obtained through gains in efficiency of resource use and, unlike traditional conservation policies which advocate restraint in consumption with consequent restrictions on personal freedom, *Factor Four* achieves its improvement by increased industrial, commercial and managerial efficiency.

The report argues that one of the main barriers to achieving such massive efficiency gains is due to established social and cultural barriers to modern resource efficiency. For example, architects and engineers are paid a percentage commission of the total cost of a new building or machine they design. It is to the benefit of the designer to make the machine more complex because the profit margin on a complicated machine is greater than on a machine with fewer parts. A complex machine may require more frequent or more costly servicing charges, so continuing the profit motivation downstream. Designers should, instead, be paid a premium for ensuring their product can be recycled at the end of its life. Advertising should focus on low running costs of a machine; electricity, gas and water charges should be loaded in favour of small efficient users instead of, as at present, discounts given to consumers using most. The report provides 50 examples of how resource productivity can be quadrupled. Three examples are provided in Box 9.4.

Society needs a sound economic and industrial basis from which to generate wealth in order to sustain our quality of life. In the past, wealth was generated by utilising the resources of the planet with little thought for the future availability of resources or of the effects on the environment caused by pollution generated by our industrialised societies. We now understand sufficient about

Table 9.3 Renewable energy cost trends in the USA, 1980–1998

Energy source	1980	1998
Photovoltic	90¢/kWh	<20¢/kWh
Bioethanol	$3.50/gal	$1.00
Wind turbines	40¢/kWh	5¢/kWh
Geothermal	9¢/kWh	4¢/kWh
Solar thermal	40¢/kWh	6¢/kWh
Biomass (energy crops)	11¢/kWh	8¢/kWh
Thermal power station for comparison		8¢/kWh

(*Source*: http://www.hawaii.gov/dbedt/ert/femp/cannonfiles/sld010.htm)

BOX 9.4

Examples of quadrupling resource productivity

Example 1   Buildings in high-latitude countries must be able to provide comfortable interior temperatures even though external temperatures in winter fall to −15°C or lower. Consumption of energy for space heating forms a major addition to the cost of living in high latitudes. Preventing heat loss from buildings through thermal insulation is vital. In 1983 Sweden adopted a thermal insulation standard of less than 60kWh/sq m-y. By comparison, in Germany, the typical heat loss from houses was 200kWh/sq m-y. By adopting Swedish standards, the German construction industry could achieve an almost fourfold improvement thereby meeting the target of *Factor Four*. Spurred into meeting the Swedish target, German architects designed and built the *Passivhaus* with an energy requirement for space heating of only 15kWh/sq m-y. First costs of energy-efficient houses are greater than for traditional designs, but the German experience has shown that with savings achieved from mass production combined with substantially lowered running costs, the lifetime costs of energy efficient houses are less than traditional designs. In addition, the new designs offer higher comfort levels, are quieter and more secure.

Example 2   Lighting buildings and streets consumes up to 20% of all electricity generated, with an additional 5% necessary to remove the heat created when the lamps are in use. Approximately half of all lamps in use today are of the 'old-fashioned' incandescent type, invented in the 1930s. Compact fluorescent lamps were introduced in 1981. A modern 18-watt compact fluorescent lamp produces as much light as a 75-watt incandescent lamp, a fourfold improvement and lasts ten times as long. Over its lifetime, the 18-watt lamp can save one tonne of carbon dioxide output, 4kg sulpher dioxide and 1kg nitrogen oxide. It saves 200 litres of fuel oil that would otherwise have been consumed in an oil-fired thermal power station. When combined with a daylight-sensing switch, the lamp can be made to switch off automatically when required. In some situations, converting from incandescent to compact fluorescent lamps has been sufficient to avoid expanding existing power stations, or to prevent 'load shedding', so common in developing countries where generating capacity is insufficient to meet peak demand for electricity.

Example 3   Water used for irrigation accounts for about 65% of fresh water consumption. The quality of water used in irrigation need not meet the purity standards for drinking purposes: it can be discoloured, or contain higher levels of bacteria but should not contain excessive concentrations of potentially harmful dissolved salts, such as magnesium or cobalt. The efficiency of irrigation systems is measured by the amount of water applied to a field that is actually taken up by the crop. Conventional irrigation systems based on furrow or flood irrigation have an efficiency in the range 40–60%. The rest is lost through surface runoff, deep percolation or losses via wind spray. By replacing surface or overhead irrigation systems by sub-surface pipes, water use efficiency can rise to 95%, an improvement of about 1.6 times. Many other benefits follow from using sub-surface systems. Less water is required, reducing (by up to 50%) the amount and cost of electricity used in pumping, and reducing the amount paid for using large volumes of water. Insecticides and fertilisers can be added to the water supply delivered to the root zone of the crop, helping to increase crop productivity and provide higher economic returns. Less time and energy is required to cultivate soils that are irrigated by sub-surface pipes, leading to further cost savings and helping to achieve a fourfold improvement in benefits from using sub-surface irrigation.

(*Source*: based on von Weizsäker *et al.*, 1998)

our planet to realise that resources need to be conserved and that the true meaning of the term 'economics' derived from its Greek root is 'the [wise] management of the household'. Many of the proposals made in *Factor Four* are expressions of the wise management of resources.

9.10 Conclusion

The challenge of coping with our environment is as old as the human race itself. As a species we have been confined and challenged by the environment. It has provided a challenge to our developing intellect and in the past has placed a restriction on the way in which we have evolved. It is unlikely that the relationship between humans and the environment will change substantially within the next 100 years. Mistakes in the way we have interacted with the environment have occurred and this has led to the so-called 'environmental disasters'. A feature of the human intellect is that we can learn from our mistakes and in recent decades we have become more aware than ever of the need to manage our environment so that the risk of creating insurmountable problems for future generations will, as far as we can predict, be avoided.

A large number of different environmental problems have been covered in earlier chapters of this book, but several common themes can be identified throughout. First, has been the way in which misuse of technological skills have resulted in overconsumption, pollution and degradation of specific parts of the biosphere. This has been due in part to ignorance and to economic greed. It has also been due to a lack of 'ownership' of the biosphere. Apart from the land and the resources to be found on and beneath it, few other parts of the biosphere have been seen as the responsibility either of individuals or of society as a whole. As economists are keen to point out, lack of ownership can often lead to misuse and Hardin's exposition of the *Tragedy of the Commons* (outlined in section 7.6) showed how a lack of community responsibility led to a philosophy that encouraged, or even dictated, that we exploit the biosphere. It is paradoxical that the fear of poverty drives us to exploit the biosphere reserves. A society trapped

in poverty is prevented us from spending time and money on conservation of the environment. There comes a point in the fortunes of society when, through the use of biosphere resources, sufficient wealth has been generated to pay back the environment through a form of environmental management. Society needs to recognise its indebtedness to the environment and to develop the social maturity and ethical responsibility to ensure it can meet its debt. For as long as we measure the well-being of society only in economic output we will have difficulties in recognising our environmental debt. Perhaps the greatest advance would be the adoption of an index of ecological well-being as proposed by Prescott-Allen in section 9.9.2. In a modern world obsessed by performance indicators it has become necessary to find an accurate method of evaluating the state of the environment.

While there are undoubtedly many environmental problems remaining to be discovered, there are also many areas in which we have achieved success in the way we have managed the environment. The reafforestation programme adopted by almost all mid- and high-latitude countries has meant that we have a far greater area of forest cover at the beginning of the twenty-first century than we did 100 years earlier. It is true that much of the new forest comprises 'exotic' species (not native to the area in which they have been planted) but at least the new forest has protected the soil from erosion and helps prevent runoff and flooding. The forest also absorbs atmospheric carbon dioxide and without the forest our atmosphere would have warmed even more than it has. Urban air quality is also immeasurably better in most of the old industrial nations now compared to 100 years ago.

Another major achievement for society has been the way in which our elected politicians recognise the need for international collaboration and agreement on ways to manage environmental problems. It is true that many different opinions exist in achieving the best way to manage the environment, but through the process of informed debate and negotiation very considerable achievements have been made. For example, through the work of CITES (see section 8.13), it has been possible to control the international trade of endangered species. The embargo on the trade in ivory has been particularly successful, especially when it is realised that, for some African countries, trade in ivory could generate a substantial proportion of foreign earnings.

The willingness of many of the more impoverished developing nations to accept the need to manage their natural resources has been remarkable. Section 3.12 examined the development of international treaties and agreements to which the developing nations have been willing participants. Despite this good intent on the behalf of the developing nations, the environmental problems they face in the twenty-first century are formidable. Shortage of fresh water and soil erosion have been considered in Chapters 5 and 7 respectively. Solving these problems will probably be beyond the financial ability of developing nations. Indeed, it will be to the benefit of developed nations to actively provide technical assistance and financial aid to mitigate these, and many other environmental problems, in developing nations. We must accept that the economic and industrial globalisation of the world's resources has meant that scarcity of resources and environmental problems are not only the concern of local communities but have assumed international significance. Perhaps at last, the fundamental importance of the all-embracing role of the biosphere, that forms the core of environmental geography, will be accepted by society as a whole. If that can be achieved then the prospect for the environment in the twenty-first century has a positive outlook.

Useful websites for this chapter

California Energy Commission
http://www.energy.ca.gov/electricity/
silicon_valley_consumption.html

Cold Fusion
http://www.alternativescience.com/cold_fusion.htm

Department of Natural Resources Energy,
British Columbia
http://www.gnb.ca/cnb/promos/NBPower/
Refurbishment-e.asp

Ecomanagement and Audit Scheme (EMAS)
http://europa.eu.int/comm/environment/emas/about/
summary_en.htm

Environment Agency NatRegs
http://www.environment-agency.gov.uk/netregs/
legislation/?lang=_e

EU energy supply and demand
http://europa.eu.int/comm/energy_transport/atlas/
htmlu/mardep1.html

Hawaii Energy, Resources, and Technology Division
http://www.state.hi.us/dbedt/ert/

ISO 14000
http://www.iso.ch/iso/en/prodsservices/otherpubs/
iso14000/index.html

River Deep
http://www.riverdeep.net/current/2000/11/
112200_worldwatch.jhtml

UK government white paper on Energy, February 2002
http://www.dti.gov.uk/energy/whitepaper/
ourenergyfuture.pdf

Well-being Of Nations
http://www.idrc.ca/en/ev-41564-201-1-DO_TOPIC.html

Bibliography

Abo Hassan, A. A. (1981) Rangeland management in Saudi Arabia. *Rangelands* 3(2): 51–3.

Abuzinada, A. H. and Chapterild, G. (1991) Developing a system of protected areas in Saudi Arabia. National Commission for Wildlife Conservation and Development, Riyadh. Paper presented at the Third Man and Biosphere Meeting on Mediterranean Biosphere Reserves and the First IUCN–CNPPA meeting for the Middle East and North Africa, 14–19 October 1991, Tunis.

Acocks, J. P. H. (1975) *Veld Types of South Africa* (2nd edn). Botanical Research Institute, Department of Agricultural Technical Services, South Africa.

Adger, W. N. (2000) Environmental and ecological economics in O'Riordan, T., *Environmental Science for Environmental Management* (2nd edn). Longman, Harlow.

Agassiz, L. (1840) *Etudes sur les Glaciers*. Neuchâtel.

Ahmed, S. (2001) Framework for an internet-based decision support system. PhD thesis, Department of Geography, University of Strathclyde, Glasgow.

Ahrens, C. D. (1994) *Meterology Today. An Introduction to Weather, Climate and the Environment*. West Publishing Company, Minneapolis/St Paul.

Aiello, L. C. and Collard, M. (2001) Palaeoanthropology: our newest oldest ancestor? *Nature* 410, 526–7.

Akintola, F. O., Areola, O. and Faniran, A. (1980) The elements of quality and social costs in rural water supply and utilisation. *Water Supply and Management* 4, 275–82.

Albrecht, W. A. (1971) Physical, chemical and biochemical changes in the soil community in Detwyler, T. R., *Man's Impact on Environment*. McGraw-Hill, New York.

Aldous, T. (1972) *Battle for the Environment*. Fontana, Wm Collins Sons & Co. Ltd, Glasgow.

Ali-Ibrahim, A. (1991) Excessive use of ground water resources in Saudi Arabia: impacts and policy options. *Ambio* 20(1): 34–7.

Allan, R., Lindesay, J. and Parker, D. (1996) El Niño, southern oscillation and climatic variability. CSIRO, Collingwood, Victoria, Australia.

Allen, S. W. and Leonard, J. W. (1966) *Conserving Natural Resources*. McGraw-Hill, New York.

Allison, R. (1994) Environment and water resources in the arid zone in Cooper, D. E. and Palmer, J. A. (eds), *The Environment in Question. Ethics and Global Issues*. Routledge, London.

Anderson, D. (1997) Turning back the harmful red tide. *Nature* 388: 513–14.

Anderson, G. C. (1972) The head has a stomach in Smith, R. L. (ed.), *The Ecology of Man: An Ecosystem Approach*. Harper & Row, New York.

Armstrong, S. J. and Botzler, R. G. (1993) *Environmental Ethics: Divergence and Convergence*. McGraw-Hill, New York.

Arnell, N. (2002) *Hydrology and Global Environmental Change*. Prentice Hall, Harlow.

Ba Kader, A. B. A., Al Sabbagh, A. L. T. S., Al Glenid, M. S. and Izzidien, M. Y. S. (1983). Islamic principles for the conservation of the natural environment. IUCN Environmental Policy and Law Paper No. 20.

Bahro, R. (1986) *Building the Green Movement*. GMP, London.

Barber, C. V. and Schweithelm, J. (2000) *Trial by Fire: Forest Fires and Forestry Policy in Indonesia: Era of Crisis and Reform*. World Resources Institute, Washington, DC.

Barrow, C. J. (1991) *Land Degradation. Development and Breakdown of Terrestrial Environments*. Cambridge University Press, Cambridge.

Barry, R. G. and Chorley, R. J. (1998) *Atmosphere, Weather and Climate* (6th edn). Routledge, London.

Bates, M. The Human Ecosystem in *Resources and Man*. National Academy of Sciences. National Research Council, Freeman, San Francisco.

Batterbury, S. (1998) Shifting sands, *Geographical* May: 40–5.

Bean, M. J. (1997) A policy perspective on biodiversity protection and ecosystem management in Pickett, S. T. A., Ostfeld, R. S., Shackak, M. and Likens, G. E. (eds), *The Ecological Basis of Conservation. Heterogeneity, Ecosystems and Biodiversity*. Chapman & Hall, New York.

Bedi, R. (1998) Heat kills 713 in India. *Daily Telegraph* 1 June, Issue 1102.

Benchley, P. and Gradwohl, J. (2000) *Ocean Planet: Writings and Images of the Sea*. Harry N. Abrams Inc., New York.

Benn, D. I. and Evans, D. J. A. (1998) *Glaciers and Glaciation*. Arnold, London.

Berger, A. and Loutre, M. F. (1991) Insolation values for the climate of the last ten years. *Quarternary Science Reviews* **10**: 297–317.

Berry, C. R. and Hepting, G. H. (1964) Injury to Eastern White Pine by unidentified atmospheric constituents. *Forest Science* **10**(1): 2–15.

Bewley, R. J. F., Jeffries, R. and Bradley, K. (2000) Chromium contamination: field and laboratory remediation trials. CIRIA, London.

Bibby, C. J. (1995) Recent past and future extinctions in birds in Lawton, J. H. and May, R. M. (eds), *Extinction Rates*. Oxford University Press, Oxford.

Bibby, J. S. and Mackney, D. (1977) Land use capability classification. Technical Monograph No. 1. The Soil Survey, Harpenden, Herts.

Bigg, G. R. (1996) *The Oceans and Climate*. Cambridge University Press, Cambridge.

Billings, W. D. (1969) *Plants and the Ecosystem*. Wadsworth Publishing, Belmont, CA.

Blackmore, D. J. (1995) Murray–Darling Basin Commission: a case study in integrated catchment. *Wat. Sci. Tech.* **32**(5–6), 15–25.

Blaikie, P. (1985) *The Political Economy of Soil Erosion in Developing Countries*. Longman, London.

Blaikie, P. and Brookfield, H. (1987) *Land Degradation and Society*. Methuen, London.

Blunden, J. and Reddish, A. (eds) (1996) *Energy, Resources and Environment* (2nd edn). Hodder & Stoughton, London.

Boardman, J. and Evans, R. (1997) Soil erosion in Britain: a review in Goudie, A. S. (ed.), *The Human Impact Reader. Readings and Case Studies*. Blackwell, Oxford.

Bocock, R. (1993) *Consumption*. Routledge, London.

Bolin, B., Döös, B. R., Jäger, J. and Warrick, R. A. (eds) (1986) *The Greenhouse Effect, Climate Change, and Ecosystems*. SCOPE 29. Wiley & Sons, Chichester.

Borrow, G. (1836) *Wild Wales*. Reprinted 1955. Collins, London.

Botzler, R. G. and Armstrong, S. J. (1993) *Environmental Ethics: Divergence and Convergence*. McGraw-Hill, New York.

Boughey, A. S. (1968) *Ecology of Populations*. Macmillan, New York.

Boulding, K. (1965) Earth as a spaceship. Kenneth E. Boulding Papers, Archives (Box #38), University of Colorado at Boulder Libraries. Also available at: http://csf.colorado.edu/authors/Boulding.Kenneth/spaceship-earth.html.

Bowler, J. M., Johnston, H., Olley, J. M., Prescott, J. R., Roberts, R. G., Shawcross, W. and Spooner, N. A. (2003) New ages for human occupation and climatic change at Lake Mungo, Australia. *Nature* **421**: 837–40.

Bradbury, I., Boyle, J. and Morse, A. (2002) *Scientific Principles for Physical Geographers*. Prentice Hall, Harlow.

Bradshaw, M. and Weaver, R. (1995) *Foundations of Physical Geography*. W. C. Brown, Dubuque, IA.

Braidwood, R. J. (1971) The agricultural revolution in Ehrlich, P. R., Holdren, J. P. and Holm, R. W. (eds), *Man and the Ecosphere*. Scientific American Books, W. H. Freeman and Co., San Francisco.

Brammer, H. (1990) Floods in Bangladesh I: geographical background to the 1987 and 1988 floods. *The Geographical Journal* **156**(1): 12–22.

Brokensha, D. W., Warren, D. M. and Werner, O. (1980) *Indigenous Knowledge Systems and Development*. University of America Press, Washington, DC.

Bronowski, J. (1973) *The Ascent of Man*. British Broadcasting Corporation, London.

Brown, L. (1997) Facing the prospect of food shortage in World Watch Institute, *State of the World 1997*. Earthscan, London.

Brunet, M. *et al.* (2002) A new hominid from the Upper Miocene of Chad, Central Africa. *Nature* 11 July, **418**: 145–51.

Brunhes, J. (1920) *Géographie Humaine de la France*. Hanotaux, Histoire de la France, Paris.

Brundtland, G. H. (1987) *Our Common Future*. World Commission on Environment and Development, Oxford University Press, Oxford.

Bryant, D., Burke, L., McManus, J. and Spalding, M. (1998) *Reefs at Risk: A Map-based Indicator of Threats to the World's Coral Reefs*. World Resources Institute, Washington, DC.

Bryant, D., Neilson, D. and Tangley, L. (1997) *The Last Frontier Forests: Ecosystems and Economics on the Edge*. World Resources Institute, Washington, DC.

Bryson, R. A. (1971) 'All other factors being constant . . .' – theories of global climatic change in Detwyler, T. R. (ed.), *Man's Impact on Environment*. McGraw-Hill, New York.

Buck, K. (1989) Brave new botany. *New Scientist* **122**(1667): 50–5.

Budge, I., Crewe, D., McKay, D. and Newton, K. (eds), (1998) *The New British Politics*. Longman, Harlow.

Burton, I., Kates, R. W. and White, G. F. (1994) *The Environment as Hazard*. Longman, Harlow.

Butzler, K. W. (1982) *Archaeology as Human Ecology.* Cambridge University Press, Cambridge.

Buzan, B. (1976) *Seabed Politics.* Praeger, New York.

Carson, R. (1963) *Silent Spring.* Houghton Mifflin, Boston, MA.

Caughley, G. and Sinclair, A. R. E. (1994) *Wildlife Ecology and Management.* Blackwell Scientific, Oxford.

Chapman, R. N. (1931) *Animal Ecology with Especial Reference to Insects.* McGraw-Hill, New York.

Chepstow-Lusty, A. J., Bennett, K. D., Fjeldsa, J., Kendall, A., Galiano, W. and Tupayachi Herrera, A. (1998) Tracing 4,000 years of environmental history in the Cuzco area, Peru, from the pollen record. *Mountain Research and Development* **18.2**: 159–72.

Child, G. and Grainger, J. (1990) *A System Plan for Protected Areas for Wildlife Conservation and Sustainable Rural Development in Saudi Arabia.* NCWCD, Riyadh.

Chisholm, A. and Dumsday, R. (1987) *Land Degradation. Problems and Policies.* Cambridge University Press, Cambridge.

Chivian, E. (1993) *Critical Condition: Human Health and the Environment.* MIT Press, Cambridge, MA.

Christianson, G. E. (1999) *Greenhouse. The 200-year Story of Global Warming.* Constable, London.

Cicin-Sain, B. and Knecht, R. (1998) *Integrated Coastal and Ocean Management. Concepts and Practices.* Island Press, Washington, DC.

Coffey, M. (1978) The dust storms. *Natural History* (New York) **87**: 72–83.

Cohen, J. B. and Ruston, A. G. (1911) *Smoke, A Study of Town Air.* Edward Arnold & Co., London.

Cohen, J. E. (1997) Conservation and human population growth: what are the linkages? in Pickett, S. T. A., Ostfeld, R. S., Shackak, M. and Likens, G. E. (eds), *The Ecological Basis of Conservation. Heterogeneity, Ecosystems and Biodiversity.* Chapman & Hall, New York.

Cohen, M. N. and Armelagos, G. J. (1984) *Paleopathology at the Origins of Agriculture.* Academic Press, Orlando.

Common, A. S. and Perrings, C. (1992) Towards an ecological economics of sustainability. *Ecological Economics* **6**: 7–34.

Common Ground (2002) A new role for renewable energy. Common ground. A triannual report on Germany's environment. Federal Ministry for the Environment, Nature Conservation and Nuclear Safety. 02/2002, Berlin.

Commoner, B. (1971) Evaluating the biosphere in Detwyler, T. R., *Man's Impact on Environment.* McGraw-Hill, New York.

Conway, W. (1988) Can technology aid species preservation? in Wilson, E. O. and Peter, F. M. (eds), *Biodiversity.* National Academy Press, Washington, DC.

Cooke, R. H. and Doornkamp, J. C. (1990) *Geomorphology on Environmental Management* (2nd edn). Clarendon Press, Oxford.

Cooper, D. E. (1992) The idea of environment in Cooper, D. E. and Palmer, J. A. (eds), *The Environment in Question. Ethics and Global Issues.* Routledge, London.

Cooper, D. E. and Palmer, J. A. (eds) (1992) *The Environment in Question. Ethics and Global Issues.* Routledge, London.

Cooper, N. S. and Carling, R. C. J. (eds) (1996) *Ecologists and Ethical Judgements.* Chapman & Hall, London.

Couper-Johnston, R. (2000) *El Niño: The Weather Phenomenon that Changed the World.* Hodder & Stoughton, London.

Cousteau, J. Y. (1981) *The Cousteau Almanac. An Inventory of Life on Our Water Planet.* Jacques-Ives Cousteau and the staff of the Cousteau Society, Doubleday, Garden City, NY.

Cowles, H. C. (1899) The ecological relations of the vegetation of the sand dunes of Lake Michigan. *Bot. Gaz.* **27**: 95–117, 167–202, 281–308, 361–91.

Critchley, W. R. S., Reij, C. and Wilcocks, T. (1994) Indigenous soil and water conservation: a review of the state of knowledge and prospects for building on tradition. *Land Degradation and Rehabilitation* **5**: 293–314.

Crosby, A. (1986) *Ecological Imperialism: The Biological Expansion of Europe, 900–1900.* Cambridge University Press, Cambridge.

Crutzen, P. J. and Golitsyn, G. S. (1992) Linkages between global warming, ozone depletion, acid deposition and other aspects of global environmental change in Mintzer, I. M. (ed.), *Confronting Climate Change. Risks, Implications and Responses.* Cambridge University Press, Cambridge.

Cunningham, W. P. and Saigo, B. W. (2001) *Environmental Science: A Global Concern.* McGraw-Hill, Boston.

Cutter, S. L. and Renwick, W. H. (1999) *Exploitation, Conservation, Preservation. A Geographic Perspective on Natural Resource Use* (3rd edn). Wiley & Sons Inc., New York.

Dahl, B. and Blanck, H. (1996) Toxic effects of the antifouling agent Irgaol 1051 on periphyton communities in coastal water microcosms. *Marine Pollution Bulletin* **32**: 342–50.

Daily, G. (ed.) (1997) *Nature's Services: Societal Dependence on Natural Ecosystems*. Island Press, Washington, DC.

Daily Telegraph (1996) Russians to sink rotting N-sub fleet. *Daily Telegraph* 29 April, Issue 372.

Daily Telegraph (1996) More than 80 killed as big freeze grips Europe. *Daily Telegraph* 31 December, Issue 585.

Darling, F. F. and Dasmann, R. F. (1972) The ecosystem view of human society in Smith, R. L. (ed.), *The Ecology of Man: An Ecosystem Approach*. Harper & Row, New York.

Davidson, D. A. and Jones, G. E. (1986) A land resources information system (LRIS) for land use planning. *Applied Geography* 6: 255–65.

Deegan, J. (1987) Looking back at Love Canal. *Environmental Science and Technology* 21: 328–31.

DEFRA (2001) *Climate Change. UK Programme – Summary*. Department of Environment, Food and Rural Affairs. HMSO, London.

Department of the Environment (DoE) (1991) *The Potential Effects of Climate Change in the United Kingdom*. HMSO, London.

Department of the Environment (DoE) (1995) *The Habitats Directive. How it will Apply to Great Britain*. Department of the Environment, HMSO, London.

Department of the Environment (DoE) (1996) *Review of the Potential Effects of Climate Change in the United Kingdom*. UK Climate Change Impacts Review Group, HMSO, London.

Department of the Environment (DoE) (1998) *UK Climate Change Programme – A Consultation Paper*. Department of the Environment, Transport and the Regions, HMSO, London.

Department of Transport (1993) *The Braer Incident, Shetland Islands, January 1993*. Department of Transport, Marine Emergencies Organisation, Southampton.

Des Jardins, J. R. (1993) *Environmental Ethics. An Introduction to Environmental Philosophy*. Wadsworth Publishing, Belmont, CA.

Detwyler, T. R. (ed.) (1971) *Man's Impact on Environment*. McGraw-Hill, New York.

Diamond, J. (1997) *Guns, Germs and Steel: The Fates of Human Societies*. Norton, New York.

Diamond, J. (2002) Evolution, consequences and future of plant and animal domestication. *Nature* **418**: 700–7.

Dixon, T. (1990) An investigation by von Franeker (1985), quoted by Dixon in Marine Forum for Environmental Issues, North Sea Report.

Dobson, A. (1996) *Green Political Thought* (2nd edn). Routledge, London.

Dobson, A. P. (1995) *Conservation and Biodiversity*. Scientific American Library, W. H. Freeman and Co., New York.

Dodds, F. (1992) *Earth summit '92*. United Nations Association, London.

Dolan, R. and Goodell, H. G. (1986) Sinking cities. *American Scientist* 74: 38–47.

Douguédroit, A. (1997) On relationships between climate variability and change, and societies, pp. 21–41 in Yoshino, M., Domrös, M., Douguédroit, A., Paszyński, J. and Nkemdirim, L. C. *Climates and Societies – A Climatological Perspective*. Kluwer Academic Publishers, Dordrecht, The Netherlands.

Draz, O. (1985) The *hema* system of range reserves in the Arabian Peninsula, its possibilities in range improvement and conservation projects in the Near East in McNeely, J. A. and Pitt, D. (eds), *Culture and Conservation: The Human Dimension in Environmental Planning*. Croom Helm, London.

Duckham, A. N. and Roberts, J. G. W. (eds) (1976) *Food Production and Consumption. The Efficiency of Human Food Chains and Nutrient Cycles*. North Holland, Amsterdam.

Dury, G. H. (1981) *Environmental Systems*. Heinemann, London.

Eagles, P. F. J. (1984) *The Planning and Management of Environmentally Sensitive Areas*. Longman, London.

Earle, S. (1999) Foreword in Thorne-Miller, B., *The Living Ocean. Understanding and Protecting Marine Biodiversity*. Island Press, Washington, DC.

Eckersley, R. (1992) *Environmentalism and Political Theory: Towards an Ecocentric Approach*. UCL Press, London.

Eden, M. (1996a) Land degradation: environmental, social and policy issues in Eden, M. and Parry, J. T. (1996) *Land Degradation in the Tropics. Environmental Policy and Issues*. Pinter, London.

Eden, M. (1996b) Forest degradation in the tropics: environmental and management issues in Eden, M. and Parry, J. T. (1996) *Land Degradation in the Tropics. Environmental Policy and Issues*. Pinter, London.

Eder, K. (1996) *The Social Construction of Nature*. Sage, London.

Ehrlich, A. H. and Ehrlich, P. R. (1987) *Earth*. Franklin Watts, New York.

Ehrlich, P. R. (1965) *The Population Bomb*. Ballantine, New York.

Ehrlich, P. R. and Ehrlich, A. H. (1982) *Extinction*. Gollancz, London.

Ehrlich, P. R. E. (1977) *Ecoscience: Population, Resources, Environment*. Freeman Books, San Francisco.

Ehrlich, P. R., Ehrlich, A. H. and Holdren, J. P. (1973) *Human Ecology: Problems and Solutions*. W. H. Freeman and Co., San Francisco.

Ehrlich, P. R., Holdren, J. P. and Holm, R. W. (eds) (1971) *Man and the Ecosphere*. Scientific American Books, W. H. Freeman and Co., San Francisco.

Ekins, P. (1993) 'Limits to growth' and 'sustainable development': grappling with ecological realities. *Ecological Economics* 8(3): 269–88.

Ekins, P., Hillman, M. and Hutchinson, R. (1992) *Wealth Beyond Measure: An Atlas of New Economics*. Gaia, London.

Elton, C. (1927) *Animal Ecology*. Macmillan, New York.

Environmental Development Consultants (1997) *Compliance Costings for Approximation of EU Environmental Legislation in the CEECs*. Environmental Development Consultants, Dublin.

Evans, D. (1992) *A History of Nature Conservation in Britain*. Routledge, London.

Faith, W. L. A. (1972) *Air Pollution*. Wiley Interscience, New York.

Farman, J. (1987) What hope for the ozone layer now? *New Scientist* 116(1586): 50–4.

Farman, J., Gardiner, B. G. and Shanklin, J. D. (1985) Large losses in total ozone in Antarctica reveal seasonal ClO_x/NO_x interaction. *Nature* 315: 207–10.

Fawcett, C. B. (1917) The natural divisions of England. *Geographical Journal* 49: 124–41.

Federal Ministry of the Environment (1997) Towards sustainable development in Germany. Report of the Government of the Federal Republic of Germany, Federal Ministry of the Environment, Nature Conservation and Nuclear Safety, Bonn.

Fellmann, J., Getis, A. and Getis, J. (1995) *Human Geography. Landscapes of Human Activity*. W. C. Brown, Dubuque, IA.

Ferraro, P. J. and Kiss, A. (2002) Direct payments to conserve biodiversity. *Science* 298: 1718–19.

Field, B. C. and Field, M. K. (2002) *Environmental Economics: An Introduction* (3rd edn). McGraw-Hill/Irwin, New York.

Fitter, R. and Fitter, M. (1987) *The Road to Extinction*. IUCN, Gland, Switzerland.

FitzPatrick, E. A. (1971) *Pedology. A Systematic Approach to Soil Science*. Oliver & Boyd, Edinburgh.

Flavin, C. and Dunn, S. (1998) Responding to the threat of climate change in *State of the World 1998*. Worldwatch Institute, Earthscan, London.

Flenley, J. R. (1979) *The Equatorial Rain Forest. A Geological History*. Butterworth, London.

Flynn, A. and Lowe, P. (1992) The greening of the Tories: the Conservative Party and the environment in Rudig, W. (ed.), *Green Politics Two*. Edinburgh University Press, Edinburgh.

Folland, C. K., Karl, T. R. and Vinnikov, K. Y. A. (1990) Observed climate variations and change in Houghton, J. T., Jenkins, G. J. and Ephraums, J. J. (eds), *Climate Change. The IPCC Scientific Assessment*. World Meteorological Organization/United Nations Environment Programme, Cambridge University Press, Cambridge.

Folsome, C. E. (1979) *The Origin of Life. A Warm Little Pond*. W. H. Freeman and Co., San Francisco.

Food and Agriculture Organization (1992) *World Food Supplies and the Prevalence of Chronic Undernutrition in Developing Regions*. FAO, Rome.

Food and Agriculture Organization (1999) Report of the Food and Agriculture Organization of the United Nations (FAO) for the Third Conference of Parties of the Convention to Combat Desertification (CCD). Recife, Brazil 16–29 November. FAO, Rome.

French, P. W. (1997) *Coastal and Estuarine Management*. Routledge, London and New York.

Fuller, K. S. (1988) Debt-for-nature-swaps: a new conservation tool. *Economics of Environmental Protection* 65: 39–44.

Furley, P. A. and Newey, W. W. (1983) *Geography of the Biosphere*. Butterworths, London.

Garcia, S. M. and I. De Leiva Moreno (2000) Trends in world fisheries and their resources: 1974–1999 in *State of Fisheries and Aquaculture 2000*. FAO, Rome.

Gardner, G. (1997) *State of the World*. Earthscan, London.

Garrod, G. (2000) *Economic Evaluation of the Environment – Methods and Case Studies*. Edward Elgar, Cheltenham, England.

Gaston, K. J. (ed.) (1996) *Biodiversity. A Biology of Numbers and Difference*. Blackwell Science, Oxford.

Gause, G. F. (1934) *The Struggle for Existence*. Williams and Wilkins, Baltimore. Reprinted 1964, Hafner, New York.

Gerasimov, I. P. (1976) Problems of natural environment transformation in Soviet constructive geography. *Progress in Geography* 9: 75–99.

Gerrard, S. (2000) Environmental risk management in O'Riordan, T. (ed.), *Environmental Science for Environmental Management* (2nd edn). Longman, Harlow.

GESAMP (1981) Scientific aspects of marine pollution. Reports and Studies No. 14. Joint Group of Experts on the Scientific Aspects of Marine Pollution. World Meteorological Organization, Geneva.

Getis, A., Getis, J. and Fellman, J. D. (1996) *An Introduction to Geography* (5th edn). W. C. Brown, Dubuque, IA.

Gischler, C. E. (1979) *Water Resources in the Arab Middle East and North Africa*. Middle East and North Africa Studies Press Ltd, Cambridge.

Glacken, C. J. (1967) *Traces on the Rhodian Shore: Nature and Culture in Western Thought from Ancient Times to the End of the Eighteenth Century*. University of California Press, Berkeley.

Glantz, M. H., Rubinstein, A. Z. and Zonn, I. (1993) Tragedy in the Aral Sea Basin. Looking back to plan ahead? *Global Environmental Change* 3: 174–98.

Gleik, P. H. (ed.) (1993) *Water in Crisis. A Guide to the World's Fresh Water Resources*. Oxford University Press, New York.

Gleik, P. H. (1998) *The World's Water. The Biennial Report on Freshwater Resources*. Island Press, Washington, DC.

Goddard, J. (2002) A growing concern. *Geographical* 74(3): 38–41.

Goldsmith, E., Allen, R., Allaby, M., Davoll, J. and Lawrence, S. (1972) A blueprint for survival. *The Ecologist* 2(1): 1–44.

Gossling, S. (1999) Ecotourism: a means to safeguard biodiversity and ecosystem function? *Ecological Economics* 29(2): 303–20.

Goudie, A. S. (1992) *Environmental Change* (3rd edn), Clarendon Press, Oxford.

Goudie, A. S. (ed.) (1997) *The Human Impact Reader. Readings and Case Studies*. Blackwell, Oxford.

Goudie, A. S. (2000) *The Human Impact on the Natural Environment* (5th edn). Blackwell, Oxford.

Goudie, A. S. and Viles, H. (1997) *The Earth Transformed. An Introduction to Human Impacts on the Environment*. Blackwell, Oxford.

Grainger, A. (1993) Rates of deforestation in the humid tropics: estimates and measurements. *Geographical Journal* 159: 33–44.

Grant, A. and Jickells, T. (2000) Marine and estuarine pollution in O'Riordan, T. *Environmental Science for Environmental Management* (2nd edn). Longman, Harlow.

Grant, W. (1995) *Pressure Groups, Politics and Democracy in Britain* (2nd edn). Philip Allan, New York.

Graedel, T. E. and Crutzen, P. J. (1993) *Atmospheric Change: An Earth System Perspective*. W. H. Freeman, New York.

Gregory-Knight, C. (1980) Ethnoscience and the African farmer: rationale and strategy in Brokensha, D. W., Warren, D. M. and Werner, O. (1980) *Indigenous Knowledge Systems and Development*. University of America Press, Lanham, MD.

Grenon, M. and Batisse, M. (eds) (1989) *Futures for the Mediterranean Basin: The Blue Plan*. Oxford University Press, New York.

Greuter, W. (1995) Extinctions in Mediterranean areas in Lawton, J. H. and May, R. M. (eds), *Extinction Rates*. Oxford University Press, Oxford.

Grimston, M. (1990) A critique of 'green' science. *Atom* 408: 17–20.

Gruen, L. (1997) Revaluing nature in Warren, K. J. (ed.) (1997) *Ecofeminism. Women, Culture, Nature*. Indiana University Press, Bloomington.

Hadley Centre (2002) Stabilisation and commitment to future climate change. Scientific results from the Hadley Centre, October. Meteorological Office, Bracknell, UK.

Haggett, P. (2001) *Geography. A Global Synthesis*. Prentice Hall, Harlow.

Hall, D. R. (1990) This changing world. Geographical dimensions of change. *Geography* 70: 239–44.

Hansen, J., Johnson, D., Lacis, A., Lebedeff, S., Lee, P., Rind, D. and Russell, G. (1981) Climatic impact of increasing atmospheric carbon dioxide. *Science* 213: 957–66.

Hardin, G. (1968) The tragedy of the commons. *Science* 162: 1243–8.

Hardy, J. D. and Dubois, E. F. (1940) Differences between men and women in their response to heat and cold. *Proc. Nat. Acad. Sci.* 26: 389–98.

Harper, J. L. and Hawksworth, D. L. (1995) *Biodiversity. Measurement and Estimation*. Chapman & Hall, London.

Harvey, L. D. D. (2000) *Climate and Global Environmental Change*. Longman, Harlow.

Haslam, S. M. (1994) *River Pollution: An Ecological Perspective*. Wiley & Sons, Chichester.

Hawken, P., Lovins, A. and Lovins, H. L. (1999) *Natural Capitalism: Creating the Next Industrial Revolution*. Little, Brown and Company, New York.

Heggestad, H. E. and Darley, E. F. (1969) Plants as indicators of the air pollutants ozone and PAN in Air pollution. Proceedings of the First European Congress on the Influence of Air Pollution on Plants and Animals. Wageningen, The Netherlands.

Herweg, K. and Stillhardt, B. (1999) The variability of soil erosion in the highlands of Ethiopia and Eritrea. Soil Conservation Research Report 42. Centre for Development and Environment, Institute of Geography, University of Berne, Switzerland.

Hofmann, D. J. (1996) Recovery of the Antarctic ozone hole. *Nature* 384: 222–3.

Holliday, R. H. (1976) The efficiency of solar energy conversion by the whole crop in Duckham, A. N. and Roberts, J. G. W. (eds), *Food Production and Consumption. The Efficiency of Human Food Chains and Nutrient Cycles*. North Holland, Amsterdam.

Hollister, C. and Nadis, S. (1998) Burial of radioactive waste under the sea. *Scientific American* **278**(1): 40–5.

Honey, M. (1999) *Ecotourism and Sustainable Development. Who Owns Paradise?* Island Press, Washington, DC.

Houghton, J. T., Jenkins, G. J. and Ephraums, J. J. (eds) (1990) *Climate Change: The IPCC Scientific Assessment*. World Meteorological Organization/United Nations Environment Programme, Cambridge University Press, Cambridge.

Howe, E. D. (1978) *Desalting Plans and Progress: An Evaluation of the State of the Art and Future Research and Development Requirements*. Fluor Engineering and Constructors, Inc., Irvine, CA.

Howe, G. M. (1976) *Man, Environment and Disease in Britain: A Medical Geography of Britain Through the Ages*. Penguin, Harmondsworth.

Howells, G. (1990) *Acid Rain and Acid Waters*. Ellis Horwood, London.

Huang, S., Pollack, H. N. and Shen, P.-Y. (2000) Temperature trends over the past five centuries reconstructed from borehole temperatures. *Nature* **403**: 756–8.

Huckle, J. and Martin, A. (2001) *Environments in a Changing World*. Prentice Hall, Harlow.

Hughes, J. M. R. and Goodall, B. (1992) Marine pollution in Mannion, A. M. and Bowlby, S. R., *Environmental Issues in the 1990s*. Wiley & Sons, Chichester.

Hulme, M. and Kelly, M. (1993) Exploring the links between desertification and climate change. *Environment* **35**(6): 5–11, 39–45.

Humborg, C., Ittekkot, V., Cociasu, A. and Bodungen, B. V. (1997) Effect of Danube River Dam on Black Sea biochemistry and ecosystem structure. *Nature* **386**: 385–8.

Hutchings, J. A. (2000) Collapse and recovery of marine fishes. *Nature* **406**: 882–5.

Hutchinson, G. E. (1957) Concluding remarks. *Cold Spring Harbor Symposia on Quantitative Biology* **22**: 415–27.

Huxley, J. (1943) *TVA, Adventure in Planning*. The Architecture Press, Cheam, Surrey.

Inhaber, H. (1976) *Environmental Indices*. Wiley & Sons, New York.

International Human Genome Sequencing Consortium (2001) Initial sequencing and analysis of the human genome, *Nature* **409**: 860–921.

IPCC (1999) *Climate Change 2001: The Scientific Basis*. Cambridge University Press, Cambridge.

IPCC (2001) *Summary for Policymakers. Report of Working Group 1*. Cambridge University Press, Cambridge.

Isachenko, A. G. (1974) On the so-called anthropogenic landscapes. *Soviet Geography* **15**: 467–75.

IUCN Species Survival Commission (2000). http://www.redlist.org/info/tables.html.

IUCN/UNEP (1986a) *Review of the Protected Area System in the Indo-Malayan Realm*. IUCN, Gland, Switzerland.

IUCN/UNEP (1986b) *Review of the Protected Area System in the Afrotropical Realm*. IUCN, Gland, Switzerland.

IUCN/UNEP/WWF (1980) *World Conservation Strategy: Living Resource Conservation for Sustainable Development*. Gland, Switzerland.

Jaber, J. O., Probert, S. D. and O. Badr (1997) Water scarcity: a fundamental crisis for Jordan. *Applied Energy* **57**(2/3): 107–27.

Jackson, A. R. W. and Jackson, J. M. (2000) *Environmental Science. The Natural Environment and Human Impact* (2nd edn). Prentice Hall, Harlow.

Jacobson, T. and Adams, R. M. (1958) Salt and silt in ancient Mesopotamian agriculture. *Science* **128**: 1251–8.

Jarvis, P. J. (2000) *Ecological Principles and Environmental Issues*. Prentice Hall, Harlow.

Jeleff, S. (1999) *Oceans*. Council of Europe Publishing, Strasbourg.

Johansson, T. B., Bodlund, B. and Williams, R. H. (eds) (1989) *Electricity: Efficient End-use and New Generation Technologies and their Planning Implications*. Lund University Press, Lund.

Johnson, D. L. and Lewis, L. A. (1995) *Land Degradation: Creation and Destruction*. Blackwell, Cambridge.

Jones, G. E. (1979) *Vegetation Productivity*. Longman, London.

Jones, G. E. and Ahmed, S. (2000) The impact of coastal flooding on conservation areas: a study of the Clyde Estuary, Scotland. *J. Coastal Cons.* **6**: 171–80.

Jones, G. E. and Hollier, G. P. (1997) *Resources, Society and Environmental Management*. Paul Chapman, London.

Jones, M. and Brown, T. (2000) Agricultural origins: the evidence of modern and ancient DNA. *The Holocene* **10**(6): 769–76.

Jordan, A. and O'Riordan, T. (2000) Environmental politics and policy processes in O'Riordan, T. (ed.), *Environmental Science for Environmental Management* (4th edn). Prentice Hall, Harlow.

Judson, S. (1968) Erosion of the land – or what's happening to our continents? *American Scientist* **56**: 356–74.

Karl, T., Knight, R. and Baker, B. (2000) The record breaking global temperatures of 1997 and 1998: evidence for an increase in the rate of global warming? *Geophysical Research Letters* **27**(5): 719–22.

Keeton, W. T. (1980) *Biological Science* (3rd edn). W. W. Norton & Co., New York.

Kellogg, C. E. (1961) *Soil Interpretation in the Soil Survey*. Soil Conserv. Agric. U.S. Dep. Agriculture, Washington D.C.

Kemp, D. D. (1994) *Global Environmental Issues: A Climatological Approach* (2nd edn). Routledge, London.

Kennet, W. (1974) The politics of conservation in Warren, A. and Goldsmith, F. B. *Conservation in Practice*. Wiley & Sons, London.

Key, M. H. (2001) Fast track to fusion energy. *Nature* 412: 775–6.

King, A. and Schneider, B. (1991) *The First Global Revolution: From the Problematique to the Resolutique*. The Council of the Club of Rome, New York, Pantheon.

King, J. (1959) *The Conservation Fight, from Theodore Roosevelt to the Tennessee Valley Authority*. Public Affairs Press, Washington DC.

Klein, B. C. (1989) Effects of forest fragmentation on dung and carrion beetle communities in central Amazonia. *Ecology* 70: 1715–25.

Klingebiel, A. A. and Montgomery, P. H. (1961) *Land Capability Classification*. US Dep. Agric., Soil Conserv. Agric. Handbook No. 210.

Kloppenburg, J. and Burrows, B. (1996) Biotechnology to the rescue? Twelve reasons why biotechnology is incompatible with sustainable agriculture. *The Ecologist* 26(2): 61–7.

Knight, J. (2001) Biology's last taboo. *Nature* 413: 12–15.

Kormondy, E. J. (1969) *Concepts of Ecology*. Prentice-Hall, Englewood Cliffs, New Jersey.

Kowolak, M. (1993) Common threads: research lessons from acid rain, ozone depletion and global warming. *Environment* 35(6): 12–20, 35–8.

Lack, D. (1972) *The Life of the Robin*. Fontana New Naturalist, Collins, London.

Lanza, F. (1988) *Feasibility of Disposal of High-Level Radioactive Waste into the Seabed. Vol. 8, Review of Processes Near a Buried Waste Canister*. Nuclear Energy Agency, OECD, Paris.

Lawton, J. H. and May, R. M. (eds) (1995) *Extinction Rates*. Oxford University Press, Oxford.

Leakey, M. D. (1979) *Olduvai Gorge: My Search for Early Man*. Collins, London.

Leakey, R. and Lewin, R. (1992) *Origins Reconsidered. In Search of What Makes us Human*. Little, Brown and Co., London.

Leicester Report (1945) *Report on Air Pollution*. HMSO, London.

Leighton, P. A. (1971) Geographical aspects of air pollution in Detwyler, T. R. (ed.), *Man's Impact on Environment*. McGraw-Hill, New York.

Leopold, A. (1949) *A Sand County Almanac*. Oxford University Press, New York. (Enlarged edn (1966). OUP, New York.)

Leroux, M. (1998) *Dynamic Analysis of Weather and Climate*. Wiley & Sons, Chichester.

Loaiciga, H. A. and Renehan, S. (1997) Municipal water use and water rates driven by severe drought: a case study. *Journal of the American Water Resources Association* 33(6): 1313–26.

Lomborg, B. (2001) *The Sceptical Environmentalist. Measuring the Real State of the World*. Cambridge University Press, Cambridge.

Lovelock, J. (1989) *The Ages of Gaia: A Blueprint of Our Living Earth*. Oxford University Press, Oxford.

Lowi, M. R. (1993) *Water and Power*. Routledge, London.

Lukes, S. (1974) *Power: A Radical View*. Macmillan, Reading.

Lutz, W., Sanderson, W. and Scherbov, S. (2001) The end of world population growth. *Nature* 412: 543–5.

Macdonald, A. and Kay, D. (1988) *Water Resources. Issues and Strategies*. Longman, London.

Mannion, A. M. (1995) *Agriculture and Environmental Change*. Wiley & Sons, Chichester.

Mannion, A. M. (1992a) Environmental change: lessons from the past in Mannion, A. M. and Bowlby, S. R., *Environmental Issues in the 1990s*. Wiley & Sons, Chichester.

Mannion, A. M. (1992b) Biotechnology and genetic engineering: new environmental issues in Mannion, A. M. and Bowlby, S. R., *Environmental Issues in the 1990s*. Wiley & Sons, Chichester.

Mannion, A. M. (1992c) Acidification and eutrophication in Mannion, A. M. and Bowlby, S. R., *Environmental Issues in the 1990s*. Wiley & Sons, Chichester.

Mannion, A. M. and Bowlby, S. R. (1992) *Environmental Issues in the 1990s*. Wiley & Sons, Chichester.

Martin, P. S. and Wright, H. E. (1967) *Pleistocene Extinctions*. Yale University Press, New Haven.

Marshal, T. (2000) The drowning wave. *New Scientist* 2259: 27–30.

Master, L. (1990) The imperilled state of North American aquatic animals. *The Nature Conservancy Biodiversity Network News* 3: 1–8.

Matthews, E. R., Payne, M., Rohweder, M. and Murray, S. (2000) *Pilot Analysis of Global Ecosystems: Forest Ecosystems*. World Resources Institute, Washington, DC.

Maunder, J. (1962) A human classification of climate. *Weather* 17: 3–12.

May, R. M., Lawton, J. H. and Stork, N. E. (1995) Assessing extinction rates in Lawton, J. H. and May,

R. M. (eds), *Extinction Rates*. Oxford University Press, Oxford.

McIlveen, R. (1992) *Fundamentals of Weather and Climate*. Chapman & Hall, London.

McKinney, M. L. and Schoch, R. M. (1998) *Environmental Science. Systems and Solutions*. Jones and Bartlett, Sudbury, MA.

McNeely, J. A., Miller, K. R., Reid, W. V., Mittermeier, R. A. and Werner, T. B. (1990) *Conserving the World's Biological Diversity*. IUCN, Gland, Switzerland.

Meadows, D. H., Meadows, D. L., Randers, J. and Behrens III, W. W. (1972) *The Limits to Growth: A Report to the Club of Rome's Project on the Predicament of Mankind*. Potomac Associates, New York.

Mee, L. D. (1992) The Black Sea crisis: a need for concerted international action. *Ambio* **21**: 278–86.

Mellanby, K. (1967) *Pesticides and Pollution*. William Collins, London.

Meredith, M. P., Wade, I. P., McDonagh, E. L. and Heywood, K. J. (2000) Managing the oceans in O'Riordan, T. (2000) *Environmental Science for Environmental Management* (2nd edn). Prentice Hall, Harlow.

Meteorological Office (2001) *Climate Change Science. Some results from the Hadley Centre*. Met Office, Hadley Centre, Bracknell, England.

Micklin, P. P. (1988) Desiccation of the Aral Sea: a water management disaster in the Soviet Union. *Science* **241**: 1170–5.

Middleton, N. (1999) *The Global Casino. An Introduction to Environmental Issues* (2nd edn). Arnold, London.

Ministerie van Volkshuivesting (1992) *Environmental Policy in the Netherlands*. Ministerie van Volkshuivesting, Den Haag.

Mintzer, I. M. (1992) Living in a warmer world in Mintzer, I. M. (ed.), *Confronting Climate Change. Risks, Implications and Responses*. Cambridge University Press, Cambridge.

Mitchell, B. (2002) *Resource and Environmental Management* (2nd edn). Prentice Hall, Harlow.

Monroe, J. S. and Wicander, R. (1994) *The Changing Earth. Exploring Geology and Evolution*. West Publishing Company, Minneapolis/St Paul.

Morgan, R. P. C. (1995) *Soil Erosion and Conservation* (2nd edn). Longman, London.

Morse, S. and Stocking, M. (1995) *People and the Environment*. UCL Press, London.

Münchau, W. (1993) Floods on the way to being America's worst disaster. *The Times*, 17 July, 14.

Munn, R. E. (1996) Global change: both a scientific and a political issue in Munn, R. E., La Rivière, J. W. M. and van Lookeren Campagne, N. (eds), *Policy Making in an Era of Global Environmental Change*. Kluwer, Dordrecht.

Murakami, M., Hirose, K., Yurimoto, H., Nakashima, S. and Takafuji, N. (2002) Water in earth's lower mantle. *Science* **295**: 1885–7.

Myers, N. (1984) *The Primary Source. Tropical Forests and our Future*. Norton, New York.

Myers, N. (1988) Threatened biotas: 'hotspots' in tropical forests. *Environmentalist* **8**(3): 1–20.

Naess, A. (1993) The deep ecological movement: some philosophical aspects in Armstrong, S. J. and Botzler, R. G., *Environmental Ethics: Divergence and Convergence*. McGraw-Hill, New York.

Nader, R. (1973) *The Consumer and Corporate Accountability*. Harcourt Brace, Stamford, CT.

NARMAP (1996) *The Directory of Belizian Protected Areas and Sites of Nature Conservation Interest* (2nd edn). NARMAP, Belize City.

Nature (1990) 'Big green' update. *Nature* **348**, 15 November.

Neal, E. (1958) *The Badger*. Penguin, Harmondsworth.

Neal, P. (1992) The ocean environment: marine development, problems and pollution in Cooper, D. E. and Palmer, J. A. (eds), *The Environment in Question. Ethics and Global Issues*. Routledge, London.

Nebel, B. J. and Wright, R. T. (1998) *Environmental Science* (6th edn). Prentice-Hall, New Jersey.

Nelson, B. (1968) *Galapagos: Islands of Birds*. Longman, London.

Nesterov, A. I. (1974) The development of the anthropogenic landscape science at the Geography Faculty of Voronezh University. *Soviet Geography* **15**: 463–6.

New Scientist (1993) Nuclear dump could pollute water. *New Scientist* **138**(1874): 9.

New Scientist (1997) Sellafield leaves its mark on the frozen north. *New Scientist* **154**(2081): 14.

New Scientist (1999) Fasten your seat belt. *New Scientist* **164**(2218): 5.

New Scientist (2000) All for one! *New Scientist* **167**(2246): 30.

New Scientist (2002) Scientists condemn river reversal plan. *New Scientist* **176**(2366): 12.

Newson, M. (1997) *Land, Water and Development. Sustainable Management of River Basin systems*. Routledge, London.

Nicholson, M. (1970) *The Environmental Revolution. A Guide for the New Masters of the World*. Penguin, Harmondsworth.

Niklas, K. J. and Tiffney, B. H. (1995) The quantification of plant biodiversity through time in Hawksworth, D. L. (ed.), *Biodiversity. Measurement and Estimation*. Chapman & Hall, London.

Nkemdirim, L. C. (1997) Climate and life in the Caribbean Basin in Yoshino, M., Domrös, M., Douguédroit, A., Paszyński, J. and Nkemdirim, L. C. (1997) *Climates and Societies – A Climatological Perspective*. Kluwer Academic Publishers, Dordrecht, The Netherlands.

Nørgård, J. S. (1989) Low electricity appliances – options for the future in Johansson, T. B., Bodlund, B. and Williams, R. H. (eds), *Electricity: Efficient End-use and New Generation Technologies and their Planning Implications*. Lund University Press, Lund.

O'Connor, M. (2000) Pathways for environmental evaluation: a walk in the (hanging) gardens of Babylon. *Ecological Economics* **34**(2): 175.

O'Connell, M. (1996) Legislation in Spellerberg, I. F., *Conservation Biology*. Longman, Harlow.

Odum, E. P. (1971) *Ecology*. Holt, Rinehart and Winston, London.

OECD (1976) *Water Management in Japan. Study on Economic and Policy Instruments for Water Management*. Organisation for Economic Co-operation and Development, Paris.

OECD (1998a) *Water Consumption and Sustainable Water Resources Management*. Organisation for Economic Co-operation and Development, Paris.

OECD (1998b) *Water Management: Performance and Challenges in OECD Countries*. Organisation for Economic Co-operation and Development, Paris.

Open University (1977) *Oceanography. Introduction to the Oceans*. Open University Press, Milton Keynes.

Opie, J. (1993) *Ogallala: Water for a Dry Land*. University of Nebraska, Nebraska.

O'Riordan, T. (1981) *Environmentalism* (2nd edn). Pion, London.

O'Riordan, T. (2000) Environmental science on the move in O'Riordan, T. (ed.), *Environmental Science for Environmental Management* (2nd edn). Prentice Hall, Harlow.

O'Riordan, T. and Jordan, A. (2000) Managing the global commons in O'Riordan, T. (ed.), *Environmental Science for Environmental Management* (2nd edn). Prentice Hall, Harlow.

OTA (US Congress Office of Technology Assessment) (1987) *Technologies to Maintain Biological Diversity*. US Government Printing Office, Washington, DC.

Ott, W. (1998) The Kyoto Protocol: finished and unfinished business. *Environment* **40**(7): 45–9.

Otzen, U. (1993) Reflections on the principles of sustainable agricultural development. *Environmental Conservation* **20**: 310–16.

Packard, V. (1960) *The Waste Makers*. Penguin, Harmondsworth.

Pahl-Wostl, C. (1995) *The Dynamic Nature of Ecosystems: Chaos and Order Entwined*. Wiley & Sons, New York.

Palmer, J. (1992) Destruction of the rain forests: principles or practices? in Cooper, D. E. and Palmer, J. A. (eds), *The Environment in Question*. Routledge, New York.

Park, K. (2001) *The Environment. Principles and Applications* (2nd edn). Routledge, London.

Parker, J. and Frost, S. (2000) Environmental health aspects of coastal bathing water standards in the UK. *Environmental Management and Health* **11**(5): 447–54.

Parry, M. L. and Swaminathan, M. S. (1992) Effects of climate change on food production in Mintzer, I. M. (ed.), *Confronting Climate Change. Risks, Implications and Responses*. Cambridge University Press, Cambridge.

Pearce, D. and Moran, D. (1994) *The Economic Value of Biodiversity*. Earthscan, London.

Pears, N. V. (1985) *Basic Biogeography* (2nd edn). Longman, Harlow.

Pennington, W. (1969) *The History of British Vegetation*. The English Universities Press Ltd, London.

Pepper, D. (1984) *The Roots of Modern Environmentalism*. Croom Helm, London.

Pepper, D. (1996) *Modern Environmentalism. An Introduction*. Routledge, London.

Phillips, M. and Mighall, T. (2000) *Society and Exploitation Through Nature*. Prentice Hall, Harlow.

Pickering, K. T. and Owen, L. A. (1994) *An Introduction to Global Environmental Issues*. Routledge, London.

Pickett, S. T. A., Ostfeld, R. S., Shackak, M. and Likens, G. E. (eds), *The Ecological Basis of Conservation. Heterogeneity, Ecosystems and Biodiversity*. Chapman & Hall, New York.

Pietila, H. (1990) The daughters of earth: women's culture as a basis for sustainable development in Engel, J. R. and Engel, J. G. (eds), *Ethics of Environment and Development: Global Challenge and the International Response*. Belhaven, London.

Pimentel, D. (1976) Land degradation: effects on food and energy resources. *Science* **194**: 149–55.

Pimm, S. L., Russell, G. R., Gittleman, J. L. and Brooks, T. M. (1995) The future of biodiversity. *Science* **269**: 347–50.

Plumwood, V. (1992) Women, humanity and nature in Sayers, S. and Osborne, P. (eds), *Socialism, Feminism and Philosophy: A Radical Philosophy Reader*. Routledge, London.

Porritt, J. and Winner, D. (1988) *The Coming of the Greens*. Fontana, London.

Postel, S. (1992) *Last Oasis: Facing Water Crisis.* W. W. Norton and Co., New York.

Postel, S. (1993) Water and agriculture in Gleik, P. H. (ed.), *Water in Crisis: A Guide to the World's Fresh Water Resources.* Oxford University Press, New York.

Postel, S. (1999) *Pillar of Sand.* Norton, New York.

Powell, J. C. and Craighill, A. (2000) Waste management in O'Riordan, T. (ed.), *Environmental Science for Environmental Management* (2nd edn). Prentice Hall, Harlow.

Prescott-Allen, R. (1999) *The Wellbeing of Nations. Reviewing the Wellbeing and Sustainability of 180 Nations.* Earthscan, London.

Prescott-Allen, R. (2001) *The Wellbeing of Nations. A Country-by-Country Index of Quality of Life and the Environment.* IDRC/Island Press, Washington, DC.

Press, F. and Siever, R. (2001) *Understanding Earth.* W. H. Freeman and Co., New York.

Pressey, R. L. (1996) Protected areas: where should they be and why should they be there? in Spellerberg, I. F., *Conservation Biology.* Longman, Harlow.

Price, E. O. and King, J. A. (1971) Domestication and adaptation in Detwyler, T. R. (ed.), *Man's Impact on Environment.* McGraw-Hill, New York.

Pringle, C. M. (1997) Expanding scientific research programs to address conservation challenges in freshwater ecosystems in Pickett, S. T. A., Osterfeld, R. S., Shackak, M. and Likens, G. E. (eds), *The Ecological Basis of Conservation. Heterogeneity, Ecosystems and Biodiversity.* Chapman & Hall, New York.

Prins, G. (1993) *Threats without Enemies.* Earthscan, London.

Programme for Belize (1996) *Rio Bravo Conservation and Management Area. Management Plan* (3rd edn). Programme for Belize, Belize City.

Prosser, R. F. (1992) The ethics of tourism in Cooper, D. E. and Palmer, J. A., *The Environment in Question.* Routledge, London.

Quarrie, J. (1992) *Earth Summit 1992.* The Regency Press Corporation, Wickford, Essex.

Quayle, R. and Doehring, F. (1981) Heat stress: a comparison of indices. *Weatherwise* 34: 1210–24.

Raskin, P. (1997) Water futures: assessment of long-range patterns and problems, *Comprehensive Assessment of the Freshwater Resources of the World.* Stockholm Environment Institute, Stockholm.

Ratcliffe, D. A. (1980) *The Peregrine Falcon.* Poyser, Carlton.

Raup, D. M. (1986) Biological extinction in earth history. *Science* 231: 1528–33.

Raven, P. H. (1995) The importance of biodiversity. Keynote speech in Symposium on Biodiversity Along the Central California Coast, 3–5 March, San Francisco.

RCEP (2000) *Royal Commission on Environmental Pollution.* HMSO, London. Summary document available at http://www.recep.org.uk/pdf/ar-summary.pdf.

Rees, J. (1991) Equity and environmental property. *Geography* 76(4): 292–303.

Regan, T. (1993) The case for animal rights in Armstrong, S. J. and Botzler, R. G., *Environmental Ethics: Divergence and Convergence.* McGraw-Hill, New York.

Reid, L. (1962) *The Sociology of Nature.* Penguin, Harmondsworth.

Reid, W. V. (1998) Biodiversity hotspots. *Trends in Ecology and Evolution* 13: 275–80.

Rettie, D. F. (1995) *Our National Park System.* University of Illinois Press, Urbana and Chicago.

Richards, P. W. (1952) *The Tropical Rain Forest. An Ecological Study.* Cambridge University Press, Cambridge.

Robin, G. de Q. (1986) Predicting the rise in sea level caused by warming of the atmosphere in Bolin, B., Döös, B. R., Jäger, J. and Warrick, R. A. (eds), *The Greenhouse Effect, Climate Change, and Ecosystems.* SCOPE 29. Wiley & Sons, Chichester.

Rocheleau, D., Thomas-Slayter, B. and Wangari, E. (1996) *Feminist Political Ecology. Global Issues and Local Experiences.* Routledge, London.

Rolston, H. (1988) *Environmental Ethics. Duties to and Values in the Natural World.* Temple University Press, Philadelphia.

Rosegrant, M. W. (1997) *Water Resources in the Twenty-First Century: Challenges and Implications for Action.* Food, Agriculture and Environment Discussion Paper 20. International Food Policy Research Institute, Washington, DC.

Rowland, F. S. (1989) Chlorofluorocarbons and the depletion of stratospheric ozone. *American Scientist* 77: 36–45.

Royal Commission on Environmental Pollution (1998) Setting environmental standards. Twenty-first Report (Cm 4053). Royal Commission on Environmental Pollution, London.

Rudig, W. (ed.) (1992) *Green Politics Two.* Edinburgh University Press, Edinburgh.

Ruffner, J. A. and Blair, F. E. (1984) *The Weather Almanac* (4th edn). Gale, Detroit.

Saiko, T. (2001) *Environmental Crises: Geographical Case Studies in Post-socialist Eurasia.* Prentice Hall, Harlow.

Saiko, T. and Zonn, I. (1994) Deserting a dying sea. *Geographical* 66: 12–15.

Santibáñez, F. (1997) Possible impact on agriculture due to climatic change and variability in South America in Yoshino, M., Domrös, M., Douguédroit, A., Paszyński, J. and Nkemdirim, L. C. (1997) *Climates and Societies – A Climatological Perspective*. Kluwer Academic Publishers, Dordrecht, The Netherlands.

Sauer, C. (1952) Agricultural origins and dispersals. *Amer. Geogr. Soc. Bowman Memorial Lectures* Series 2, New York.

Schmid-Schonbein, O. (1992) Debt-for-nature swaps: funding conservation in third world countries. Swiss Bank Corporation magazine. Issue No. 6.

Schumacher, E. (1973) *Small is Beautiful*. Blond and Briggs: London.

Schneider, S. H. (1989) The greenhouse effect: science and policy. *Science* **243**: 771–81.

Scientific American (1970) The oxygen cycle in *The Biosphere*. Scientific American Books, W. H. Freeman and Co., San Francisco.

Scottish Executive (2002) *Key Scottish Environment Statistics*. The Stationery Office, Edinburgh.

Scragg, A. (1999) *Environmental Biotechnology*. Longman, Harlow.

Sears, P. B. (1972) The process of environmental change by man in Smith, R. L. (ed.), *The Ecology of Man: An Ecosystem Approach*. Harper & Row, New York.

Seidman, S. (1998) *Contested Knowledge: Social Theory in the Post-modern Era*. Blackwell, Oxford.

Serageldin, I. (1995) *Towards Sustainable Management of Water Resources*. The World Bank, Washington, DC.

Sheail, J. (1998) *Nature conservation in Britain. The formative years*. The Stationery Office, London.

Shiklomanov, I. A. (1993) World fresh water resources in Gleik, P. H. (ed.), *Water in Crisis: A Guide to the World's Fresh Water Resources*. Oxford University Press, New York.

Shimwell, D. (1971) *Description and Classification of Vegetation*. Sidgwick & Jackson, London.

Siever, R. (1975) The Earth in Folsome, C. E. (ed.), *Life: Origin and Evolution*. Scientific American Books, W. H. Freeman and Co., San Francisco.

Simmons, I. G. (1996) *Changing the Face of the Earth: Culture, Environment, History* (2nd edn). Blackwell, Oxford.

Simon, J. L. (1996) *The Ultimate Resource 2*. Princetown University Press, Princetown, New Jersey.

Singer, P. (1979) *Practical Ethics*. Cambridge University Press, Cambridge.

Singer, P. (1993) Equality for animals in Armstrong, S. J. and Botzler, R. G., *Environmental Ethics: Divergence and Convergence*. McGraw-Hill, New York.

Sioli, H. (1997) The effects of deforestation in Amazonia in Goudie, A. (ed.), *The Human Impact Reader. Readings and Case Studies*. Blackwell, Oxford.

Slater, D. (1997) *Consumer Culture and Modernity*. Polity Press, Cambridge.

Smith, C. T. (1971) The drainage basin as an historical basis for human activity in Chorley, R. J. (ed.), *Introduction to Geographical Hydrology*. Methuen & Co., London.

Smith, K. (1975) *Principles of Applied Climatology*. McGraw-Hill, Maidenhead.

Smith, K. (2001) *Environmental Hazards: Assessing Risk and Reducing Disaster*. Routledge, London.

Smith, R. L. (ed.) (1972) *The Ecology of Man: An Ecosystem Approach*. Harper & Row, New York.

SOED (1991) Environmental studies, 5–14. Curriculum and assessment in Scotland. A policy for the 90s. Working Paper No. 13. The Scottish Office Education Department, Edinburgh.

SOFA (1997) *The State of Food and Agriculture 1997*. Food and Agriculture Organization, Rome.

Solomon, M. E. (1969) *Population Dynamics*. The Institute of Biology Studies in Biology No. 18. Edward Arnold, London.

Soulé, M. E. and Wilcox, B. A. (eds) (1986) *Conservation Biology: The Science of Scarcity and Diversity*. Sinauer Associates, Sunderland, MA.

Soussan, J. G. (1992) Sustainable development in Mannion, A. M. and Bowlby, S. R. (eds), *Environmental Issues in the 1990s*. Wiley & Sons, Chichester.

Sparks, B. W. (1969) *Geomorphology*. Longmans, Green and Co. Ltd, London.

Spellerberg, I. F. (ed.) (1996) *Conservation Biology*. Longman, Harlow.

Spray, J. G., Kelley, S. P. and Rowley, D. B. (1998) Evidence for a late Triassic multiple impact on earth. *Nature* **392**: 171–3.

Stanley, S. M. (1987) *Extinction*. Scientific American Library, New York.

Stebbins, G. L. (1971) *Processes of Organic Evolution*. Prentice-Hall, New Jersey.

Stewart, S. A. (2002) A 20km diameter multi-ringed impact structure in the North Sea. *Nature* **418**: 520–3.

Stocking, M. (1984) Erosion and soil productivity, a review. FAO soil conservation programme land and water development division consultants', Working Paper, 1, Rome.

Stocking, M. (2000) Soil erosion and land degradation in O'Riordan T., *Environmental Science for Environmental Management* (2nd edn). Prentice Hall, Harlow.

Stoker, H. and Seager, S. L. (1972) *Environmental Chemistry: Air and Water Pollution*. Scott, Foresman and Co., Glenview, IL.

Strahler, A. H. and Strahler, A. N. (1991) *Modern Physical Geography* (4th edn). Wiley & Sons, New York.

Strahler, A. N. and Strahler, A. H. (1974) *Introduction to Environmental Science*. Hamilton Publishing Company, Santa Barbara, CA.

Sullivan, A. (1985) *Greening the Tories: New Politics on the Environment*. Centre for Policy Studies, London.

Summerfield, M. (1991) *Global Geomorphology*. Longman, Harlow.

Swanson, T. M. (1995) Why does biodiversity decline? The analysis of forces for global change in Swanson, T. M. (ed.), *The Economics and Ecology of Biodiversity Decline*. Cambridge University Press, Cambridge.

Sweeting, J. E. N., Bruner, A. G. and Rosenfeld, A. B. (1999) The green host effect: an integrated approach to sustainable tourism and resort development. Conservation International Policy Paper, Washington, DC.

Sydney Water (1995) *Demand Management Strategy*. Sydney Water Corporation, Sydney, Australia.

Tansley, A. G. (1923) *Practical Plant Ecology. A Guide for Beginners in Field Study of Plant Communities*. Completely revised edition by Willis, A. J. (1973) Allen & Unwin, London.

Tansley, A. G. (1946) *Introduction to Plant Ecology*. Allen & Unwin, London.

Taylor, J. (2001) *Petra and the Lost Kingdom of the Nabataeans*. I. B. Tauris, London.

Taylor, J. A. (1967) *Weather and Agriculture*. Pergamon Press, Oxford.

Terjung, W. H. (1966) Physiologic climates of the coterminous United States: a bioclimate classification based on man. *Ann. Ass. Amer. Geogrs.* **56**: 141–79.

Terjung, W. H. (1968) World patterns of the distribution of the monthly comfort index. *Int. J. Biometeor.* **12**: 119–51.

Thomas, D. S. G. and Middleton, N. J. (1997) Salinization: new perspectives on a major desertification issue in Goudie, A. S. (ed.), *The Human Impact Reader. Readings and Case Studies*. Blackwell, Oxford.

Thomas, J. A. and Morris, M. G. (1995) Rates and patterns of extinction among British invertebrates in Lawton, J. H. and May, R. M. (eds), *Extinction Rates*. Oxford University Press, Oxford.

Thomas, W. A. (1972) *Indicators of Environmental Quality*. Plenum Press, New York.

Thomas, W. L. (ed.) (1956) *Man's Role in Changing the Face of the Earth*. University of Chicago Press, Chicago.

Thompson, R. D. (1992) The changing atmosphere and its impact on planet earth in Mannion, A. M. and Bowlby, S. R., *Environmental Issues in the 1990s*. Wiley & Sons, Chichester.

Thorne-Miller, B. (1999) *The Living Ocean. Understanding and Protecting Marine Biodiversity*. Island Press, Washington, DC.

Tickell, C. (1993) The human species: a suicidal success? *Geographical Journal* **159**: 219–26.

Tobin, G. A. and Montz, B. E. (1997) *Natural Hazards. Explanation and Integration*. The Guilford Press, New York.

Treiber, M. and Helbing, D. (2001) Microsimulations of freeway traffic including control measures. *Automatisierungstechnik* **49**: 478–84.

Trewavas, A. (1999) Much food, many problems. *Nature* **402**: 231–2.

Trusted, J. (1992) The problem of absolute poverty: what are our moral obligations to the destitute? in Cooper, D. E. and Palmer, J. A. (eds), *The Environment in Question. Ethics and Global Issues*. Routledge, London.

Tudge, C. (1988) *Food Crops for the Future*. Blackwell, Oxford.

Tyndall Centre (2002) The effect. Tyndall Centre for Climate Change Research, School of Environmental Sciences, University of East Anglia, Norwich, England.

Tyson, P. D. (1963) Some climatic factors affecting pollution in South Africa. *South African Geog. J.* **XLV**: 277–84.

UKCIP02 (2002) Climate change scenarios for the United Kingdom. The UKCIP02 Briefing Report. Tyndall Centre for Climate Change Research, University of East Anglia, Norwich.

United Kingdom Government (1994) *Biodiversity. The UK Action Plan*. Command No. 2428, HMSO, London.

United Nations (1973) *Report of the United Nations Conference on Human Environment, Stockholm 1972*. United Nations, New York.

United Nations (1982) *Coastal Area Management and Development*. United Nations Department of International Economic and Social Affairs, Pergamon Press, Oxford.

United Nations (1987) *Our Common Future* (The Brundtland Report). World Commission on Environment and Development, Oxford University Press, Oxford.

United Nations (1992) *The Draft Rio Declaration on Environment and Development*. United Nations, New York.

United Nations (1993) *The Global Partnership for Environment and Development. A Guide to Agenda 21*. Post-Rio de Janeiro Conference edition. United Nations, New York.

United Nations (1996) *Negotiating Committee hears Details of Efforts to Combat Desertification and Land Degradation*. Press Release, ENV/DEV/380 United Nations, New York.

United Nations Development Programme (UNDP) (2000) *World Resources 2000–2001*. World Resources Institute, Washington, DC.

United Nations Environment Programme (UNEP) (1999) *Global Environmental Outlook 2000*. Earthscan, London.

United Nations Population Division (1998) *World Population Prospects: The 1998 Revision*. UN Population Division, New York.

Van Kooten, G. C., Bulte, E. H. and Sinclair, A. R. E. (2000) Introduction: conserving biological diversity in Van Kooten, G. C., Bulte, E. H. and Sinclair, A. R. E. (eds), *Conserving Nature's Diversity. Insights from Biology, Ethics and Economics*. Ashgate, Aldershot, UK.

Van Loon, G. W. and Duffy, S. J. (2000) *Environmental Chemistry: A Global Perspective*. Oxford University Press, Oxford.

Vernberg, F. J. and Vernberg, W. B. (1970) *The Animal and the Environment*. Holt Rinehart and Winston, New York.

Veryard, R. G. (1958) Some climatological aspects of air pollution. *Smokeless Air* **106**: 277–80.

Vice, R. B., Guy, H. P. and Ferguson, G. E. (1969) Sediment movement in areas of suburban highway construction. Scott Run Basin, Fairfax County, Virginia, 1961–64. U.S. Geological Survey Water Supply Paper. 1591-E.

von Weizsäker, E., Lovins, A. B. and Lovins, L. H. (1998) *Factor Four. Doubling Wealth, Halving Resource Use*. Earthscan, London.

Wagaar, J. A. (1970) Growth versus the quality of life. *Science* **168**(6): 1179–84.

Walker, G. (2000) Catch the wave. *New Scientist* **2269**: 40–3.

Walling, D. E. and Quine, T. A. (1991) Recent rates of soil loss from areas of arable cultivation in the UK. IAHS Publication 203, 123–31.

Walter, H. (1973) *Vegetation of the Earth in Relation to Climate and the Eco-physiological Conditions*. The English Universities Press Ltd, London.

Warren, A. and Goldsmith, F. B. (1974) *Conservation in Practice*. Wiley & Sons, London.

Warren, K. J. (ed.) (1997) *Ecofeminism. Women, Culture, Nature*. Indiana University Press, Bloomington.

Warren, M. (1999) No relief in sight as 140 are killed by Moscow heat. *Daily Telegraph*, 3 July, Issue 1499.

Wasson, R. (1987) Detection and measurement of land degradation processes in Chisholm, A. and Dumsday, R., *Land Degradation. Problems and Policies*. Cambridge University Press, Cambridge.

Weale, A. (1998) Environmental policy in Budge, I., Crewe, D., McKay, D. and Newton, K. (eds), *The New British Politics*. Longman, Harlow.

Webb, N. R. and Haskins, L. E. (1980) An ecological survey of heathlands in the Poole Basin, Dorset, England in 1978. *Biological Conservation* **17**: 281–96.

Wellens, J. and Millington, A. C. (1992) Desertification in Mannion, A. M. and Bowlby, S. R., *Environmental Issues in the 1990s*. Wiley & Sons, Chichester.

White, L. Jnr (1967) The historical roots of the ecological crisis. *Science* **155**(3767): 1203–7.

Whittow, J. (1979) *Disasters: The Anatomy of Environmental Hazards*. University of Georgia Press, GA.

Wilcove, D. S. and Bean, M. J. (1994) The big kill: biodiversity in America's lakes and rivers. Environmental Defense Fund, Washington, DC.

Wilcove, D. S., McLellan, C. H. and Dobson, A. P. (1986) Habitat fragmentation in the temperate zone in Soulé, M. E. (ed.), *Conservation Biology: The Science of Scarcity and Diversity*. Sinauer Associates, Sunderland, MA.

Williams, P. H., Gaston, K. J. and Humphries, C. J. (1994) Do conservationists and molecular biologists value differences between organisms in the same way? *Biodiv. Lett.* **2**: 67–78.

Willis, A. J. (1973) *Introduction to Plant Ecology*. Completely revised edition of Tansley's *Introduction to Plant Ecology*. Allen & Unwin, London.

Wilson, E. O. and Peter, F. M. (eds) (1988) *Biodiversity*. National Academy Press, Washington, DC.

Wischmeier, W. H. and Smith, D. D. (1965) Predicting rainfall erosion losses from cropland east of the Rocky Mountains. USDA Agricultural Handbook 282. US Government Printing Office, Washington, DC.

Wood, A., Stedman-Edwards, P. and Mang, J. (eds) (2000a) *The Root Causes of Biodiversity Loss*. Earthscan, London.

Wood, B. (2002) Who are we? *New Scientist* **176**(2366): 44–7.

Wood, S., Sebastian, K. and Scherr, S. (2000b) Pilot analysis of global ecosystems (PAGE): agroecosystems technical report. World Resources Institute and International Food Policy Research Institute, Washington, DC.

Worcester, R. (1998) Greening the millennium: public attitudes to the environment. *Public Opinion Quarterly* **69**: 73–85.

World Bank (1994) *A Strategy for Managing Water in the Middle East and North Africa*. Washington, DC.

World Bank (1999) *World Development Indicators 1999*. The World Bank, Washington, DC.

World Bank (2000) *World Development Indicators 2000*. The World Bank, Washington, DC.

World Commission on Environment and Development (WCED) (1987) *Our Common Future* (The Brundtland Report). Oxford University Press, Oxford.

World Conservation Monitoring Centre (1992) *Global Biodiversity: Status of the Earth's Living Resources.* Chapman & Hall, London.

World Meteorological Organization (1997) *Comprehensive Assessment of the Freshwater Resources of the World*. WMO, Geneva.

World Resources Institute (1986) *World Resources, 1996. An Assessment of the Resources Base that Supports the Global Economy*. World Resources Institute, Washington, DC.

World Resources Institute (1996) *World Resources, 1996–1997*. World Resources Institute, Washington, DC.

World Resources Institute (1999) *World Resources, 1998–1999*. World Resources Institute, Washington, DC.

World Resources Institute (2000) *World Resources, 2000–2001*. World Resources Institute, Washington, DC.

Worldwatch Institute (1997) *State of the World 1997*. Earthscan, London.

Worldwatch Institute (1998) Fastest mass extinction in earth history. www.enn.com/features/1998/09/091698/fea0916.asp.

Worldwatch Institute (2000) http://www.riverdeep.net/current/2000/11/112200_worldwatch.jhtml.

Worldwide Fund for Nature (1999) *The Living Planet Index*. WWF, Gland, Switzerland.

Worster, D. (1985) *Nature's Economy: A History of Ecological Ideas*. Cambridge University Press, Cambridge.

Worthington, E. B. (ed.) (1977) *Arid Land Irrigation in Developing Countries: Environmental Problems and Effects*. Pergamon, Oxford.

Wynne, B. and Meyer, S. (1993) How science fails the environment. *New Scientist* 5 June, 33–5.

Yoshino, M., Domrös, M., Douguédroit, A., Paszyński, J. and Nkemdirim, L. C. (1997) *Climates and Societies – A Climatological Perspective*. Kluwer Academic Publishers, Dordrecht, The Netherlands.

Ziswiler, J. (1971) *Extinct and Vanishing Animals*. Springer-Verlag, New York.

Zukovskij, P. M. (1962) *Cultivated Plants and their Wild Relatives*. Translated by Hudson, P. S. Commonwealth Agricultural Bureau, Farnham Royal, Buckinghamshire, England.

Useful websites

Website addresses may be discontinued or change over time. The following list was current at time of going to press.

Association for Biodiversity Information
http://www.abi.org

Belize Tourism
http://www.belizetourism.org

Biodiversity: an overview
http://www.wcmc.org.uk/infoserv/biogen/biogen.html

Biodiversity and World Map
(an older site but provides a good introduction to the topic)
http://www.nhm.ac.uk/science/projects/worldmap

Conservation of Arctic Flora and Fauna
http://www.grida.no/caff/

Conservation International
http://www.conservation.org
http://www.conservation.org/WEB/NEWS/PRESSREL/01-0123.htm

Convention on Biological Diversity
http://www.biodiv.org

Convention on International Trade in Endangered Species of Wild Fauna and Flora
http://www.wcmc.org.uk/CITES/

Ecology and biodiversity
http://conbio.rice.edu/vl/

Ecotourism.cc – the Ecotourism Portal
http://www.ecotourism.cc

Ecotourism Information Centre
http://www.life.csu.edu.au/ecotour/EcoTrHme.html

Ecotravel Centre
http://www.ecotour.org

Electronic Telegraph
(search for specific oil tanker disasters)
http://www.telegraph.co.uk

Environmental News Network
http://www.enn.com/enn-news-archive/2000/03/031320000/medtourist_10789.asp

Environmental web resources
http://www.herts.ac.uk/lis/subjects/natsci/

Eurobarometer 55.2, leading national trends in science and technology
http://europa.eu.int/dg10/epo/eb/eb55/eb5552_sctech_national_eu.pdf

European Federation of Green Parties
http://www.europeangreens.org

Genetic engineering and its dangers
http://online.sfsu.edu/%7Erone/GEessays/gedanger.htm

Genetic Engineering News
http://www.genennews.com/

Global change (climate change metasite)
http://www.globalchange.org

Greenpeace Memorial Site for Contaminated Workers in Cubatao
http://www.greenpeace.org/pressreleases/toxics/1999jan12.html

International Aral Sea Rehabilitation (IFAS)
http://www.ifas-almaty.kz/ENG/

International Centre for Genetic Engineering
http://www.icgeb.trieste.it/

International Ecotourism Society
http://www.ecotourism.org

International Maritime Organisation
http://www.imo.org

King and Schneider, the First Global Revolution
http://www/kipawa.com/philosophy/first_global_revolution.htm

Kyoto Climate Agreement
http://www.sciam.com/forum/121597Kyoto/

Mother Earth Monday
http://www.dailyrevolution.org/monday/aralsea.html

Murray–Darling Basin Commission
http://www.mdbc.gov.au/index.htm

NASA Landsat satellite data
http://geo.arc.nasa.gov/sge/landsat/landsat.html

National Biodiversity Network
http://www.nbn.org.uk

National biodiversity profiles
http://www.wcmc.org.uk/nbp/

Plan of control of natural disasters, Cubatao (Brazil) (in Spanish but translation facilities provided)
http://www.habitat.aq.upm.es/bpal/onu/bp033.html

Regional environmental centre for the Black Sea
http://www.bsein.mhi.iuf.net/

Royal Commission on Environmental Pollution
http://www.rcep.org.uk/pdf/ar-summary.pdf

Stockholm International Water Institute
http://www.siwi.org/

Sustainable development information service
http://www.wri.org/trends/water2.html

Tennessee Valley Authority
http://tva.gov

UK biodiversity
http://ibs.uel.ac.uk/ibs/other/ukbiodiv/ukbiodiv.htm

United Nations, Rio + 10 Conference
http://www.johannesburgsummit.org/

University of Minnesota, Department of Statistics
http://www.geom.umn.edu/education/calc-init/population/

Warwick University Department of Engineering (water harvesting portal)
http://www.eng.warwick.ac.uk/DTU/rainwaterharvesting/links.htm

World Meteorological Organization
http://www.wmo.ch

World Resources Institute
http://www.wri.org/

Worldwatch Institute
www.enn.com/features/1998/09/091698/fea0916.asp

Zero Population Growth
http://www.zpg.org/

Index

Royal Society for Protection of Birds (RSPB), 43
runoff, 113, **114**

Sahel, **185**
salinisation, **49**, 168, 169, 174
salinity, of oceans, 16, 140–1
satisficing, 36
scientific revolution, 32
sea-level rise scenarios, **103**
Sellafield, **151**
settled agriculture, 172
shifting cultivation, 172
short-wave energy, 79
Sierra Club, 43
slime moulds, 20
small and medium-sized (SMS) businesses, 225
smog, 62
social organisation, 32
soft technologists, **40**, 41
Soil Conservation Service (USA), 177, 183
soil erosion, 53–4, 168–9, 174
    control of, 177
    rates, 172, 175–6
soil nutrient depletion, 168
solar energy, 16, 79, 138
species abundance curve, **196**
species protection, 209
Species Survival Unit (SSU), 214
stabilised population curve, **10**
Stockholm Conference on the Human Environment, 157
stockpiling, 42
stratosphere, 80, **81**
stewardship, 224
subsidies, for agriculture, 178
sulphur dioxide, 37
surface microlayer, 140
Surfers Against Sewage (SAS), **154**

sustainable development, 39, 238, 247–9
sustainability *see* sustainable development

tanker disasters, **148**, 148
technocentrism/technocentrists, 39, **40**, 59, 238
technology, 3, 6, 14, 53, 54, 56, 72, 178, 187, 228, 232–6
temperature inversion, 97
Tennessee Valley Authority (TVA), 115, **116**, 177
Thallophyta, 21
thermocline, 140
thermosphere, 80, **81**
threatened species, categories, 214
    distribution, **216**
Torrey Canyon, **144**
total weed killers, 199, 200
Tracheophytes, 22–4
traditional medicine, 30
Tragedy of the Commons, 54, 174, 251
transhumance, 184
tributyltin (TBT), **144**
trophic layer, 56
tropical rain forest, 190
tropopause, 80, **81**
troposphere, 80–2, **81**, 98, 100
Tyndall Centre for Climate Change Research, 100

United Nations Conference on Environment and Development (UNCED), 70, 74, 115, 158, 184, 187, 219, 222, 237
United Nations Environment Programme (UNEP), 74, 213
universal soil loss equation (USLE), 175

user pays, 128
utility, of the land, 178

vertebrates, 27

water
    management, 115, 132
    mining, 123, 131
    pollution, 128
    rights, 121
    scarcity, 127
    stress, 127
    symposium, 108
    table, 111
    vapour, 15, 79, 110
water deficit, 134
Water Resources Act 1963 (UK), 223
weather, 82–3, **83**, 86, 94
well-being index (WI), 248
well-being stress index (WSI), 248
wilderness areas, 35, 192
wind chill, 89
World Commission on Environment and Development, 74, 237
World Conservation Monitoring Centre (WCMS), 213
World Conservation Strategy (WCS), 207
World Resources Institute, 219
World Summit on Sustainable Development, 12, 75, 115
World Trade Organization (WTO), 238, **239**
World Water Forum, 108
World Wide Fund for Nature (WWF), 200, 213, **217**
worldview, 5, 243–51

Yellowstone National Park, 207

zapovedniks, 192